Springer Climate

Series Editor
John Dodson (iD), Chinese Academy of Sciences, Institute of Earth Environment,
Xian, Shaanxi, China

Springer Climate is an interdisciplinary book series dedicated to climate research. This includes climatology, climate change impacts, climate change management, climate change policy, regional climate studies, climate monitoring and modeling, palaeoclimatology etc. The series publishes high quality research for scientists, researchers, students and policy makers. An author/editor questionnaire, instructions for authors and a book proposal form can be obtained from the Publishing Editor.

Now indexed in Scopus® !

More information about this series at http://www.springer.com/series/11741

Wolfgang Buchholz • Anil Markandya •
Dirk Rübbelke • Stefan Vögele
Editors

Ancillary Benefits of Climate Policy

New Theoretical Developments and Empirical Findings

Editors
Wolfgang Buchholz
Department of Economics
and Econometrics
University of Regensburg
Regensburg, Germany

Anil Markandya
Basque Centre for Climate Change (BC3)
Leioa, Vizcaya, Spain

Dirk Rübbelke
Faculty of Economics
Technische Universität Bergakademie
Freiberg
Freiberg, Germany

Stefan Vögele
Institute of Energy and Climate Research
Forschungszentrum Jülich
Jülich, Germany

ISSN 2352-0698   ISSN 2352-0701   (electronic)
Springer Climate
ISBN 978-3-030-30977-0   ISBN 978-3-030-30978-7   (eBook)
https://doi.org/10.1007/978-3-030-30978-7

© Springer Nature Switzerland AG 2020
This work is subject to copyright. All rights are reserved by the Publisher, whether the whole or part of the material is concerned, specifically the rights of translation, reprinting, reuse of illustrations, recitation, broadcasting, reproduction on microfilms or in any other physical way, and transmission or information storage and retrieval, electronic adaptation, computer software, or by similar or dissimilar methodology now known or hereafter developed.
The use of general descriptive names, registered names, trademarks, service marks, etc. in this publication does not imply, even in the absence of a specific statement, that such names are exempt from the relevant protective laws and regulations and therefore free for general use.
The publisher, the authors, and the editors are safe to assume that the advice and information in this book are believed to be true and accurate at the date of publication. Neither the publisher nor the authors or the editors give a warranty, express or implied, with respect to the material contained herein or for any errors or omissions that may have been made. The publisher remains neutral with regard to jurisdictional claims in published maps and institutional affiliations.

This Springer imprint is published by the registered company Springer Nature Switzerland AG.
The registered company address is: Gewerbestrasse 11, 6330 Cham, Switzerland

Foreword

Do we need ambitious climate policies to create a good future for humanity? We clearly do—such goes the premise that has given rise to rounds and rounds of global negotiations spearheaded by the United Nations. This process has now entered its fifth decade since its genesis: the First World Climate Conference was held in February 1979, then organized by the UN's meteorological organization. Lately, global climate diplomacy has developed a momentum that many could not foresee. It produces regulations and voluntary commitments aimed at all the world's countries, and both the EU and Germany have already made concrete pledges to contribute. The Fridays For Future youth demonstrations have been mobilizing young people in more than 120 nations. Climate change is at the top of the political agenda, and many industry players have ceased to resist the development and begun instead to promote predictable, efficient climate policies.

But the countermovement is gaining traction, too. The USA, the world's largest economy, has recently announced its withdrawal from the Paris Climate Agreement next year. Brazil appears to be wavering on the issue of climate protection, while populist parties openly question the entire concept in many countries. Should we do nothing instead? Some believe that idleness is indeed the best approach, and their voices grow louder by the day. The Internet with its ubiquitous echo chambers and climate change denial blogs amplifies them effectively. Politicians have no choice but to take these factors seriously: climate policy is about much more than the climate itself. It is about democratic legitimacy and, not infrequently, about political survival.

The onus is on those in charge to navigate the ongoing struggle among the various factions of global civil society and make the right promises. Those promises must be ambitious enough that they can inspire voters and stakeholders to embrace change. At the same time, they must be realistic enough to prevent a crisis of legitimacy in the climate protection movement: their benefits must be clearly defined and put in perspective to their costs; feasible future approaches must be presented in accessible ways. Situated at the intersection between science and politics, climate research carries out crucial groundwork to facilitate progress.

This book is a valuable contribution to its efforts. It provides a realistic overview of the ancillary benefits (or co-benefits) of climate protection: the positive "side effects" of climate policy beyond their immediate chain of effects (lower greenhouse gas emissions—less climate change—less damage to the climate). In the heat of political discourse, these effects are often used as a pretext for careless, even false promises: transforming our energy system will bring about better jobs, greener parks, cleaner air, greater democracy, and so on. Such hollow words are easily debunked by opponents of climate protection and ultimately contribute little to the cause. The following selection of research papers by authors from four continents provides a scientific assessment of these co-benefits. Their findings establish a solid basis for convincing arguments, which may well encourage policymakers to dare take ambitious action.

Explaining the benefits of climate policy can be tricky: after all, its purpose is to prevent or lessen something that is negative, poorly understood, and difficult to quantify. Demands for ambitious measures can easily take a negative tone, as arguments emphasize the "threat" of climate change. But society is increasingly aware that the excessive exploitation of global resources, such as the atmosphere, spells trouble for prosperity in the twenty-first century. In turn, the fair and efficient use of global resources secures prosperity. But in the twenty-first century, the notion of prosperity must be different to that of the preceding two centuries. It is this hypothetical new idea of prosperity that will allow us to quantify the co-benefits of climate policy adequately.

This work skillfully highlights the importance of the Sustainable Development Goals in this context. Ending poverty, securing food, and giving everyone access to healthcare and education are demands and guiding principles that fuel political debate and motivate voters and politicians alike. Research into the co-benefits of climate policy has inspired important discussions in welfare states such as Germany, showing clearly that society is interested in fair income distribution, healthy ecosystems, and sustainable consumerism. Ultimately, this book employs specific scientific perspectives to discuss a very simple, fundamental topic: climate policy as a guarantor of a high standard of living.

Mercator Research Institute on Global
Commons and Climate Change (MCC)
gGmbH, Berlin, Germany

Ottmar Edenhofer

Potsdam Institute for Climate Impact
Research, Potsdam, Germany

Technical University Berlin, Faculty VI
and VII, Berlin, Germany

Preface

While concerns about accelerating global warming preoccupy the public—as one can observe from the recent school strikes as protest against too few climate protection efforts—the Paris Agreement on Climate Change is starting to take effect in several areas of our planet. Yet, while in Europe, levels of carbon dioxide emissions are declining, in order to reach the 2° goal and even more the 1.5° goal of the agreement, global climate protection efforts have to be strengthened considerably. This becomes even more urgent as concentrations of methane, which is another very potent greenhouse gas besides carbon dioxide, have increased in an unexpectedly strong way in recent years.

Owing to the vital need for action to protect the climate, there have been intensive efforts to find additional incentives to raise protection levels. Apart from the generation of incentives via policy instruments like emission trading schemes, climate policy has other effects, so-called ancillary effects, which tend to raise the attractiveness of climate protection. The aim of this volume is to deal with these effects and the various associated benefits for economy and society.

The fact that besides its underlying objective a policy measure may have other impacts has regularly been present in the research of the editors of this volume, even beyond the context of environmental protection. We have addressed such issues in different ways, and in doing so, we have included the implications of different degrees of "publicness" or the extent to which the ancillary effects take the form of a public good. Inspired by this research and the high relevance of these issues for climate policy, we had the idea to compile a book including recent research concerned with joint production effects. Although the chapters in the book focus on the joint effects of climate policy, it is straightforward to apply the models and techniques used in this book also to joint production in other fields, for instance to research on philanthropy, performing arts, green goods consumption, terrorism, and military alliances.

In the process of drafting and publishing this book, many people were involved to whom we owe our thanks. We are in particular grateful to Theresa Stahlke who supported us in organizing the submission and review processes. She also gave valuable comments on several parts of this book. We would also like to thank

Johannes Glaeser at Springer Nature, who was involved in the publication process in a very helpful and constructive way. Finally, we thank Lisa Broska, Anja Brumme, Kristina Govorukha, Philip Mayer, and Anja Zenker for valuable suggestions on parts of the book. Of course, none of these people are responsible for any errors or omissions in this book.

Regensburg, Germany	Wolfgang Buchholz
Leioa, Spain	Anil Markandya
Freiberg, Germany	Dirk Rübbelke
Jülich, Germany	Stefan Vögele

Contents

Analysis of Ancillary Benefits of Climate Policy 1
Wolfgang Buchholz, Anil Markandya, Dirk Rübbelke, and Stefan Vögele

Part I Ancillary Benefits and Development Co-effects

**Can the Paris Agreement Support Achieving the Sustainable
Development Goals?** 15
Lorenza Campagnolo and Enrica De Cian

Co-benefits Under the Market Mechanisms of the Paris Agreement 51
Axel Michaelowa, Aglaja Espelage, and Stephan Hoch

Technological Transition and Carbon Constraints Under Uncertainty 69
Alexander Golub

**Part II Conceptual and Theoretical Approaches for the Analysis
of Ancillary Benefits**

**Sustainable International Cooperation with Ancillary Benefits
of Climate Policy** 91
Nobuyuki Takashima

**Impure Public Good Models as a Tool to Analyze the Provision
of Ancillary and Primary Benefits** 109
Anja Brumme, Wolfgang Buchholz, and Dirk Rübbelke

Private Ancillary Benefits in a Joint Production Framework 125
Claudia Schwirplies

Impure Public Goods and the Aggregative Game Approach 141
Anja Brumme, Wolfgang Buchholz, and Dirk Rübbelke

**Multi-criteria Approaches to Ancillary Effects: The Example
of E-Mobility** 157
Stefan Vögele, Christopher Ball, and Wilhelm Kuckshinrichs

ix

Part III Ancillary Benefits in Different Sectors and in Adaptation to Climate Change

Ancillary Benefits of Adaptation: An Overview 181
Elisa Sainz de Murieta

Economic Assessment of Co-benefits of Adaptation to Climate Change... 197
Christiane Reif and Daniel Osberghaus

Ancillary Benefits of Carbon Capture and Storage 213
Asbjørn Torvanger

Health Co-benefits of Climate Mitigation Policies: Why Is It So Hard to Convince Policy-Makers of Them and What Can Be Done to Change That? 227
Anil Markandya and Jon Sampedro

Financing Forest Protection with Integrated REDD+ Markets in Brazil 243
Ronaldo Seroa da Motta, Pedro Moura Costa, Mariano Cenamo, Pedro Soares, Virgílio Viana, Victor Salviati, Paula Bernasconi, Alice Thuault, and Plinio Ribeiro

Ancillary Benefits of Climate Policies in the Shipping Sector 257
Emmanouil Doundoulakis and Spiros Papaefthimiou

Part IV Climate Actions in Urban Areas and Their Ancillary Benefits

Co-benefits of Climate Change Mitigation and Pollution Reduction in China's Urban Areas 279
Liu Jie, Pan Jiahua, Liu Ziwei, and Jiao Shanshan

Climate Co-benefits in Rapidly Urbanizing Emerging Economies: Scientific and Policy Imperatives 301
Mahendra Sethi

Protocol of an Interdisciplinary and Multidimensional Assessment of Pollution Reduction Measures in Urban Areas: MobilAir Project 325
Sandrine Mathy, Hélène Bouscasse, Sonia Chardonnel, Aïna Chalabaëv, Stephan Gabet, Carole Treibich, and Rémy Slama

Analysis of Ancillary Benefits of Climate Policy

Wolfgang Buchholz, Anil Markandya, Dirk Rübbelke, and Stefan Vögele

The main pillars of climate policy are adaptation to and mitigation of climate change. The primary objective of adaptation policies—in turn—is to make adjustments in ecological, social, or economic systems that prevent adverse impacts (like losses in agricultural yields) of climate change or take advantage of this change (like using an ice-free Northwest Passage, providing a shipping shortcut between the Pacific and Atlantic Oceans). In the case of policies to mitigate climate change, the primary aim is basically climate protection by reducing the emissions of greenhouse gases.

A large strand of scientific literature focuses on the analysis of these primary objectives of climate policy by assessing their associated benefits and on investigating the tools to bring about an efficient combination of adaptation and mitigation.

Yet, since the early 1990s at least (see e.g. Ayres and Walter 1991; Pearce 1992), another category of benefits accruing from climate policy has also aroused the interests of researchers, namely the ancillary benefits, which are brought about in

---

W. Buchholz
University of Regensburg, Regensburg, Germany
e-mail: wolfgang.buchholz@wiwi.uni-regensburg.de

A. Markandya
Basque Centre for Climate Change (BC3), Leioa, Spain
e-mail: anil.markandya@bc3research.org

D. Rübbelke (✉)
Technische Universität Bergakademie Freiberg, Freiberg, Germany
e-mail: dirk.ruebbelke@vwl.tu-freiberg.de

S. Vögele
Institute of Energy and Climate Research – Systems Analysis and Technology Evaluation (IEK-STE), Forschungszentrum Jülich GmbH, Jülich, Germany
e-mail: s.voegele@fz-juelich.de

© Springer Nature Switzerland AG 2020
W. Buchholz et al. (eds.), *Ancillary Benefits of Climate Policy*, Springer Climate,
https://doi.org/10.1007/978-3-030-30978-7_1

addition to the primary benefits of mitigation or adaptation.[1] So, if a mitigation policy reduces greenhouse gas (GHG) emissions via a decrease in the burning of fossil fuels, this will generally be accompanied by the reduction of other pollutants like particulate matter and sulphur dioxide. As such extra benefits were not primarily intended, they are frequently ignored in scientific studies on climate policy, although such co-effects tend to be of significant size.

A large number of empirical studies on ancillary benefits include assessments of benefits enjoyed from the reduction of those local or regional air pollutants that regularly accompany GHG emissions (e.g. Burtraw et al. 2003; Rive and Rübbelke 2010; Markandya and Wilkinson 2007; Markandya et al. 2018). Among the reasons for the high importance of air pollution reduction benefits in the ancillary benefits literature is the high level of associated positive health effects and—quite simply— the fact that this ancillary benefit category is easier to assess than most others. Nemet et al. (2010) survey studies considering air-quality-related ancillary benefits and find that a range of studies assessed them to be of a similar order of magnitude to greenhouse gas abatement cost estimates. Bain et al. (2016) analyse different types of ancillary effects and although they confirm that these benefits could motivate action on climate change, they assess that ancillary benefits associated with pollution reduction are among those types of benefits that are the weakest motivators of action.

For that reason alone, it makes a lot of sense not to take into account only pollution-related benefits, but to consider the broad "umbrella" of multiple benefits.[2] Other (not air-pollution-related) ancillary benefit categories include, amongst others, traffic-related benefits (e.g. from reduced noise, traffic congestion, road surface damage),[3] benefits enjoyed from the prevention of soil erosion and biodiversity loss, from positive effects exerted on the labour market and on competitiveness, and from a rise in domestic energy security.

Co-effects of climate policy may not be beneficial throughout, i.e. there may also be ancillary costs. An example where such ancillary costs arise is the following: Wind turbines, on the one hand, substitute renewable energy for fossil fuels and thereby reduce GHG emissions and provide primary benefits, but on the other hand, they also threaten the wildlife as their rotor blades may hit and kill birds or bats and thus cause ancillary cost. Krupnick et al. (2000: 71) give another example: A switch to hydroelectric power could create many negative externalities to river ecosystems involving ancillary costs. Other authors (e.g. Halsnaes 2002: 61) point

---

[1] There are several terms conveying the idea of ancillary benefits, e.g. co-benefits, secondary benefits or spillover benefits. As Markandya and Rübbelke (2004: 489) explain, "*[t]he main difference is the relative emphasis given to the climate change mitigation benefits versus the other benefits. For some policies these 'other benefits' may be as important as the GHG reduction benefits, in which case the term 'co-benefits' is more appropriate*".

[2] That is why Mayrhofer and Gupta (2016b) call co-benefits an "'umbrella' concept".

[3] Younger et al. (2008) discuss "built environment strategies" (focusing not only on the component transportation, but also regarding buildings and land use) that affect climate change and health outcomes. Such a strategy in the transport sector would e.g. be the promotion of telecommuting.

to the possibility that certain climate protection activities may trigger a decrease in employment.

During the 1990s, ancillary benefits studies were mainly of empirical nature and assessed the magnitude of such benefits. Ekins (1996) reviews the early estimates of the size of ancillary benefits. Some papers also discussed policy implications, e.g. Heintz and Tol (1996) consider the implications in the context of climate finance. Yet, the examination of policy and strategic implications of ancillary benefits and the research on the theory behind this concept have received quite little consideration in the literature and this does not only hold for this period of the 1990s. As Mayrhofer and Gupta (2016b) find out more recently, articles on the theory behind the co-benefits concept are underrepresented (merely 5% of the literature they reviewed) and only 20% of the papers address ancillary benefits in a qualitative way.

In 2000, the OECD made efforts to promote research on ancillary benefits and published a seminal proceedings book including contributions by leading experts in this field (OECD 2000). The objective of this publication project was to serve as a reference to those scientists *"seeking to develop integrated policies to meet a range of policy objectives simultaneously"* (OECD 2000). The scientific literature on ancillary benefits has significantly grown since then and benefitted from the concepts conveyed by the OECD book project.

In the past 20 years, scientists from different disciplines not only prepared numerous empirical studies considering different ancillary effects in different world regions but also subjected theoretical aspects to closer inspection. In their study on carbon offsetting and climate-friendly consumption, Schwirplies and Ziegler (2016) interpret warm-glow (see Andreoni 1990) as a co-effect of climate protection. In doing so, they address both theoretical as well as empirical aspects. Buchholz and Sandler (2017) also take into account "psychological" co-effects and investigate their influence on the effects of climate-friendly leadership of a country. Buchholz et al. (2014) consider social esteem as a co-benefit of public good provision (like the provision of climate protection) where the level of social esteem and thus of the co-benefits is influenced by a non-governmental agency. Bahn and Leach (2008) and Pittel and Rübbelke (2017) model ancillary effects in dynamic settings. Harlan and Ruddell (2011) consider both ancillary benefits of policies to mitigate and to adapt to climate change. For an experimental approach to investigate both global and local co-benefits of climate change mitigation efforts, see Löschel et al. (2018).

Strategic implications of ancillary benefits at the international level also increasingly caught researchers' attention (see e.g. Finus and Rübbelke 2013). Ancillary benefits have in many cases properties of a private good for climate-protecting countries, which may (favourably) affect strategic behaviour of countries in global public good provision (Pittel and Rübbelke 2008). Furthermore, if one takes "private" ancillary benefits of climate change mitigation into account, then the effects of climate finance and transfers may differ from those that would arise in the case where only pure public effects of mitigation are considered. So the famous Warr neutrality (Warr 1983) of income redistribution tends not to hold anymore. Altemeyer-Bartscher et al. (2014) consider international side-payments to raise global climate protection levels while taking into account the positive impact that ancillary effects may exert on the scope for financing such side-payments. Pittel and Rübbelke (2013) investigate in particular the potential effects of adaptation transfers where one could regard induced fairness improvements as a co-effect of such transfers, which in turn tend to influence countries' behaviour in international negotiations.

Flachsland et al. (2009) consider distributional aspects related to ancillary benefits that arise by linking cap-and-trade systems. They argue that outsourcing of climate protection activities to other regions will bring about a loss of ancillary benefits in those regions that outsource their abatement activities. Similarly, Krook-Riekkola et al. (2011: 4992) point out for the case of climate policy in Sweden that "*an increase in emission reductions abroad also implies a lost opportunity of achieving important welfare gains from the reductions of a number of regional and local environmental pollutants*".

There are also instances of global co-effect generation by global environmental policies, i.e. a generation of effects that do not exclusively benefit the region hosting an environmental protection project. Velders et al. (2012) give the example of the Montreal Protocol that controls ozone-depleting substances. In doing so, it helps to stop the depletion of the stratospheric ozone layer and with it protects the planet against dangerous solar ultraviolet radiation. As most ozone-depleting substances are also potent GHGs, the Protocol serves a twin goal, i.e. combating the global threats of stratospheric ozone layer depletion and of climate change. Thus, it produces two global characteristics jointly. Velders et al. (2012) discuss how the substantial climate benefits of the Montreal Protocol can be preserved in the future.

In accordance with Lancaster (1971: 1), who stresses that "*[g]ood theory should be as universal as possible*", Ürge-Vorsatz et al. (2014) recommend the analysis of co-effects in a multiple-objective/multiple-impact framework. Hence, one should capture all relevant effects in order to incorporate related benefits and costs into decision-making schemes. According to Ürge-Vorsatz et al. (2014), this can be achieved not only by standard cost–benefit analysis but also by integrated assessment models or by a multi-criteria analysis. Creutzig et al. (2012) conduct a multi-criteria analysis of co-benefits enjoyed from decarbonising urban transport in European cities. They consider this kind of assessment to be a useful complement to cost–benefit analysis of climate change mitigation. Dubash et al. (2013) apply the multi-criteria technique to co-benefits of Indian climate policy, while de Bruin et al. (2009) examine options to adapt to climate change in The Netherlands by this method. West et al. (2013) investigate co-effects of global climate change mitigation on future air quality and human health in an integrated assessment framework. Oxley et al. (2012) use such a framework to examine co-benefits in the UK's road transport sector. Cost–benefit analysis of ancillary benefits is extensively conducted,[4] e.g. by Bollen et al. (2009) who—however—suggest in their paper that climate change mitigation is an ancillary benefit of air pollution policy (as they

---

[4]Longo et al. (2012) use the contingent valuation method to elicit the WTP for ancillary and global benefits of climate change mitigation policies.

find in their analysis framework higher benefits from local air pollution control than from greenhouse gas control). Li (2006) uses a multi-period cost–benefit framework in order to examine the economic and social welfare implications of local health benefits of climate change mitigation in Thailand.

From a different methodological perspective, Mayrhofer and Gupta (2016a) argue that in co-benefit research, *"the institutional context and choice architectures in which the idea of co-benefits is embedded"* should get more attention in the future as these aspects have been largely ignored so far. In order to overcome these short-comings, these authors make use of the framework of discursive institutionalism and draw on the concept of storylines.

Even though the description above is far from complete, it gives an idea of the multiplicity and variety of research topics in the context of ancillary benefits and of developments in this scientific field.

In this book, drawing on these past contributions, some more recent research efforts on ancillary benefits of climate policy will be presented.

The book is organised such that in the *first part* we address ancillary benefits derived from development co-effects. The first two chapters in this part are concerned with the links between ancillary benefits and sustainable development in particular against the background of the Paris Agreement on Climate Change. The third chapter takes a more theoretical stand in its investigation of effects of climate policy on long-term economic growth.

Campagnolo and De Cian (2019) raise the question whether the Paris Agreement on Climate Change can support the attainment of the Sustainable Development Goals (SDGs). In general, there can be synergies and trade-offs between climate policy and sustainable development. The authors evaluate the impact of the Paris Agreement implementation on a set of SDG indicators by 2030 using a computable general equilibrium model. Their methodology of combining empirically based relationships between macroeconomic variables and sustainability indicators makes it possible to explore the sustainability consequences of mitigation policy on 16 SDGs.

Michaelowa et al. (2019) focus in their analysis on sustainable development co-benefits of activities for climate change mitigation under the market mechanisms of the Paris Agreement. In the past, there have been criticisms regarding limited co-benefits of projects initiated via the Clean Development Mechanism (CDM) of the Kyoto Protocol. In response to this, the CDM regulators provided a voluntary tool for sustainable development benefit assessment. For the new market mechanisms under Article 6 of the Paris Agreement, the rules currently negotiated foresee a continuation of the voluntary approach. Nevertheless, the rising political concern regarding sustainable development co-benefits is likely to trigger efforts by buyers of emission credits to ensure significant co-benefits of the underlying activities.

In his chapter, Golub (2019) explores the transition of low- or middle-income economies from carbon intensive technological structures to cleaner structures. In this context, development traps are a hazard but as the author argues, timely climate policy may facilitate the transition and positive ancillary development effects could be induced. The analysis takes into account that different forms of uncertainties may arise.

Conceptual and theoretical approaches are in the focus of the *second part* of this book.

In a game-theoretic framework, Takashima (2019) presents new theoretical findings concerning the effects of ancillary benefits on international environmental agreements reached by two types of asymmetric countries. As this analysis shows, ancillary benefits are a key factor for enhancing the feasibility of sustained, full participation in an international environmental agreement.

Brumme et al. (2019a) elaborate on the adequacy of the impure public good approach developed by Cornes and Sandler (1984) for analysing ancillary and primary benefits. They describe how this theoretical approach evolved and outline its key concepts and techniques. As the authors argue, the analysis of climate policy in this framework is fruitful, but also quite intricate. Therefore, they stress the importance of developing new tools to facilitate the analysis and the interpretation of results. They see one way for doing so in the application of the aggregative game approach developed by Cornes and Hartley (see e.g. Cornes and Hartley 2007; Cornes 2016; Dickson 2017; Cornes et al. 2019).

Schwirplies (2019) refers to the impure public good framework in her review of literature on climate policy co-effects like monetary savings, internal satisfaction, health benefits, and fairness enhancement. In her study, she examines whether actors from the public and private sector should lay more emphasis on ancillary benefits when promoting climate protection measures.

Brumme et al. (2019b) employ the aggregative game approach in order to trace back the impure public good model to the conventional pure public good model and thereby illustrate a way to facilitate the application of the impure public good approach to the analysis of climate policy. In this chapter, differences between the impure and the pure public good model become evident, e.g. the emergence of non-contributors in the Cournot–Nash equilibrium is less likely in the impure than in the pure public good model.

Vögele et al. (2019) use a multi-criteria approach to analyse ancillary benefits of decarbonising the transport sector. They employ the example of e-mobility and consider a broad range of relevant beneficial and adverse factors. From their findings, they draw the conclusion that due to ancillary effects certain stakeholders will exert resistance if they are pushed towards e-mobility.

In the *third part* of this book, we take stock of some novel developments in the research on ancillary benefits like the examination of ancillary benefits of adaptation to climate change or of carbon capture and storage. Furthermore, scientific literature on climate policy effects in specific sectors like forestry, health and international shipping is surveyed.

The first two chapters in this part are concerned with ancillary benefits and costs of adaptation to climate change.

Sainz de Murieta (2019), after a brief discussion of ancillary benefits of mitigation, examines different groups of ancillary effects of adaptation to climate change. In her analysis, she distinguishes between three categories of ancillary benefits:

economic, social and environmental co-benefits of adaptation. Furthermore, she takes into account that adaptation can also involve ancillary costs. The author points out that co-benefits of adaptation are underrepresented in the scientific literature compared to those of mitigation. Furthermore, she argues that there is little evidence about the extent to which ancillary benefits in general are incorporated into decision-making processes.

Reif and Osberghaus (2019) provide a synthesis of the literature on beneficial and harmful impacts of measures to adapt to climate change. The authors discuss the ancillary effects of adaptation in five different fields, i.e. mitigation, ecological systems, economic development, decisions under uncertainty, and disaster resilience. The difficulties to assess co-benefits and conflicts of adaptation, which arise due to the complex interlinkages between the domains, uncertainties and unquantifiable values, are illustrated.

Torvanger (2019) reviews the literature on ancillary benefits of carbon capture and storage (CCS). The author categorises different key groups of ancillary benefits of CCS. Among these are the use of carbon dioxide for synthetic fuels, chemicals, plastics, and building materials. Learning from CCS applied to industry and fossil-based power sources can enable two negative emission technologies: bioenergy combined with CCS, and "Direct Air Capture".

Markandya and Sampedro (2019) assess present action on reducing greenhouse gases to be sub-optimally low. They argue that the presence of ancillary benefits has not played a major role in driving the discussion on raising climate protection levels. In their chapter, they take a closer look at the reasons for this and offer some suggestions on ways to give health co-benefits a greater role in climate policy.

Seroa da Motta et al. (2019) review estimates of the ancillary benefits from land use mitigation options in Brazil. Furthermore, they present an integrated REDD+ framework that catalyses the transfer of financial resources to the land use sector, while ensuring that non-REDD+ options continue to receive financial resources for the innovation and decarbonisation of activities in different economic sectors (energy, industry and transport). As the authors argue, ancillary benefits will promote national and regional economies in supporting water flows regarding urban and rural supply and hydroelectricity.

Doundoulakis and Papaefthimiou (2019) review ancillary benefits of climate policies in the shipping sector. A special feature of the international shipping sector is that emissions cannot be attributed to an individual country, which is due to the global nature of activity and the complexity of shipping operations. The chapter considers different regulations concerning air pollution caused by the shipping sector and discusses potential effects. While doing so, they illustrate and categorise potential co-benefits that arise from the current regulatory framework of the international shipping sector.

The *fourth part* of this volume is particularly concerned with ancillary benefits in urban areas.

Liu et al. (2019) see the combat of global warming and regional air pollution control as two major challenges for China's environmental policy. They conduct an empirical analysis of co-benefits of carbon dioxide emissions mitigation and air pollution control in 30 provincial capital cities in China. From the results of this study, they derive policy recommendations for haze and carbon reduction in China's urban areas.

Sethi (2019) addresses the low performance of cities in emerging economies concerning parameters of social development, equity, functional autonomy and financial capacity. As he argues, the co-benefits approach has proved to be a key analysis mechanism and he stresses the importance of assessing the magnitude of such benefits in emerging economies. The author discusses assessment tools, lessons learned in the past from research on ancillary benefits and remaining knowledge gaps. In case studies, the applicability of urban co-benefits concepts in India, China, Brazil and Turkey are tested.

Mathy et al. (2019) present and describe a specific project that strives for identifying measures to reduce significant atmospheric pollution in cities and its impacts. Synergies between short-term public health issues related to air pollution and the reduction of greenhouse gas emissions play an important role in this context. As the authors argue, the key tasks of the project are to better comprehend population's exposure to pollution, improve the understanding of the determinants of mobility behaviour and support public decision-making.

References

Altemeyer-Bartscher M, Markandya A, Rübbelke D (2014) International side-payments to improve global public good provision when transfers are refinanced through a tax on local and global externalities. Int Econ J 28:71–93

Andreoni J (1990) Impure altruism and donations to public goods: a theory of warm-glow giving. Econ J 100:464

Ayres RU, Walter J (1991) The greenhouse effect: damages, costs and abatement. Environ Resour Econ 1:237–270

Bahn O, Leach A (2008) The secondary benefits of climate change mitigation: an overlapping generations approach. Comput Manag Sci 5:233–257

Bain PG, Milfont TL, Kashima Y, Bilewicz M, Doron G, Garðarsdóttir RB, Gouveia VV, Guan Y, Johansson L-O, Pasquali C, Corral-Verdugo V, Aragones JI, Utsugi A, Demarque C, Otto S, Park J, Soland M, Steg L, González R, Lebedeva N, Madsen OJ, Wagner C, Akotia CS, Kurz T, Saiz JL, Schultz PW, Einarsdóttir G, Saviolidis NM (2016) Co-benefits of addressing climate change can motivate action around the world. Nat Clim Chang 6:154–157

Bollen J, van der Zwaan B, Brink C, Eerens H (2009) Local air pollution and global climate change: a combined cost-benefit analysis. Resour Energy Econ 31:161–181

Brumme A, Buchholz W, Rübbelke D (2019a) Impure public good models as a tool to analyze the provision of ancillary and primary benefits. In: Buchholz W, Markandya A, Rübbelke D, Vögele S (eds) Ancillary benefits of climate policy—new theoretical developments and empirical findings. Springer, Cham

Brumme A, Buchholz W, Rübbelke D (2019b) Impure public goods and the aggregative game approach. In: Buchholz W, Markandya A, Rübbelke D, Vögele S (eds) Ancillary benefits of climate policy—new theoretical developments and empirical findings. Springer, Cham

Buchholz W, Sandler T (2017) Successful leadership in global public good provision: incorporating behavioural approaches. Environ Resour Econ 67:591–607

Buchholz W, Falkinger J, Rübbelke D (2014) Non-governmental public norm enforcement in large societies as a two-stage game of voluntary public good provision. J Public Econ Theory 16:899–916

Burtraw D, Krupnick A, Palmer K, Paul A, Toman M, Bloyd C (2003) Ancillary benefits of reduced air pollution in the US from moderate greenhouse gas mitigation policies in the electricity sector. J Environ Econ Manag 45:650–673

Campagnolo L, De Cian E (2019) Can the Paris agreement support achieving the sustainable development goals? In: Buchholz W, Markandya A, Rübbelke D, Vögele S (eds) Ancillary benefits of climate policy—new theoretical developments and empirical findings. Springer, Cham

Cornes R (2016) Aggregative environmental games. Environ Resour Econ 63:339–365

Cornes R, Hartley R (2007) Aggregative public good games. J Public Econ Theory 9:201–219

Cornes R, Sandler T (1984) Easy riders, joint production, and public goods. Econ J 94:580–598

Cornes R, Hartley R, Tamura Y (2019) Two-aggregate games: demonstration using a production–appropriation model. Scand J Econ 121:353–378

Creutzig F, Mühlhoff R, Römer J (2012) Decarbonizing urban transport in European cities: four cases show possibly high co-benefits. Environ Res Lett 7:44042

de Bruin K, Dellink RB, Ruijs A, Bolwidt L, van Buuren A, Graveland J, de GRS, Kuikman PJ, Reinhard S, Roetter RP, Tassone VC, Verhagen A, van Ierland EC (2009) Adapting to climate change in the Netherlands: an inventory of climate adaptation options and ranking of alternatives. Clim Chang 95:23–45

Dickson A (2017) Multiple-aggregate games. In: Buchholz W, Rübbelke D (eds) The theory of externalities and public goods: essays in memory of Richard C. Cornes. Springer, Cham, pp 29–59

Doundoulakis E, Papaefthimiou S (2019) Ancillary benefits of climate policies in the shipping sector. In: Buchholz W, Markandya A, Rübbelke D, Vögele S (eds) Ancillary benefits of climate policy—new theoretical developments and empirical findings. Springer, Cham

Dubash NK, Raghunandan D, Sant G, Sreenivas A (2013) Indian climate change policy: exploring a co-benefits based approach. Econ Polit Wkly 48:47–61

Ekins P (1996) The secondary benefits of CO_2 abatement: how much emission reduction do they justify? Ecol Econ 16:13–24

Finus M, Rübbelke D (2013) Public good provision and ancillary benefits: the case of climate agreements. Environ Resour Econ 56:211–226

Flachsland C, Marschinski R, Edenhofer O (2009) To link or not to link: benefits and disadvantages of linking cap-and-trade systems. Clim Pol 9:358–372

Golub A (2019) Technological transition and carbon constraints under uncertainty. In: Buchholz W, Markandya A, Rübbelke D, Vögele S (eds) Ancillary benefits of climate policy—new theoretical developments and empirical findings. Springer, Cham

Halsnaes K (2002) A review of the literature on climate change and sustainable development. In: Markandya A, Halsnaes K (eds) Climate change and sustainable development: prospects for developing countries. Earthscan, London, pp 49–72

Harlan SL, Ruddell DM (2011) Climate change and health in cities: impacts of heat and air pollution and potential co-benefits from mitigation and adaptation. Curr Opin Environ Sustain 3:126–134

Heintz RJ, Tol RS (1996) Secondary benefits of climate control policies: implications for the global environment facility. CSERGE GEC working paper 96-17, London

Krook-Riekkola A, Ahlgren EO, Söderholm P (2011) Ancillary benefits of climate policy in a small open economy: the case of Sweden. Energy Policy 39:4985–4998

Krupnick A, Burtraw D, Markandya A (2000) The ancillary benefits and costs of climate change mitigation: a conceptual framework. In: OECD (ed) Ancillary benefits and costs of greenhouse gas mitigation. OECD, Paris, pp 53–93

Lancaster K (1971) Consumer demand: a new approach. Columbia University Press, New York and London

Li JC (2006) A multi-period analysis of a carbon tax including local health feedback: an application to Thailand. Environ Dev Econ 11:317–342

Liu J, Pan J, Liu Z, Jiao S (2019) Co-benefits of climate change mitigation and pollution reduction in China's urban areas. In: Buchholz W, Markandya A, Rübbelke D, Vögele S (eds) Ancillary benefits of climate policy—new theoretical developments and empirical findings. Springer, Cham

Longo A, Hoyos D, Markandya A (2012) Willingness to pay for ancillary benefits of climate change mitigation. Environ Resour Econ 51:119–140

Löschel A, Pei J, Sturm B, Wang R, Buchholz W, Zhao Z (2018) The demand for global and local environmental protection—experimental evidence from climate change mitigation in Beijing. ZEW discussion papers 18-017, Mannheim

Markandya A, Rübbelke D (2004) Ancillary benefits of climate policy. Jahrb Natl Stat 224:488–503

Markandya A, Sampedro J (2019) Health co-benefits of climate mitigation policies—why is it so hard to convince policy-makers of them and what can be done to change that? In: Buchholz W, Markandya A, Rübbelke D, Vögele S (eds) Ancillary benefits of climate policy—new theoretical developments and empirical findings. Springer, Cham

Markandya A, Wilkinson P (2007) Electricity generation and health. Lancet 370:979–990

Markandya A, Sampedro J, Smith SJ, van Dingenen R, Pizarro-Irizar C, Arto I, González-Eguino M (2018) Health co-benefits from air pollution and mitigation costs of the Paris agreement: a modelling study. Lancet Planet Health 2:e126–e133

Mathy S, Bouscasse H, Chardonnel S, Chalabaëv A, Gabet S, Treibich C, Slama R (2019) Protocol for an interdisciplinary and multidimensional assessment of pollution reduction measures in urban areas. In: Buchholz W, Markandya A, Rübbelke D, Vögele S (eds) Ancillary benefits of climate policy—new theoretical developments and empirical findings. Springer, Cham

Mayrhofer J, Gupta J (2016a) The politics of co-benefits in India's energy sector. Environ Plann C Gov Policy 34:1344–1363

Mayrhofer J, Gupta J (2016b) The science and politics of co-benefits in climate policy. Environ Sci Pol 57:22–30

Michaelowa A, Espelage A, Hoch S (2019) Co-benefits under the market mechanisms of the Paris agreement. In: Buchholz W, Markandya A, Rübbelke D, Vögele S (eds) Ancillary benefits of climate policy—new theoretical developments and empirical findings. Springer, Cham

Nemet GF, Holloway T, Meier P (2010) Implications of incorporating air-quality co-benefits into climate change policymaking. Environ Res Lett 5:14007

OECD (2000) Ancillary benefits and costs of greenhouse gas mitigation. OECD, Paris

Oxley T, Elshkaki A, Kwiatkowski L, Castillo A, Scarbrough T, ApSimon H (2012) Pollution abatement from road transport: cross-sectoral implications, climate co-benefits and behavioural change. Environ Sci Pol 19–20:16–32

Pearce D (1992) Secondary benefits of greenhouse gas control. CSERGE working paper 92-12, London

Pittel K, Rübbelke D (2008) Climate policy and ancillary benefits: a survey and integration into the modelling of international negotiations on climate change. Ecol Econ 68:210–220

Pittel K, Rübbelke D (2013) International climate finance and its influence on fairness and policy. World Econ 36:419–436

Pittel K, Rübbelke D (2017) Thinking local but acting global? The interplay between local and global internalization of externalities. In: Buchholz W, Rübbelke D (eds) The theory of externalities and public goods: essays in memory of Richard C. Cornes. Springer, Cham, pp 271–297

Reif C, Osberghaus D (2019) Economic assessment of co-benefits of adaptation to climate change. In: Buchholz W, Markandya A, Rübbelke D, Vögele S (eds) Ancillary benefits of climate policy—new theoretical developments and empirical findings. Springer, Cham

Rive N, Rübbelke D (2010) International environmental policy and poverty alleviation. Rev World Econ 146:515–543

Sainz de Murieta E (2019) Ancillary benefits of adaptation: an overview. In: Buchholz W, Markandya A, Rübbelke D, Vögele S (eds) Ancillary benefits of climate policy—new theoretical developments and empirical findings. Springer, Cham

Schwirplies C (2019) Private ancillary benefits in a joint production framework. In: Buchholz W, Markandya A, Rübbelke D, Vögele S (eds) Ancillary benefits of climate policy—new theoretical developments and empirical findings. Springer, Cham

Schwirplies C, Ziegler A (2016) Offset carbon emissions or pay a price premium for avoiding them? A cross-country analysis of motives for climate protection activities. Appl Econ 48:746–758

Seroa da Motta R, Moura Costa P, Cenamo M, Soares P, Viana V, Salviati V, Bernasconi P, Thuault A, Ribeiro P (2019) Financing forest protection with integrated REDD+ markets in Brazil. In: Buchholz W, Markandya A, Rübbelke D, Vögele S (eds) Ancillary benefits of climate policy—new theoretical developments and empirical findings. Springer, Cham

Sethi M (2019) Climate co-benefits in rapidly urbanizing emerging economies: scientific and policy imperatives. In: Buchholz W, Markandya A, Rübbelke D, Vögele S (eds) Ancillary benefits of climate policy—new theoretical developments and empirical findings. Springer, Cham

Takashima N (2019) Sustainable international cooperation with ancillary benefits of climate policy. In: Buchholz W, Markandya A, Rübbelke D, Vögele S (eds) Ancillary benefits of climate policy—new theoretical developments and empirical findings. Springer, Cham

Torvanger A (2019) Ancillary benefits of carbon capture and storage. In: Buchholz W, Markandya A, Rübbelke D, Vögele S (eds) Ancillary benefits of climate policy—new theoretical developments and empirical findings. Springer, Cham

Ürge-Vorsatz D, Herrero ST, Dubash NK, Lecocq F (2014) Measuring the co-benefits of climate change mitigation. Annu Rev Environ Resour 39:549–582

Velders GJM, Ravishankara AR, Miller MK, Molina MJ, Alcamo J, Daniel JS, Fahey DW, Montzka SA, Reimann S (2012) Climate change. Preserving Montreal protocol climate benefits by limiting HFCs. Science 335:922–923

Vögele S, Ball C, Kuckshinrichs W (2019) Multi-criteria approaches to ancillary effects: the example of e-mobility. In: Buchholz W, Markandya A, Rübbelke D, Vögele S (eds) Ancillary benefits of climate policy—new theoretical developments and empirical findings. Springer, Cham

Warr PG (1983) The private provision of a public good is independent of the distribution of income. Econ Lett 13:207–211

West JJ, Smith SJ, Silva RA, Naik V, Zhang Y, Adelman Z, Fry MM, Anenberg S, Horowitz LW, Lamarque J-F (2013) Co-benefits of global greenhouse gas mitigation for future air quality and human health. Nat Clim Chang 3:885–889

Younger M, Morrow-Almeida HR, Vindigni SM, Dannenberg AL (2008) The built environment, climate change, and health: opportunities for co-benefits. Am J Prev Med 35:517–526

Part I
Ancillary Benefits and Development Co-effects

Can the Paris Agreement Support Achieving the Sustainable Development Goals?

Lorenza Campagnolo and Enrica De Cian

1 Introduction

With the advent of the United Nations' 2030 Agenda and the Paris Agreement in 2015 (United Nations (UN) 2015), a growing number of studies have been exploring the synergies and trade-offs between climate policy and sustainable development. Synergies and trade-offs can go in both directions. On the one hand, the mitigation literature in the context of the new scenario framework of the Shared Socioeconomic Pathways (SSPs) and Representative Concentration Pathways (RCPs, O'Neill et al. 2017; van Vuuren et al. 2014) highlights how deep decarbonization (Rogelj et al. 2019) can be achieved more easily under sustainable scenarios, such as the SSP1 narrative, which poses lower challenges to mitigation and adaptation. On the other hand, climate mitigation policies can generate a wide range of non-climate ancillary benefits or obstacles in achieving the Sustainable Development Goals (SDGs, Roy et al. 2019). Aligning mitigation policies with SDGs is key for ensuring social acceptability of the required structural transformation and for fostering the more ambitious action required to contain global warming below 1.5 °C in 2100.

This chapter contributes to the emerging literature on the synergies and trade-offs between mitigation and sustainable development by evaluating the impact of the Paris Agreement implementation on a set of SDG indicators by 2030 using a Computable General Equilibrium (CGE) model. A macroeconomic framework provides a system perspective analysis, highlighting the aggregate impacts of mitigation policy on multiple sustainable development dimensions at the same time, while

---

L. Campagnolo (✉) · E. De Cian
Ca'Foscari University of Venice, Venice, Italy

Euro-Mediterranean Center on Climate Change, Venice, Italy

RFF-CMCC European Institute on Economics and the Environment, Venice, Italy
e-mail: lorenza.campagnolo@unive.it

© Springer Nature Switzerland AG 2020
W. Buchholz et al. (eds.), *Ancillary Benefits of Climate Policy*, Springer Climate,
https://doi.org/10.1007/978-3-030-30978-7_2

taking into account the general equilibrium adjustments induced by price changes. Ex-ante assessments, such as those based on simulation or numerical models, make it possible to explore the implications of mitigation policies of different ambition, broadening the evidence beyond the policies actually implemented in the past. They can examine synergies and trade-offs into the future, and provide a benchmark for policy evaluation and design while accounting for policy and socioeconomic uncertainty. This chapter develops projections of selected SDG indicators in a reference and mitigation policy scenario, contributing to expand the existing literature on mitigation pathways in the context of sustainable development. The major limitation of the few existing integrated assessment approaches available to date is the focus on economic and technological indicators, a choice that is driven by the limited ability of quantitative models to represent the social dimensions of sustainable development (McCollum et al. 2018b; von Stechow et al. 2016). The method presented in this chapter combines regression analysis to estimate empirically-based relationships between 16 economic, social, and environmental SDG indicators and the key socioeconomic variables represented in the CGE model. Gender inequality is the only goal left unexplored (SDG5).

The remainder of the chapter is organised as follows: Section 2 synthesises the most recent literature on the mitigation co-benefits and side effects on sustainable development. Section 3 describes the ex-ante approach used to assess the SDG implications of the Paris Agreement. Section 4 discusses the advantages and limitations of our methodology in the sustainability assessment of policy implementation and concludes, highlighting some directions for future research concerning the co-benefits of mitigation and adaptation policies.

2 Mitigation Policy and Sustainable Development: Recent Contributions from the Literature

The SDGs define broad and ambitious development targets for both developed and developing countries encompassing all sustainability dimensions (economic, social, and environmental), including minimising climate change impacts (SDG13), with the ambition of informing pathways towards inclusive green growth. The tight linkage among the economic, social, and environmental dimensions is reflected in the connections across different goals integrated into the broader framework. Given the multiple interactions among different SDGs, integrated approaches, such as those based on Integrated Assessment Models (IAMs) or integrated energy-economy climate models, can quantify the synergies and trade-offs between target-specific policies, such as mitigation, and all other goals with a system perspective (von Stechow et al. 2016, 2015).

Despite the growing number of efforts, current integrated modelling research remains confined to sectoral studies offering a limited view on the possible co-effects and focusing on a narrow set of specific objectives. Most of the literature, recently reviewed in the IPCC Special Report 1.5, has focused on food security and hunger (SDG2), air pollution and health (SDG3), clean energy for all (SDG7), water security (SDG6). Only McCollum et al. (2018a) conduct a systematic review of the literature to evaluate the nature and strength of interactions between SDG7 and all other SDGs. The review relies on forward-looking, quantitative scenario studies focusing on multiple objectives. SDG7 is connected to the implementation of mitigation policies through the specific targets on access to modern energy services, increased share of renewables and improved energy intensity. Since these targets are basic requirements of any mitigation policy, McCollum et al. (2018a) indirectly shed some light on the interaction between mitigation policy and SDGs. It is interesting to note that, the model-based literature reviewed in the paper is not able to identify contributions assessing social indicators (such as poverty SDG1, education SDG4, gender equality SDG5, reduced inequalities SDG10). In order to provide some evidence on these dimensions, McCollum et al. (2018a) select historical, empirical, or case-study papers.

The social indicators for which most evidence is found are SDG2 and SDG3. Regarding SDG3, good health and well-being, most literature focuses on reduced air pollution (Rao et al. 2016; Markandya et al. 2018) and diminished impacts of climate change and environmental degradation (Ebi et al. 2018). Mitigation policy stimulates the development and the diffusion of renewable technologies that appear decisive in improving energy access especially in remote and not connected areas (McCollum et al. 2018a). Regarding SDG2 (undernutrition reduction), the literature on the impacts of uncontrolled emission growth and temperature rise on agricultural production and on undernutrition prevalence is wide (Hasegawa et al. 2016; Nelson et al. 2010; Lloyd et al. 2011). Achieving mitigation targets helps reducing these side effects, but at the same time can generate some trade-offs pushing large-scale deployment of bio-energy, competition for land, and increased food prices. These are trade-offs that can be mitigated by decarbonization strategies oriented more towards demand-side actions (Grubler et al. 2018) or through the adoption of complementary distributional policies. The literature on the link between mitigation and poverty (SDG1) and inequality (SDG10) reduction is also quite scattered. On the one hand, as in the case of SDG2, poor people are the most exposed to climate change impacts that can be 70% higher for the bottom 40% of the population than for the average (Hallegatte and Rozenberg 2017). Therefore, mitigation can have a pro-poor and equalising effect. On the other hand, emission cuts, by setting a price on carbon, can have regressive implications if an adequate revenue recycling scheme supporting the poorest layers of the population is not predisposed (Hassett et al. 2009; Metcalf 1999). The social dimension of SDG7, achieving universal energy access, can also be hindered by a mitigation policy that increases energy prices in fossil fuel-intensive countries and burdens poor households. At the same time, the efficiency improvements, especially of renewable energy technologies, combined with pro-poor incentives can reduce this tarde-off (Dagnachew et al. 2018; Jakob and Steckel 2014). Direct effects of mitigation policy on SDG4 (quality of education) and SDG16 (preserve peace) have not been explored, though the literature on the link between global warming and conflicts is growing (Hsiang et al. 2011).

With respect to the economic indicators (SDGs 8, 9, 17), a broad literature on the interaction between technology and environmental externalities (Carraro et al. 2010) highlights the positive impacts of climate policy on innovation and technology diffusion (SDG8, decent work and economic growth). With respect to employment opportunities the evidence is mixed. Green jobs are mostly high-skill, entail higher wages, and tend to be concentrated in high-tech areas (Vona et al. 2018b). Although there are distributional implications, impacts on overall employment seem to be modest (Vona et al. 2018a). Despite the multiple channels through which mitigation policy can stimulate growth (Hallegatte et al. 2012), the IAM-based mitigation literature highlights the macroeconomic costs of stringent mitigation actions, mostly due to early retirement of capital, higher energy costs for producers and consumers, terms of trade effects (Paltsev and Capros 2013). The regional distribution of impacts on economic performance can also be expected to be uneven, mostly due to terms-of-trade effects, which would penalise net exporters and work in favour of net energy importers. In developing countries prioritising poverty-related issues, emission costs could divert funds necessary to development policies.

Even mitigation with a compensatory scheme by industrialised countries can lead to a "climate finance curse", sluggish investments and technological change in energy-intensive sectors and, ultimately, slower economic growth (Jakob and Steckel 2014). Regarding SDG17, the IPCC 1.5 report highlights that the diffusion of new technologies related to decarbonization strategies requires transnational capacity building and knowledge sharing and could contribute to international partnership (Roy et al. 2019). Impacts on industry, innovation, and infrastructure (SDG9) are mixed and sector-specific, with a tendency to penalise energy-intensive sectors and infrastructure. Transforming the industrial sector towards a renewable-based and more efficient system aligns with the goal of upgrading energy infrastructure and making the energy industry more sustainable (McCollum et al. 2018b).

With respect to the environmental indicators, there is strong positive interaction between mitigation and SDG11, sustainable cities and infrastructure. This is driven by the multiple co-benefits of the behavioural and technological transformations mitigation policy might induce. According to Reis et al. (2018), meeting the 1.5 °C policy target may limit spikes of pollutant concentration (except PM2.5) above the safe thresholds in all countries. Furthermore, mitigation commitments might stimulate the development of renewable energy technologies and energy-efficient urban infrastructure solutions boosting urban environmental sustainability by further improving air quality, reducing noise and energy expenditure (McCollum et al. 2018a).

A strong positive interaction with high agreement and confidence is also found with water availability and quality (SDG6), natural resource protection (SDG12) through the reduced depletion of several natural resources, life below water (SDG14) through the reduced risk of ocean acidification, life on land (SDG 15) through reduced deforestation, though some weak trade-offs are also found especially for SDG 14 and 15 (McCollum et al. 2018a). The scaling up of renewable energy would lower the water demand for energy (e.g. for cooling power plants), though some specific options (e.g. hydropower) could induce trade-offs through competition for water use. A mitigation pathway that more strongly relies on bio-energy might have higher requirements in terms of water for irrigation, reducing availability for other sectors.

To conclude, the existing literature seems to suggest that the degree of competition between mitigation objectives and sustainable development depends on the type of transition pathway adopted. While energy supply or land and ocean mitigation options tend to entail a larger number of trade-offs and risks, demand-side measures can significantly reduce the risks associated with mitigation policies, as they tend to bring about a larger set of co-benefits. Yet, actual synergies and trade-off will be unevenly distributed across regions and nations (Roy et al. 2019).

3 An Ex-Ante Assessment of the Paris Agreement

3.1 Framework Description

The Aggregated Sustainable Development goals Index (ASDI) framework developed in this chapter aims at offering a comprehensive assessment of current well-being and future sustainability based on 27 indicators related to 16 Sustainable Development Goals.[1] As describe in Fig. 1, ASDI combines an empirical, regression approach based on historical data (grey) with a modelling, future-oriented framework (black) to offer an internally-consistent set-up that makes it possible to analyse future patterns of sustainability indicators and their inter-linkages.

The **selection of the SDG indicators** was informed by the work of the UN Inter-agency Expert Group on SDG Indicators (United Nations (UN) 2017a), which listed 232 indicators to be used in assessing SDGs, and follows these guidelines: (1) relevance for the SDG they refer to, (2) connection with one of the SDG Targets, (3) sufficient data coverage for each country, (4) linkage to the macroeconomic variables that are output of the model. These are the main constraints on indicator selection of any systemic and multi-approach analysis of Agenda 2030 (von Stechow et al. 2016), including the ASDI framework here described. On the one side, the global perspective of the proposed modelling exercise requires the broadest coverage of indicators, dismissing some promising indicators for which sufficient data coverage is not yet available for a large number of countries. On the other side, given the goal of generating future projections of the selected sustainable indicators, we have to exclude indicators that could not be linked to any of the model variable outcomes or not showing a significant correlation with them. For this reason, at the moment, our analysis does not cover SDG5 (gender equality). We were not able to find a robust relation linking a gender-related indicator to an endogenous

---

[1] SDG5 on gender inequality is not explored.

Fig. 1 ASDI framework

variable generated by the model. Table 1 lists the selected indicators and classifies them in the sustainability pillar they pertain to: economy (ECO), society (SOC), and environment (ENV). Among them, 16 are computed using model results, 7 requires regression analysis to be linked to them (SDG1, SDG2, SDG3a, SDG3b, SDG4, SDG7a, SDG10), and the remaining 4 are kept constant at historical levels (SDG14, SDG15a, SDG15c, SDG16).

The **collection of historical data** of indicators relies on several international databases (World Development Indicators (World Bank (WB) 2018), UN database (United Nations (UN) 2018), and World Income Inequality Database (WIID3.4) (United Nations (UN) 2017b)) and covers all available countries for the period 1990–2015. Historical data are used for initializing indicators in the base year of the model (2007) and for estimating the basic relationships between model's variables and indicators in the regression analysis phase.

The **regression analysis** phase makes it possible to obtain projections of those indicators not directly generated by the model: poverty headcount ratio (SDG1), undernutrition prevalence (SDG2), physician density (SDG3a), Healthy Life Expectancy (HALE) (SDG3b), literacy rate (SDG4), Palma ratio[2] (SDG10), and electricity access (SDG7a). Using independent cross-country panel regressions (reported in Annex I), we identify the historical correlation between indicators and some socioeconomic variables.[3] The selection of the relevant explanatory variables for each indicator is based on the existing literature. Regarding SDG1, poverty prevalence has a negative correlation with unequal income distribution and a positive one with average income per capita level (Ravallion 2001, 1997; Ravallion and Chen 1997). Undernourishment prevalence (SDG2) is reduced when economic conditions (Headey 2013; Heltberg 2009; Fumagalli et al. 2013) as well as food production (Headey 2013) improve, and when inequality goes down (Heltberg 2009). Physician density (SDG3a) has a positive relation with total health expenditure per capita and a negative one with private health expenditure share. The healthy life expectancy (SDG3b) increases with the level of population education (Gulis 2000), urbanisation (Bergh and Nilsson 2010), physician density (or more in general public expenditure in health) (Kabir 2008), electricity access (Youssef et al.

---

[2] The Palma Ratio is defined as the ratio of the top 10% of population's share of Gross National Income (GNI), divided by the poorest 40% of the population's share of GNI (Cobham et al. 2016).

[3] Our future sustainability scenarios are built under the assumption that the estimated relationships will hold also into the future up to 2030.

Table 1 ASDI indicators

SDG	ASDI indicator	Pillar	SDG	ASDI indicator	Pillar	SDG	ASDI indicator	Pillar
SDG1	Population below $1.90 (PPP) per day (%)	SOC	SDG8a	Annual GDP per capita growth (%)	ECO	SDG13a	Concentration of GHG emissions from AFOLU(tCO_2e/sq.km)	ENV
SDG2	Prevalence of undernourishment (%)	SOC	SDG8b	GDP per person employed ($PPP2011)	ECO	SDG13b	Compliance to Conditional NDCs (%)	ENV
SDG3a	Physician density (per 1000 population)	SOC	SDG8c	Employment-to-population ratio (%)	ECO	SDG13c	Gap from equitable and sustainable GHG emissions per capita in 2030 (tCO_2eq)	ENV
SDG3b	Healthy Life Expectancy (HALE) at birth (years)	SOC	SDG9a	Manufacturing value added (% of GDP)	ECO	SDG14	Marine protected areas (% of territorial waters)	ENV
SDG4	Youth literacy rate (% of population 15–24 years)	SOC	SDG9b	Emission intensity in industry and energy sector ($kgCO_2e$/$)	ENV	SDG15a	Terrestrial protected areas (% of total land area)	ENV
SDG6	Annual freshwater withdrawals (% of internal renewable water)	ENV	SDG9c	Share of domestic expenditure on Research and Development (% of GDP)	ECO	SDG15b	Forest area (% of land area)	ENV
SDG7a	Renewable electricity (% of total population)	ENV	SDG10	Palma ratio	ECO	SDG15c	Endangered and vulnerable species (% of total species)	ENV
SDG7b	Primary energy intensity (MJ/$PPP07)	ENV	SDG11	CO_2 intensity of residential and transport sectors over energy volumes (tCO_2/toe)	ENV	SDG16	Corruption perception index	SOC
SDG7c	Access to electricity (% of total population)	SOC	SDG12	Material productivity ($PPP2011/kg)	SOC	SDG17	Government gross debt (% of GDP)	ECO

2015), and drops in the case of a high level of undernourishment prevalence (Black et al. 2008). Regarding the literacy rate (SDG4), we consider a simple regression with education expenditure per capita and urbanisation, both fostering education attainment. The literature on electricity access (SDG7c) is wide and identifies GDP per capita (Chen et al. 2007), electricity supply, urbanisation (Lahimer et al. 2013), corruption control (Javadi et al. 2013) as favouring factors. Inequality works in the opposite side. Among the explanatory variables of our inequality measure, i.e. Palma ratio, we included public education expenditure per capita, sectoral Value Added (VA) share in agriculture and industry, corruption control and unemployment (Ferreira and Ravallion 2009; Ferreira et al. 2010).

The **modelling framework** used to develop SDG projections is the ICES model, (Eboli et al. 2010; Delpiazzo et al. 2017), a global CGE model based on the GTAP model (Corong et al. 2017) and running over the period 2007–2030 with recursive dynamics. The baseline scenario assumes no mitigation policies are implemented until 2030, while the mitigation policy scenario simulates the implementation of the conditional Nationally Determined Contributions (NDCs) submitted to the UNFCCC in the context of the Paris Agreement. By comparing the performance of the SDG targets in the two scenarios, the approach can quantitatively evaluate the implications of mitigation policy on sustainable development. Model features, baseline, and policy scenario assumptions are described in detail in the Annex II.

The **post-processing module** computes the values of the SDG indicators up to 2030 using the output of ICES. For the indicators not directly generated by the model, the estimated relationships from historical data with the regression analysis are used in an out-of-sample estimation procedure and combined with output variables of the model. All indicator values are then normalised between [0,100] using a benchmarking procedure that identifies sustainable and unsustainable thresholds for each indicator relying on the SDG targets and best practices.[4] SDG indicators are then aggregated into SDG-specific indices (simple average of the underlying indicators) and into an Aggregate Sustainable Development Index (ASDI), a simple average of the SDG indices that reaches the score 100 whether all goals are met.[5]

3.2 Regional Performance in Achieving SDGs: A 2007 Snapshot

The selected SDG indicators, normalised and aggregated as described in the previous section and in Annex I make it possible to quantify country well-being and sustainability measured in terms of proximity to all SDGs. The approach can be applied to historical as well as to future, simulated data, enabling a comparison and measurement of changes in sustainability patterns over time and scenarios.

---

[4]A more detailed description of this step and a table with benchmarks can be found in Annex I.

[5]A more detailed description of this step can be found in Annex I.

Can the Paris Agreement Support Achieving the Sustainable Development Goals?

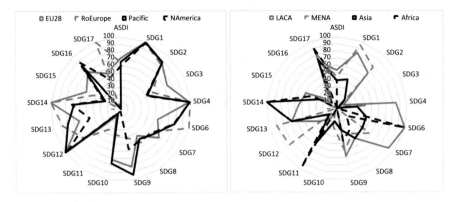

Fig. 2 Aggregate Sustainable Development Index (ASDI) and SDGs scores in 2007, top (left) and bottom (right) performers. Different grey shading represents the eight regions. Lightest grey is EU28 on the left and LACA on the right

Figure 2 synthesises collected historical indicator values at global level and shows the performance in each SDG and in the overall ASDI index of eight regional aggregates.[6] The graph on the left shows the score of top performing regions (EU28, Rest of Europe, Pacific, and North America) in 2007. All of them are still far from achieving SDGs (score of 100). The EU28 is the front-runner, with a score of 70.5, Rest of Europe and Pacific regions closely follow (both at 65.4), and North America is the last in the overall ranking (59.3). On the right are four regional aggregates lagging behind in the sustainability pathway: Latina America (LACA region, 53.4) is close to the top performing group, whereas the gap widens for Middle East and North Africa (MENA region, 43.7), Asia (39.1), and Africa (37.7).

The two radar graphs immediately visualise the noticeable difference between the two groups of countries. The top performers (left panel) are particularly close to achieving many SDGs related to the social pillar, i.e. SDG1, SDG4, SDG10, and SDG16. The graph of bottom performers (right panel) shows a more uneven regional distribution, with few isolated spikes for the SDGs mostly related to the environmental pillar, i.e. SDG14 and SDG6.

Looking more closely at regional differences, all top performers nearly meet SDG1 and SDG4, close to zero prevalence of extreme poverty and universal literacy rate, respectively. They have an average score of 87 (over 100 that represent the full sustainability) in SDG2, zero hunger, with around 2.6% of population undernourished. The score regarding reduced inequality (SDG10) and corruption perception (SDG16) is more heterogeneous: EU28 and the Pacific region score around 75 on equal income distribution, and North America only 4, with a Palma ratio of inequality equal to 1.95 (i.e. close to the unsustainable level of 2). Corruption

---

[6]It is worth remembering that the score in each SDG and in the ASDI index is restricted to the 27 selected indicators and not to all other dimensions encompassed by the UN Agenda 2030.

perception is low in North America and Pacific (on average 85) and really high in Rest of Europe (score 0).

Focusing on the economic indicators, top performers score uniformly around 50 in SDG8 (indicators relative to growth of GDP, level of GDP per employed, and employment ratio). Sustainability of public debt (SDG17) is fully achieved in the case of Rest of Europe (100) and it is null in the case of EU28 (0 due to the high debt GDP ratio in some EU28 countries). North America and Pacific region have a score around 50. The score of SDG9, combining two economic and one environmental indicators for industry, innovation, and sustainable infrastructure, is uneven, on average 88 for EU28 and Pacific region, and 49 for North America and Rest of Europe. Despite the similar levels of manufacturing value added indicator, and some heterogeneity regarding the share of investment in R&D (higher in the North America and Pacific region), the score of SDG9 strongly reflects the indicator on emission intensity in energy and industry sectors, which is low both in the Rest of Europe and North America (respectively 0 and 8.2 over 100).

Regarding the other environmental indicators, water withdraw (SDG16) is fully sustainable in Rest of Europe (100) and the least sustainable in EU28 (54). SDG7 in terms of energy intensity growth and renewable electricity share scores around 55 in top performing regions, except in the EU28, where it reaches 67. On the contrary, CO_2 intensity in residential and transport (SDG11) is too high for all top performers, in particular in North America (6 over 100). North American countries perform well in terms of efficient use of material, non-fossil resources (SDG12), while Rest of Europe scores worse (66.6). Marine ecosystems protection (SDG14) has also a score above average in all top performing regions, with Pacific region scoring the worst (67.5). Indicators relative to the protection of terrestrial ecosystem (SDG15) have a lower performance, with the Pacific region and North America scoring the worst (24.7 and 27.6, respectively). More differentiated is the result relative to SDG13, climate action, where Rest of Europe is leading with a score of 86.1, followed by Pacific (79.3) and EU28 (62). North America has the worst performance (46.3).[7]

As mentioned above, the snapshot of 2007 sustainability of the worst-performing regions is strongly heterogeneous. In Latin America (LACA) social indicator scores closely follow the North American ones, with slightly higher poverty levels (SDG1, score 83.4) and lower literacy rates (SDG4, 83.8). The social indicators most problematic for this region are undernutrition prevalence (SDG2, 66.1), good health (SDG3, 11), inequality (SDG10, 0), and corruption perception (SDG16, 16). Economic indicators are close to the average score (SDG8, 49.1, SDG17, 68.8, and SDG9, 68.1) and environmental SDGs range from good performances in water management, clean energy production, and climate action (respectively SDG6, 100, SDG7, 91.7, and SDG13, 66.6), to average results in water and land ecosystem

---

[7]SDG13 summarises three indicators: the concentration of emissions from agriculture, forestry and land sue (AFOLU), the distance from achieving NDC emissions, and the gap from equitable and sustainable GHG emissions per capita. In spite of being closer to sustainable and equitable emissions per capita than Rest of Europe and Pacific, the EU28 is characterised by a higher AFOLU emission concentration and results farther from achieving its NDC due to a more ambitious target.

protection (SDG14 and SDG15), to low outcomes regarding emission intensity in residential and transport (SDG11, 16.3), efficiency in using mineral resources (SDG12, 30).

The MENA region outperforms LACA in poverty and undernutrition reduction (SDG1, 94, and SDG2, 76), and equity (SDG10, 26). However, other social indicators are at critical levels, namely education (SDG4, 43.1), corruption perception (SDG16, 10.3) and in particular health (SDG3, 5.2). Economic indicators are slightly lower than those of LACA, excluding debt sustainability (SDG17) that for MENA is quite high (82.7). In the environmental sphere, particularly problematic are water management (SDG6, 0), CO_2 intensity in residential and transport (SDG11, 4.2) and protection of marine ecosystem (SDG14, 0).

Comparing the performance of social SDGs in Asia and MENA region, it is worth highlighting that poverty and undernutrition prevalence are considerably higher in Asia (SDG1, 43.1 and SDG2, 17.1), whereas other indicators pertaining health, education, inequality, and corruption perception share a similar low score (SDG3, 1.8, SDG4, 29.7, SDG10, 18.4, and SDG16, 10.3). Asian economic sustainability does not differ significantly from that of MENA, only SDG9 has a lower performance (30.4) due to the high emission intensity in energy and industry sector. Critical environmental SDGs are instead material efficiency (SDG12, 13.2) and terrestrial ecosystem protection (SDG15, 28.4), whereas water management, emission intensity in residential and transport, and marine areas protection is more sustainable than in MENA region.

In 2007, Africa is the region with the widest gap from achieving all SDGs. The less sustainable sphere is the social one. Poverty and undernutrition prevalence (SDG1 and SDG2), healthy life expectancy (HALE, SDG3), literacy rate (SDG4), inequality (SDG10), and corruption perception (SDG16) have a 0 score. The low level of GDP per person employed reduces SDG8 score (34.9), and the low emission intensity in industry and energy sectors lead to an average score in SDG9 (46). Two environmental SDGs have very low scores, SDG7(17.6) and SDG12 (13.2). In the case of SDG7, high growth of energy intensity and low renewable share are combined with an unsustainable level of access to electricity.

3.3 Regional Trends in Achieving SDGs: Baseline Scenario

As described in Annex II, a baseline scenario without any mitigation policy in place is projected starting in 2007, reproducing historical patterns up to 2010 and, then following similar trends of those observed in the recent decades. This is the so-called Middle of the Road narrative of the Shared Socioeconomic Pathways, SSP2 as described in O'Neill et al. (2017). The score in each SDG and in the overall sustainability indicator ASDI is computed for each simulation year and is compared with 2007 results. For the sake of clarity, results for 45 countries and macro-regions of the ASDI framework are grouped into eight regional aggregates.

The socioeconomic dynamics and technological changes characterising the baseline scenario (changes in population, employment, GDP growth, reduction in fossil fuel dependency, and rise in energy efficiency) are heterogeneous across regions as well as within regions, and determine convergence or divergence from achieving SDGs. Figure 3 shows the changes in sustainability indicators between 2007 and 2030 across regions. Asia, Africa, and MENA are gaining the most in 2030, namely 17.7, 10.7, and 9.6 percentage points (pp) with respect to 2007, instead LACA and the EU28 experience a reduced sustainability (respectively −0.1 and −2.3 pp). These changes bring Rest of Europe to the top of ranking (ASDI 71), followed by the EU28 (ASDI 68.2), whereas Asia shifts to a middle level of sustainability (ASDI 56.8).

Asian progress is relevant in reducing poverty (SDG1), undernutrition prevalence (SDG2), inequality (SDG10), in improving health (SDG3) and education (SDG4) (respectively, 55.4, 53.7, 67.4, 19.6, and 27.5 pp with respect to 2007). This evolution is fuelled by a moderate improvement in economic sustainability (SDG8, +13.7 pp) due to higher levels of GDP per person employed. The drawbacks for the environment emerge in particular regarding the intensity of water use (SDG6, −43.4 pp) and climate action (SDG13, −5.2 pp). In the latter case, economic growth implies higher emissions and therefore a widening gap from the NDC and

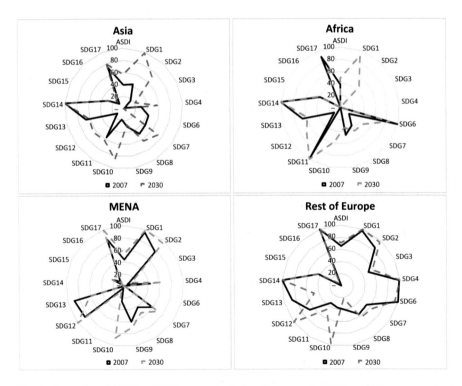

Fig. 3 Dynamics of ASDI and SDG scores, in the baseline scenario 2007 vs. 2030

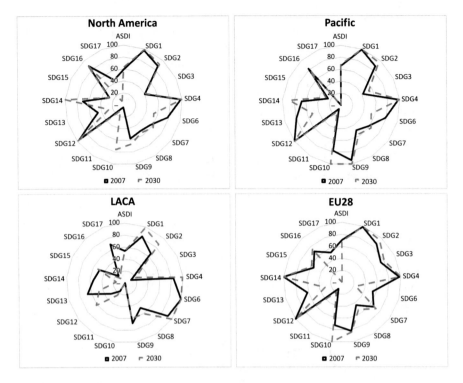

Fig. 3 (continued)

equitable and sustainable emission path. The baseline scenario exhibits exogenous improvements in efficiency, reflecting historical patterns, and this trend appears in the advancements in material productivity (SDG12, +51.3 pp), emission intensity in residential and transport (SDG11, +6 pp) and affordable and clean energy (SDG7, +26.8 pp).[8]

Despite the higher sustainability level in 2030, the African region remains at the bottom of the ranking (ASDI 48.4). Poverty, undernutrition prevalence, and inequality dramatically reduce with respect to 2007 (SDG1 +91.6 pp, SDG2 +45.9 pp, and SDG10 +57.7 pp), although progress is not enough in the case of health and education status (SDG3 and SDG4 still have a 0 score). Economic sustainability worsens in particular regarding the sustainability of public debt (SDG17 −89.4 pp). SDG8 remains stable (+9.4 pp) despite opposing changes at the indicator level (slower GDP per capita growth, but higher GDP per employed). In the environmental realm, material use productivity rises considerably (SD12 +57.8 pp) as well as the sustainability of the energy system (SDG7 +26.8 pp) due to the higher

---

[8] Asia's score in SDG7 depends on a cleaner energy system (lower growth of primary energy intensity and higher renewable electricity share), but also on the expansion of access to electricity.

share of electricity produced from renewable sources and wider access to it. Also in Africa, economic and population growth undermine the sustainable use of water resources (SDG6 −14.4 pp) and climate action (SDG13 −8.1 pp).

As mentioned above, the EU28 and LACA regions experience a reduction in sustainability by 2030. Despite the constant progress in the social SDGs, in particular inequality reduction (SDG10 +28.1 pp) and improvements in economic growth (SDG8 +8.4 pp), the sustainability of public debt deteriorates (SDG17 −53.1 pp) and some environmental indicators are negatively affected by the resource-intensive socioeconomic development foreseen in the baseline scenario. The intensity of water withdraw rises (SDG6 −30.1 pp) and the uncontrolled increase in emissions from agriculture and forest land, and of overall GHG emissions widen the distance from achieving the ambitious EU28's NDC (SDG13 −31.2 pp).

Strong improvements in energy and material efficiency (SDG9 +9.1 pp; SDG12 + 4.1 pp), the strengthening of terrestrial ecosystem protection (SDG14 +29.7 pp), the less ambitious NDC (SDG13 −19.6 pp), the high reduction of inequality (SDG10 +71.5 pp), and a lower public finance deterioration (SDG17 −38 pp) mark the divergence between North America and the EU28 between 2007 and 2030. Despite these dynamics, the EU28 sustainability score in 2030 (ASDI 68.2) remains above the North American one (ASDI 63.1).

3.4 Paris Agreement Mitigation Scenario

The Paris Agreement, adopted in 2015, initiated a new climate policy regime characterised by country-driven emission targets as part of their international effort to limit global warming beyond 2020, the so-called Nationally Determined Contributions (NDCs). The NDCs describe the mitigation efforts of the UNFCCC Parties up to 2030. They are quite heterogeneous in terms of stringency, coverage, and reference level. For example, China, India, and Chile have expressed their NDCs in terms of emission intensity. Most NDCs describe an unconditional and a conditional target: the former to be met autonomously, and the latter, more ambitious, requiring external financial and technical support.

In the policy scenario design, we focus on the conditional mitigation objectives stated in the NDCs (reported in Annex II) and on the reduction of CO_2 emissions. Our mitigation scenario starts in 2013 and assumes that each country achieves its NDC by 2030. The EU28 implements an Emission Trading System (ETS), while all other countries are assumed to implement a unilateral domestic carbon tax. Carbon tax revenues are recycled internally to households, public saving, and investments.

Our results show that the implementation of the NDCs will lead to higher sustainability for all countries, excluding MENA region, which is essentially unaffected (see Fig. 4). It is important to highlight that the change in the ASDI score induced by the mitigation policy is much smaller compared to that observed in the baseline scenario. The changes observed in the baseline scenario reflect socioeconomic and

Can the Paris Agreement Support Achieving the Sustainable Development Goals?

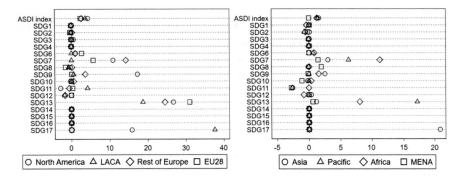

Fig. 4 Climate policy impact on SDGs in 2030 (Percentage point change relative to the baseline)

technological changes that occur between 2007 and 2030, whereas in the case of mitigation policy we only evaluate the effect of the policy on the 2030 score.

North America experiences the highest benefit from the mitigation policy (ASDI +4.3 pp with respect to 2030 baseline scenario), followed by LACA (ASDI +3.6 pp), Rest of Europe (ASDI +2.6 pp), and the EU28 (ASDI +2.4 pp). The EU28, Africa, and Pacific observe a change lower that 2 pp, and the MENA region has a modest reduction of −0.01 pp.

Mitigation policy most strongly affects the environmental SDGs. SDG13, on climate action, registers a rise between +31.1 pp in the EU28 and +0.6 pp in the MENA region, reflecting the achievement of the NDC targets and the convergence toward more equitable, sustainable emissions per capita. The SDG13a shows a general worsening because our mitigation policy focuses on CO_2 and leave uncontrolled other gases emitted by Agriculture, Forestry, and Other Land use (AFOLU). In addition, we assumed that Egypt,[9] part of MENA region, does not have a NDC. The country experiences a leakage effect that pushes it away from an equitable and sustainable emission path.

SDG7 is the second index for the magnitude of change induced by the policy, ranging between +14.2 pp in Rest of Europe and +0.01 pp in the LACA region. Also in this case, the SDG score depends on combined impacts on the underlying indicators. Mitigation targets stimulate substitution towards a cleaner energy mix characterised by a higher electricity share from renewables (between +25.5 pp in Africa and no change in the LACA region[10]) and lower primary energy intensity (between +21.7 pp in Rest of Europe and no change in LACA region[11]). It is worth noticing that the indicator on electricity access (social dimension in SDG7)

---

[9]Egypt and Bolivia do not have a quantitative NDC, therefore, we assume the two countries are not implementing any mitigation policy.

[10]LACA region is fully sustainable in this dimension (score 100) also in the baseline scenario, therefore, an improvement of this indicator does not translate into a higher score.

[11]Ibidem.

is not negatively affected by the implementation of Paris agreement, especially in those countries still far from achieving that target (no change in Asia and +0.4 pp in Africa). In both cases, regional average results mask country heterogeneity. Some Asian and African countries slightly slow down their progress in electricity access (e.g. Bangladesh and Uganda), while others see an acceleration, having an energy system more flexible to renewable switching (Ghana and Ethiopia). Positive implications of the policy spread also to SDG6, inducing a more sustainable water use. In the EU28, the score change is of +2.7 pp. The effects on SDG11 and SDG12 are more heterogeneous. CO_2 intensity in residential and transport sectors rises in Asia, MENA, North America and Pacific (SDG11 −2.5 pp), and material productivity shrinks in LACA, Rest of Europe, and Africa.

The economic SDGs show conflicting results. The carbon tax revenue improves government accounts and debt sustainability (SDG17) in LACA (37.6), Asia (20.9), and North America (+15.8 pp). In the other regions the change is not perceivable because the score of the indicator remains below the unsustainable level. SDG8 is the most sensitive to the costs of mitigation policy, reflecting a slowdown in GDP per capita growth in regions with ambitious climate policy (−4.5 pp in Europe) and a leakage effects where the interventions are modest (+5.9 pp in MENA region). The change of SDG9 ranges between +17.3 pp in North America and −0.1 pp in the MENA region and it is mainly due to the cut in emission intensity in the energy and industry sectors fulfilled with the mitigation targets.

Social indicators are slightly negatively affected by the costs of the mitigation policy and reflect the closure assumptions of the model. The carbon revenue is recycled partially to support household income, whereas government expenditure, a strong driver for social indicators, is left unchanged with respect to the baseline scenario. In Asia, social indicators slightly improve, on average, driven by the positive performance of India, whereas Indonesia, Bangladesh, and Rest of Asia highlight the need of additional pro-poor policies to complement mitigation interventions and limit their side effects. Africa shows a slow-down in poverty and undernutrition reduction (SDG1 −0.4 pp and SDG2 −0.6 pp). All countries in the region, excluding Mozambique, are negatively affected by the policy and its macroeconomic costs. As noted in Campagnolo and Davide (2018), inequality (SDG10) positively (negatively) reacts to ambitious (loose) mitigation targets, but the policy-induced inequality reduction is not sufficient to compensate the average GDP loss and often determines an increase of poverty prevalence.

4 Discussion and Conclusions

This chapter develops a framework for comparing historical and future sustainability performance measured by different SDG indicators. A CGE model is used to describe future baseline and policy scenarios at global scale for some key world regions to 2030. Relationships based on historical correlation patterns are used to link the macroeconomic variables projected by the model with 16 SDG indicators to derive sustainability implications. By looking at the sustainability issue in a dynamic manner, the approach here described makes it possible to track SDG indicator values across countries and trough time, shedding light on the regional distribution of synergies and trade-offs and contributing to expand the emerging literature on systemic analysis of climate policy and sustainable development.

Results highlight that mitigation policy reduces the gap toward achieving all sustainability goals by 2030 in all regions. Yet, regional results mask a complex relationship between mitigation policies and SDGs, which is highly country-specific, making it difficult to identify clear patterns, especially for some indicators. For example, the impact on environmental goals, such as SDG7 and 13, is unequivocally positive. Economic and social indicators are characterised by a higher regional diversity. Overall, results are in line with the evidence highlighted by the existing literature, pointing at synergies especially for environmental indicators. On the contrary, social dimensions are more frequently found to show trade-offs with mitigation policies, pointing at the need for additional pro-pure policy interventions (Roy et al. 2019). This analysis does not find evidence for strong trade-offs, one reason being the mitigation strategy, both in terms of stringency, which is moderate, and in terms of mix, as it does not rely on negative emission technologies and expansion of bio-energy.

Social and economic sustainability indicators tend to deteriorate in most regions, essentially for three reasons. First, the analysis focuses on mitigation policy without considering the benefits related to the reduced climate change impacts. Including the policy benefits in terms of reduced climate impacts, which tend to be regressive, could reverse the outcome of mitigation policies. Second, different carbon revenue recycling schemes can be designed to explicitly address the distributional implications of climate policy (Carattini et al. 2019). Third, additional mechanisms of co-benefits could operate through technological change, which in this framework remains exogenous.

The goal of this chapter is to describe the methodology and illustrate how it operates under a specific socioeconomic and policy scenario. Socioeconomic uncertainty deeply interacts with mitigation policy, and different baseline developments would affect results also with respect to the sustainability impacts of climate policy. The proposed framework can be easily adapted to handle multiple scenario combinations and to expand the set of baseline scenarios and mitigation policies.

Further refinements of the proposed framework include developing refined empirical estimates of the relationship between the SDG indicators and the model outcome variables, as well as exploring the role of uncertainty of these underlying relationships. The analysis is based on the central estimates, but confidence intervals could also be used. Widening the set of GHGs considered as well as the negative emissions from land use change could decrease the cost of the policy and the trade-off with social indicators. Whereas here the focus is on the interaction of mitigation policy with sustainable development, other existing policies could further modify the results. Adding the representation of climate change impacts and adaptation measures, which are not yet widely explored in the CGE and SDG literature, highlighting further channels of trade-offs and synergies, is needed in order to complete the characterisation of the inter-linkages between climate policy, impacts, and sustainable development. This analysis underestimates the benefits of mitigation because all impacts connected to global warming and all benefits deriving from a contained temperature increase in 2100 are not included. The emission pathway of the proposed baseline scenario falls between the Radiative Concentration Pathways RCP6.0 and RCP8.5, but the effects of the associated temperature increase on GDP growth as well as on other drivers (e.g. labour productivity) are not included. In this cost-effective approach, GDP is affected by emissions only through mitigation costs.

Acknowledgements This paper has received funding from the European Research Council (ERC) under the European Union's Horizon 2020 research and innovation programme under grant agreement No 756194 (ENERGYA).

Appendix 1

The current list of SDG indicators defined by the works of UN Inter-agency Expert Group on SDG Indicators (United Nations (UN) 2017a) considers 232 indicators. The well-established and in use indicators are less than half of them. The final ASDI screening considers 27 indicators covering 16 SDGs (All but SDG5—Achieve gender equality and empower all women and girls). Table 2 lists the ASDI indicators coupled with the related SDG, the sustainability pillar of pertinence, whether they derive from a regression (regression results are reported in Table 3), how they are computed and the main source of data.

In order to compare country performance in different SDG indicators and to compute some aggregate measures, it is necessary to bring all indicators to a common measurement unit, the [0,100] scale (**normalisation**). The normalisation is obtained using a benchmarking procedure that defines two threshold values for each indicator: unsustainable and sustainable levels. In choosing the threshold levels, we firstly looked at the 169 SDG targets, which are our preferred source if they provide quantitative targets. When the targets are qualitative, other sources are preferred such as policy targets in OECD or countries' best practices.

Table 4 shows the threshold values used for the normalisation process.

ASDI framework considers several **aggregation steps** in order to produce aggregate indices conveying more synthetic information to policymakers:

- SDG indices are the average value of indicators pertaining to each goal;
- The ASDI index is the average of scores among all SDGs.

Table 2 ASDI indicators

SDG	Pillar	Indicator	Est.	Formulation	Source
SDG1	SOC	Population below $1.90 (PPP) per day (%)	Yes	$\beta_0 + \beta_1 ln(GDPPPP pc_{t-1}) + \beta_2 Palma_{t-1}$	WDI
SDG2	SOC	Prevalence of undernourishment (%)	Yes	$\beta_0 + \beta_1 ln(GDPPPP pc_{t-1}) + \beta_2 ln(GDPPPP pc_{t-1})^2$ $+\beta_3 Palma_{t-1} + \beta_4 ln(Agri\,prod\,pc_{t-1})$	WDI-UN
SDG3a	SOC	Physician density (per 1000 population)	Yes	$ln(Physician_num_t)\beta_0 + \beta_1 ln(Pop_{t-1}) + \beta_2 ln(Healthexp\,pc_{t-1})$ $+\beta_3 Priv\,Healthexp\,Sh_{t-1}$	WDI
SDG3b	SOC	Healthy Life Expectancy (HALE) at birth (years)	Yes	$\beta_0 + \beta_1 ln(Physiciandens_{t-1}) + \beta_2 ln(Eduexp\,pc_{t-1})$ $+\beta_3 Ely_access_{t-1} + \beta_4 Undern\,pop_{t-1} + \beta_5 Urbansh_{t-1}$	WDI
SDG4	SOC	Youth literacy rate (% of population 15–24 years)	Yes	$\beta_0 + \beta_1 ln(Eduexp\,pc_{t-1}) + \beta_2 Urbansh_{t-1}$	WDI
SDG6	ENV	Annual freshwater withdrawals (% of internal renewable water)	No	(Total sectoral water use)$_t$/(Renewable water)$_t$ * 100	WDI
SDG7a	ENV	Renewable electricity (% of total population)	No	(Renewable electricity output)$_t$/(Electricity output)$_t$ * 100	IEA
SDG7b	ENV	Primary energy intensity	No	(Energy consumption)$_t$/(GDPPPP)$_t$	IEA
SDG7c	SOC	Access to electricity (% of total population)	Yes	$\beta_0 + \beta_1 ln(GDPPPP pc_{t-1}) + \beta_2 ln(GDPPPP pc_{t-1})^2$ $+\beta_3 ln(Elyout\,pc_{t-1}) + \beta_4 Urbansh_{t-1}$ $+\beta_5 Palma_{t-1} + \beta_6 Corrupt\,control_{t-1}$	WDI
SDG8a	ECO	Annual GDP per capita growth (%)	No	Growth $(GDPPPP)_t / pop_t$	WDI
SDG8b	ECO	GDP per person employed ($PPP2011)	No	$(GDPPPP)_t$/(Employed population)$_t$	WDI
SDG8c	ECO	Employment-to-population ratio (%)	No	(Employed population)$_t$/(Total population)$_t$ *100	WDI
SDG9a	ECO	Manufacturing value added (% of GDP)	No	(Manufacturing Value Added)$_t$/$(GDPPPPpc)_t$ *100	WDI
SDG9b	ENV	Emission intensity in industry and energy sector (kgCO$_2$e/$)	No	(GHG emissions)$_t$/(Value added)$_t$	WDI

(continued)

Table 2 (continued)

SDG	Pillar	Indicator	Est.	Formulation	Source
SDG9c	ECO	Share of domestic expenditure on Research and Development (% of GDP)	No	$(RD\ expenditure)_t/(GDP\,PPP)_t * 100$	WDI
SDG10	SOC	Palma ratio	Yes	$\beta_0 + \beta_1 ln(Eduexp\,pc_{t-1}) + \beta_2 ln(Agri\,VAsh_{t-1}) +$ $\beta_3 ln(HInd\,VAsh_{t-1}) + \beta_4 Corrupt\,control_{t-1} +$ $\beta_5 ln(Unempl_{t-1}) + \beta_6 Coninc$	WDI-WIID
SDG11	ENV	CO_2 intensity of residential and transport sectors over energy volumes (tCO_2/toe)	No	$(CO_2\ emissions)_t/(Energy\ use)_t$	IEA
SDG12	ENV	Material productivity ($PPP2011/kg)	No	$(GDP\,PPP)_t/(Material\ domestic\ consumption)_t$	WDI-SERI
SDG13a	ENV	Concentration of GHG emissions from AFOLU[a] (tCO_2e/sq.km)	No	$(AFOLU\ GHG\ emissions)_t/\ (Agri.\ and\ For.\ land)$	WDI-CAIT
SDG13b	ENV	Compliance to Conditional NDCs (%)	No	$(Emission_t - NDC\ Target_t)/NDC\ Target_t * 100$	WDI-CAIT
SDG13c	ENV	Gap from equitable and sustainable GHG emissions per capita in 2030 (tCO_2eq)[b]	No	$(GHG\ emissions\ per\ capita)_t - (Eq.\ and\ Sust.\ GHG\ per\ capita)_t$	CAIT
SDG14	ENV	Marine protected areas (% of territorial waters)	No	Constant after 2015	WDI
SDG15a	ENV	Terrestrial protected areas (% of total land area)	No	Constant after 2015	WDI
SDG15b	ENV	Forest area (% of land area)	No	$(Forest\ land\ area)_t\ /(Total\ land\ area)_t * 100$	WDI
SDG15c	ENV	Endangered and vulnerable species (% of total species)	No	Constant after 2015	WDI
SDG16	SOC	Corruption perception index	No	Constant after 2015	TI[c]
SDG17	ECO	Government gross debt (% of GDP)	No	$(Public\ debt)_t\ /(GDP\ PPP\ pc)_t * 100$	WDI-IMF

[a] AFOLU stands for agriculture, forestry and other land use

[b] The equitable and sustainable GHG emission per capita level in 2030 is computed as the ratio of the median GHG emission level in 2030 according to scenarios that will contain (with likelihood > 66%) the temperature increase below 2 °C by the end of the century, i.e. 42 $GtCO_2eq$ (United Nations Environment Programme (UNEP) 2015), and the median estimate of world population in 2030

[c] Transparency International

Can the Paris Agreement Support Achieving the Sustainable Development Goals? 35

Table 3 Regression table

	Palma ratio	ln(Poverty)	Access	Undernutrition	ln(Pysician number)	ln(HALE)	Literacy rate
L.ln(Education exp. pc)	−0.0990*					0.0206***	2.901**
	(0.030)					(0.000)	(0.006)
L.ln(Agriculture VA share)	−0.186***						
	(0.000)						
L.ln(Industrial VA share)	−0.0794*						
	(0.023)						
L.ln(Unemployment)	0.0395						
	(0.149)						
L.ln(GDP PPP pc)		−3.172***	0.794***	−57.03***			
		(0.000)	(0.000)	(0.000)			
L.ln(GDP PPP pc)sq			−0.0425***	2.799***			
			(0.000)	(0.000)			
L.Palma		0.139**	−0.0124*	1.215**			
		(0.002)	(0.023)	(0.003)			
L.ln(Electricity output pc)			0.0295+				
			(0.086)				
L.Urban population share			0.00901***			0.00386***	0.428***
			(0.000)			(0.000)	(0.001)
L.ln(Agricultural production pc)				−4.662*			
				(0.015)			
L.ln(population)					1.509***		
					(0.000)		
L.ln(Health exp. pc)					0.0738***		
					(0.000)		

(continued)

Table 3 (continued)

	Palma ratio	ln(Poverty)	Access	Undernutrition	ln(Pysician number)	ln(HALE)	Literacy rate
L.ln(Private health exp. share)					−0.000788* (0.015)		
L.ln(Physicians)						0.0136** (0.009)	
L.Electricity access share						0.00131** (0.003)	
L.Undernutrition prevalence						−0.00223* (0.017)	
Corruption control	−0.0880* (0.024)						
L.(Corruption control)			−0.0238* (0.014)				
Consumption/Income	−0.0123 (0.703)						
Constant	39.62*** (0.001)	28.33*** (0.000)	−3.551*** (0.000)	302.6*** (0.000)	5.147*** (0.000)	3.552*** (0.000)	47.61*** (0.000)
Country fixed effect	Yes	Yes	Yes	Yes	Yes	Yes	Yes
Year fixed effect	Yes	No	No	No	No	No	No
R^2	0.242	0.837	0.526	0.462	0.398	0.618	0.225
Countries	126	126	148	140	166	135	152
Observations	755	994	1812	1764	2044	761	1893

p-Values in parentheses

Robust standard errors, adjusted for clustering at the school level, are presented in parentheses

+ $p < 0.1$, * $p < 0.05$, ** $p < 0.01$, *** $p < 0.001$

Table 4 ASDI benchmarks

SDG	Indicator	Unsustainable level	Sustainable level
SDG1	Population below $1.90 (PPP) per day (%)	40	0
SDG2	Prevalence of undernourishment (%)	20	0
SDG3a	Physician density (per 1000 population)	2	3
SDG3b	Healthy Life Expectancy (HALE) at birth (years)	60	80
SDG4	Youth literacy rate (% of population 15–24 years)	85	100
SDG6	Annual freshwater withdrawals (% of internal renewable water)	30	5
SDG7a	Renewable electricity (% of total population)	5	60
SDG7b	Primary energy intensity (MJ/$PPP07)	10	3
SDG7c	Access to electricity (% of total population)	40	100
SDG8a	Annual GDP per capita growth (%)	0	7
SDG8b	GDP per person employed ($PPP2011)	5000	50,000
SDG8c	Employment-to-population ratio (%)	40	80
SDG9a	Manufacturing value added (% of GDP)	5	15
SDG9b	Emission intensity in industry and energy sector ($kgCO_2e/\$$)	2	1
SDG9c	Share of domestic expenditure on Research and Development (% of GDP)	0.5	3
SDG10	Palma ratio	2	1
SDG11	CO_2 intensity of residential and transport sectors over energy volumes (tCO_2/toe)	2.5	0.5
SDG12	Material productivity ($PPP2011/kg)	0.5	2
SDG13a	Concentration of GHG emissions from AFOLU[a] ($tCO_2e/sq.km$)	100	0
SDG13b	Compliance to Conditional NDCs (%)	0	100
SDG13c	Gap from equitable and sustainable GHG emissions per capita in 2030 (tCO_2eq)[b]	15	0
SDG14	Marine protected areas (% of territorial waters)	5	20
SDG15a	Terrestrial protected areas (% of total land area)	10	50
SDG15b	Forest area (% of land area)	5	60
SDG15c	Endangered and vulnerable species (% of total species)	20	5
SDG16	Corruption perception index	30	80
SDG17	Government gross debt (% of GDP)	100	20

[a] AFOLU stands for agriculture, forestry and other land use

[b] The equitable and sustainable GHG emission per capita level in 2030 is computed as the ratio of the median GHG emission level in 2030 according to scenarios that will contain (with likelihood > 66%) the temperature increase below 2 °C by the end of the century, i.e. 42 $GtCO_2eq$ (United Nations Environment Programme (UNEP) 2015), and the median estimate of world population in 2030

Appendix 2

Model Description

ICES is a recursive-dynamic multiregional Computable General Equilibrium (CGE) model developed to assess the impacts of climate change on the economic system and to study mitigation and adaptation policies (Eboli et al. 2010). The model's general equilibrium structure allows for the analysis of market flows within a single economy and international flows with the rest of the world. This implies going beyond the simple quantification of direct costs, to offer an economic evaluation of second and higher-order effects within specific scenarios either of climate change, climate policies or different trade and public-policy reforms in the vein of conventional CGE theory. The core structure of ICES derives from the GTAP-E model (Burniaux and Truong 2002), which in turn is an extension of the standard GTAP model (Corong et al. 2017). The General Equilibrium framework makes it possible to characterise economic interactions of agents and markets within each country (production and consumption) and across countries (international trade). Within each country the economy is characterised by a number of industries n, a representative household and the government. Industries are modelled as representative cost-minimising firms, taking input prices as given. In turn, output prices are given by average production costs. The production functions (Fig. 5) are specified via a series of nested Constant Elasticity of Substitution (CES) functions. In the first nest, a Value-Added-Energy nest ($QVAEN$) (primary factors, i.e. natural resources, land, and labour and a Capital+Energy composite), is combined with intermediates (QF), in order to generate the output. Perfect complementarity is assumed between value added and intermediates. This implies the adoption of a Leontief production function. For sector i in region r final supply (output) results from the following constrained production cost minimization problem for the producer:

$$min \quad PVAEN_{i,r} * QVAEN_{i,r} + PF_{i,r} * QF_{i,r}$$

$$s.t. \quad Y_{i,r} = min(QVAEN_{i,r}, QF_{i,r})$$

where $PVAEN$ and PF are prices of the related production factors.

The second nested-level in Fig. 5 (left hand side of the production tree) includes the value added and the energy composite ($QVAEN$). This composite stems from a CES function that combines four primary factors: land ($QLAND$), natural resources (QFE), labour (QFE) and the capital-energy bundle (QKE) using σ_{VAE} as elasticity of substitution. Primary factor demand on its turn derives from the first order conditions of the following constrained cost minimization problem for the representative firm:

$$min \quad P_{i,r}^{Land} * LAND_{1,r} + P_{i,r}^{NR} * NR_{i,r} + P_{i,r}^{L} * L_{i,r} + P_{i,r}^{KE} * KE_{i,r}$$

$$s.t. \quad QVAEN_{i,r} = (LAND_{i,r}^{\frac{\sigma_{VAE}-1}{\sigma_{VAE}}} + NR_{i,r}^{\frac{\sigma_{VAE}-1}{\sigma_{VAE}}} + L_{i,r}^{\frac{\sigma_{VAE}-1}{\sigma_{VAE}}} + KE_{i,r}^{\frac{\sigma_{VAE}-1}{\sigma_{VAE}}})^{\frac{\sigma_{VAE}-1}{\sigma_{VAE}}}$$

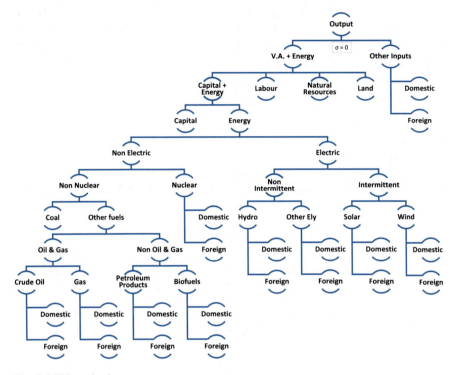

Fig. 5 ICES production tree

On its turn, the KE bundle combines capital with a set of different energy inputs. This is peculiar to GTAP-E and ICES. In fact, energy inputs are not part of the intermediates, but are associated to capital in a specific composite. The energy bundle is modelled as an aggregate of electric and non-electric energy carriers. Electricity sector differentiates between intermittent and non-intermittent sources. Wind and solar, which are intermittent sources, are separated from non-intermittent sources: hydro power and the rest of electricity produced using fossil fuel sources (coal, oil, and gas).[12] The Non-Electric bundle is a composite of nuclear and non-nuclear energy. The aggregate Non-nuclear energy combines, in a series of subsequent nests, Coal, Natural Gas, Crude Oil, Petroleum Products, and Biofuels, while Nuclear corresponds to the carrier used for electricity generation.

---

[12] ICES model further specifies renewable energy sources in electricity production, namely wind, solar and hydro-electricity, splitting them from the original electricity sector. The data collection refers to physical energy production in Mtoe (Million tons of oil equivalent) from different energy vectors and for each GTAP 8 country/region. The data source is Extended Energy Balances (both OECD and Non-OECD countries) provided by the International Energy Agency (IEA). We complemented the production in physical terms with price information (OECD-IEA 2005; Ragwitz et al. 2006; GTZ 2009, IEA country profiles and REN21).

All elasticities regarding the inter-fuel substitution bundles are those from GTAP-E (Burniaux and Truong 2002), while for the extended renewable electricity sectors we set those values considering different studies (Paltsev et al. 2005; Bosetti et al. 2006). The demand of production factors (as well as that of consumption goods) can be met by either domestic or foreign commodities which are, however, not perfectly substitute according to the "Armington" assumption. In general, inputs grouped together are more easily substitutable among themselves than with other elements outside the nest. For example, the substitutability across imported goods is higher than that between imported and domestic goods. Analogously, composite energy inputs are more substitutable with capital than with other factors. In ICES, two industries are treated in a special way and are not related to any country, viz. international transport and international investment production. International transport is a world industry, which produces the transportation services associated with the movement of goods between origin and destination regions, thereby determining the cost margin between f.o.b. and c.i.f. prices. Transport services are produced by means of factors submitted by all countries, in variable proportions. In a similar way, a hypothetical world bank collects savings from all regions and allocates investments in order to achieve equality in the absolute change of current rates of return.

Figure 6 describes the main sources and uses of regional income. In each region, a representative utility maximising household receives income, originated

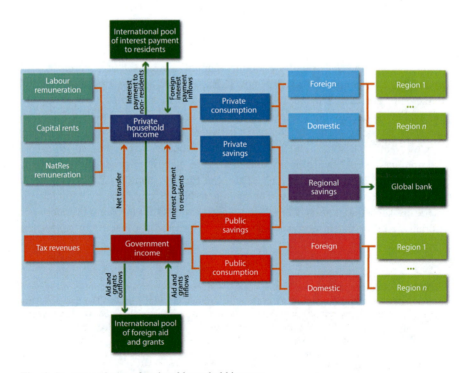

Fig. 6 Sources and uses of regional household income

by the service value of national primary factors (natural resources, land, labour, and capital) that it owns and sells to the firms. Capital and labour are perfectly mobile domestically but immobile internationally (investment is instead internationally mobile). Land and natural resources, on the other hand, are industry-specific. The regional income is used to finance aggregate household consumption and savings.

Government income equals to the total tax revenues from both private household and productive sectors, a series of international transactions among governments (foreign aid and grants) and national transfers between the government and the private (Delpiazzo et al. 2017). Both the government and the private household consume and save a fraction of their income according to a Cobb-Douglas function. The government income not spent is saved, and the sum of public and private savings determines the regional disposable saving, which enters the Global Bank as in the core ICES. Both private and public sector consumption are addressed to all commodities produced by each firm/sector. Public consumption is split into a series of alternative consumption commodities according to a Cobb-Douglas specification. However, almost all public expenditure is concentrated in the specific sector of Non-market Services, including education, defence, and health. Private consumption is analogously addressed towards alternative goods and services including energy commodities that can be produced domestically or imported. The functional specification used at this level is the Constant Difference in Elasticities (CDE) form: a non-homothetic function, which is used to account for possible differences in income elasticities for the various consumption goods.[13]

The recursive-dynamic feature is described in Fig. 7. Starting from the picture of the world economy in the benchmark year, following socioeconomic (e.g. population, primary factors stocks, and productivity) as well as policy-driven changes occurring in the economic system, agents adjust their decisions in terms of input mix (firms), consumption basket (households), and savings. The model finds a new general (worldwide and economy-wide) equilibrium in each period, while all periods are interconnected by the accumulation process of physical capital stock, net of its depreciation. Capital growth is standard along exogenous growth theory models and follows:

$$K e_r = I_r + (1 - \delta) K b_r$$

where $K e_r$ is the end of period capital stock, $K b_r$ is the beginning of period capital stock, δ is capital depreciation and I_r is endogenous investment. Once the model is solved at a given step t, the value of $K e_r$ is stored in an external file and used

---

[13]Hanoch's constant difference elasticity (CDE) demand system (Hanoch 1975) has the following formulation: $1 = \sum B_i U^{Y_i R_i} (\frac{P_i}{X})^{Y_i}$ where U denotes utility, P_i the price of commodity i, X the expenditure, B_i are distributional parameters, Y_i substitution parameters, and R_i expansion parameters. The CDE in principle does not allow to define explicitly direct utility, expenditure, or indirect utility functions. Accordingly, also explicit demand equations could not be defined. Fortunately, in a linearized equation system such as that used in GTAP, it is possible to obtain a demand function with price and expenditure elasticities.

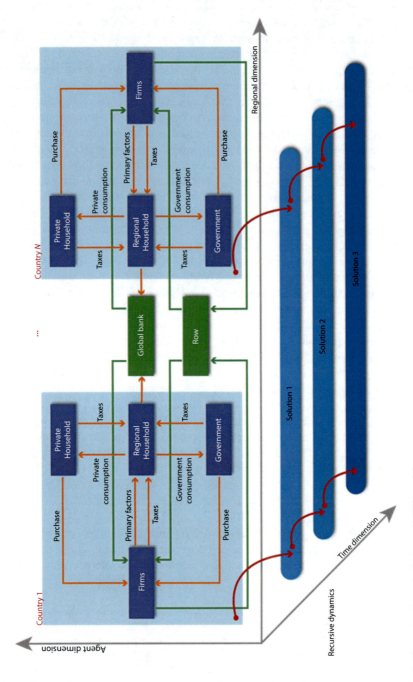

Fig. 7 Recursive-dynamic feature of ICES model

as the beginning of period capital stock of the subsequent step t+1. The matching between savings and investments only holds at the world level; a fictitious world bank collects savings from all regions and allocates investments following the rule of highest capital returns.

As with capital, at each simulation step the government net deficit at the end of the period is stored in an external file and adds up to next year debt.

Regional Aggregation

ICES is a Computable model: all the model behavioural equations are connected to the GTAP 8 database (Narayanan and McDougall 2012), which collects national social accounting matrices from all over the world and provides a snapshot of all economic flows in the benchmark year. Being based on the GTAP database, ICES has worldwide coverage. In this analysis, we consider 45 countries/regions (Fig. 8).

For sake of clarity in presenting results, we further aggregate the 45 countries/regions in eight regional aggregates following the mapping presented in Table 5.

Fig. 8 Regional aggregation ICES model

Table 5 Mapping ICES regions into macro regional aggregates

Id.	Country/region	Macro region	Id.	Country/region	Macro region
1	Australia	Pacific	24	Germany	EU28
2	New Zealand	Pacific	25	Greece	EU28
3	Japan	Pacific	26	Italy	EU28
4	South Korea	Pacific	27	Poland	EU28
5	Bangladesh	Asia	28	Spain	EU28
6	China	Asia	29	Sweden	EU28
7	India	Asia	30	UK	EU28
8	Indonesia	Asia	31	RoEU	EU28
9	RoAsia	Asia	32	RoEurope	RoEurope
10	Canada	NAmerica	33	Russia	RoEurope
11	USA	NAmerica	34	Turkey	MENA
12	Mexico	LACA	35	Egypt	MENA
13	Argentina	LACA	36	RoMENA	MENA
14	Bolivia	LACA	37	Ethiopia	Africa
15	Brazil	LACA	38	Ghana	Africa
16	Chile	LACA	39	Kenya	Africa
17	Peru	LACA	40	Mozambique	Africa
18	Venezuela	LACA	41	Nigeria	Africa
19	RoLACA	LACA	42	Uganda	Africa
20	Benelux	EU28	43	South Africa	Africa
21	Czech Republic	EU28	44	RoAfrica	Africa
22	Finland	EU28	45	RoW	RoEurope
23	France	EU28			

Reference Scenario

Our reference in designing the baseline scenario is the set of possible futures envisioned by the climate change community and known as Shared Socioeconomic Pathways (SSPs) (O'Neill et al. 2017). These are five possible futures with different mitigation/adaptation challenges and are characterised by different evolution of main socioeconomic variables. SSPs can be linked to Representative Concentration Pathways (RCPs), that envisions the GHG emission evolution and forcing and temperature rise due to specific patter of socioeconomic growth (Riahi et al. 2017). SSPs provide future patterns for population, working age population and GDP at country level. Other trends for exogenous drivers such as primary factor productivity, sector-specific efficiency, total factor productivity, and energy prices are then used in order to calibrate given endogenous variables, namely GDP, energy use, emissions and value-added shares to be coherent to the selected RCP.

Among Shared Socioeconomic Pathways (SSPs), we used as business as usual SSP2 "Middle of the Road" scenario. The main features of this scenario are:

- similar trends of recent decades, but some progresses towards achieving development goals;
- medium population growth;
- per capita income levels grow globally at a medium pace; slow income convergence across countries; some improvements in the intra-regional income distributions;
- reductions in resource and energy intensity, and slowly decreasing fossil fuel dependency.

Give the short time horizon of the proposed analysis, we focus on the SSP2 because it is the the pathway that more closely follows the historical development in terms of socioeconomic variables evolution (medium population and GDP growth). In calibrating the SSP2, we followed not only the socioeconomic trends reported in SSP database,[14] but we also adjusted energy efficiency and fuel prices in order to obtain a global emission level in line with IAM multi-model projections (Riahi et al. 2017). The literature reports a range between 61,279 and 70,005 Mt CO_2-eq/year in 2030. Our baseline global emissions are 65,140 Mt CO_2-eq/year. The projected emission range in 2100 is between 85,030 and 106,778 Mt CO_2-eq/year which corresponds to a radiative forcing between 6.561 and 7.251 W/m^2, and a temperature rise between 3.8 and 4.2 °C. These results place our baseline in between RCP6 and RCP8.5.

Mitigation Scenario

We designed a mitigation scenario mimicking Paris Agreement functioning: all parties achieve the conditional mitigation targets stated in the NDC by 2030; for regional aggregates, we computed reference and target emission levels and calculated the required regional reduction. We relies on CAIT database for computing reference historical emission levels, whereas our baseline scenario is used when NDC uses a BAU scenario as term of comparison. Due to model limitations, we impose the GHG emission targets only to CO_2 emissions. Mitigation objectives considered for each country/region are reported in Table 6. Two countries in our aggregation do not have a clear quantitative mitigation target, i.e. Egypt and Bolivia; therefore, in our simulation, we assume they have not a NDC.

The mitigation policy starts in 2013 and it is fully achieved by 2030. The European Union (EU28) opts for an Emission Trading System (ETS), while all other countries achieve their contributions unilaterally with a domestic carbon tax. China, India, and Chile have expressed their NDCs in terms of emission intensity. Carbon

---

[14]https://tntcat.iiasa.ac.at/SspDb/dsd?Action=htmlpage&page=about.

Table 6 Emission reduction target in 2030

Country	Target (%)	Target type	Country	Target (%)	Target type
Australia	−27	Emission reduction wrt 2005	Venezuela	−20	Emission reduction wrt 2030 BAU scenario
NewZealand	−30	Emission reduction wrt 2005	RoLACA	−11	Average mission reduction wrt 2030 BAU scenario
Japan	−26	Emission reduction wrt 2013	EU28	−40	Emission reduction wrt 1990
SouthKorea	−37	Emission reduction wrt 2030 BAU scenario	RoEurope	−37.2	Average mission reduction wrt 2030 BAU scenario
Bangladesh	−15	Emission reduction wrt 2030 BAU scenario	Russia	−27.5	Emission reduction wrt 1990
China	−62.5	Emission intensity reduction wrt 2005	Turkey	−21	Emission reduction wrt 2030 BAU scenario
India	−34	Emission intensity reduction wrt 2005	RoMENA	−5	Average mission reduction wrt 2030 BAU scenario
Indonesia	−41	Emission reduction wrt 2030 BAU scenario	Ethiopia	−64	Emission reduction wrt 2030 BAU scenario
RoAsia	6	Average mission reduction wrt 2030 BAU scenario	Ghana	−45	Emission reduction wrt 2030 BAU scenario
Canada	−30	Emission reduction wrt 2005	Kenya	−30	Emission reduction wrt 2030 BAU scenario
USA	−27	Emission reduction wrt 2005	Mozambique	−8	Emission reduction computed from target emission levels in 2030
Mexico	−36	Emission reduction wrt 2030 BAU scenario	Nigeria	−45	Emission reduction wrt 2030 BAU scenario
Argentina	−30	Emission reduction wrt 2030 BAU scenario	Uganda	−22	Emission reduction wrt 2030 BAU scenario
Brazil	−37	Emission reduction wrt 2005	South Africa	−22	Emission level target in 2030 is in the range 398 and 614 Mt CO_2-eq
Chile	−40	Emission intensity reduction wrt 2007	RoAfrica	−24.4	Average mission reduction wrt 2030 BAU scenario
Peru	−30	Emission reduction wrt 2030 BAU scenario	RoW	−11.3	Average mission reduction wrt 2030 BAU scenario

tax revenues are redistributed internally to government investment, public saving and transfers to households.

References

Bergh A, Nilsson T (2010) Good for living? On the relationship between globalization and life expectancy. World Dev 38(9):1191–1203

Black RE, Allen LH, Bhutta ZA, Caulfield LE, de Onis M, Ezzati M, Mathers C, Rivera J (2008) Maternal and child undernutrition: global and regional exposures and health consequences. Lancet 371(9608):243–260

Bosetti V, Carraro C, Galeotti M, Massetti E, Tavoni M (2006) A world induced technical change hybrid model. Energy J 27:13–37

Burniaux J-M, Truong TP (2002) GTAP-E: an energy-environmental version of the GTAP model. GTAP Technical Paper, 16

Campagnolo L, Davide M (2018) Can Paris deal boost SDGs achievement? An assessment of climate mitigation co-benefits or side effects on poverty and inequality. Belfer Center Discussion Paper, 2

Carattini S, Kallbekken S, Orlov A (2019) How to win public support for a global carbon tax. Nature 565:289–291

Carraro C, De Cian E, Nicita L, Massetti E, Verdolini E (2010) Environmental policy and technical change: a survey. Int Rev Environ Resour Econ 4(2):163–219

Chen S, Kuo H, Chen C (2007) The relationship between GDP and electricity consumption in 10 Asian countries. Energy Policy 35(4):2611–2621

Cobham A, Schlogl L, Sumner A (2016) Inequality and the Tails: the Palma proposition and ratio. Global Policy 7:25–36

Corong EL, Hertel TW, McDougall RA, Tsigas ME, van der Mensbrugghe D (2017) The Standard GTAP model, Version 7. J Glob Econ Anal 2:1–119

Dagnachew AG, Lucas PL, Hof AF, van Vuuren DP (2018) Trade-offs and synergies between universal electricity access and climate change mitigation in Sub-Saharan Africa. Energy Policy 114:355–366

Delpiazzo E, Parrado R, Standardi G (2017) Extending the public sector in ICES with an explicit government institution. FEEM Nota di Lavoro, 11

Ebi KL, Hasegawa T, Hayes K, Monaghan A, Paz S, Berry P (2018) Health risks of warming of 1.5 °C, 2 °C, and higher, above pre-industrial temperatures. Environ Res Letters 13

Eboli F, Parrado R, Roson R (2010) Climate change feedback on economic growth: explorations with a dynamic general equilibrium model. Environ Dev Econ 15:515–533

Ferreira FHG, Ravallion M (2009) Poverty and inequality: the global context. In: Salverda W, Nolan B, Smeeding T (eds) The Oxford handbook of economic inequality. Oxford University Press, Oxford

Ferreira FHG, Leite PG, Ravallion M (2010) Poverty reduction without economic growth?: explaining Brazil's poverty dynamics, 1985–2004. J Dev Econ 93:20–36

Fumagalli E, Mentzakis E, Suhrcke M (2013) Do political factors matter in explaining under- and overweight outcomes in developing countries? J Socio-Econ 46:48–56

Grubler A, Wilson C, Bento N, Boza-Kiss B, Krey V, McCollum DL, Rao ND, Riahi K, Rogelj J, De Stercke S, Cullen J, Frank S, Fricko O, Guo F, Gidden M, Havlík P, Huppmann D, Kiesewetter G, Rafaj P, Schoepp W, Valin H (2018) A low energy demand scenario for meeting the 1.5 C target and sustainable development goals without negative emission technologies. Nat Energy 3:515

GTZ (2009) Energy-policy framework conditions for electricity markets and renewable energies. Energy-policy framework papers, Section Energy and Transport, Deutsche Gesellschaft fur Technische Zusammenarbeit (GTZ) GmbH

Gulis G (2000) Life expectancy as an indicator of environmental health. Eur J Epidemiol 16:161–165

Hallegatte S, Rozenberg J (2017) Climate change through a poverty lens. Nat Clim Change 7:250–256

Hallegatte S, Heal G, Fay M, Treguer D (2012) From growth to green growth. A framework. World Bank Policy Research working paper, WP5872

Hanoch G (1975) Production and demand models with direct or indirect implicit additivity. Econometrica 43:395–419

Hasegawa T, Fujimori S, Takahashi K, Yokohata T, Masui T (2016) Economic implications of climate change impacts on human health through undernourishment. Clim Change 136:189–202

Hassett KA, Mathur A, Metcalf GE (2009) The incidence of a U.S. carbon tax: a lifetime and regional analysis. Energy J 30:157–179

Headey DD (2013) Developmental drivers of nutritional change: a cross-country analysis. World Dev 42:76–88

Heltberg R (2009) Malnutrition, poverty, and economic growth. Health Econ 18:77–88

Hsiang SM, Meng KC, Cane MA (2011) Civil conflicts are associated with the global climate. Nature 476:438–441

Jakob M, Steckel JK (2014) Climate change mitigation could harm development in poor countries. WIREs Clim Change 5:161–168

Javadi FS, Rismanchi B, Sarraf M, Afshar O, Saidur R, Ping HW, Rahim NA (2013) Global policy of rural electrification. Renew Sust Energy Rev 19:402–416

Kabir M (2008) Determinants of life expectancy in developing countries. J Dev Areas 41:185–204

Lahimer AA, Alghoul MA, Yousif F, Razykov TM, Amin N, Sopian K (2013) Research and development aspects on decentralised electrification options for rural household. Renew Sust Energy Rev 24:314–324

Lloyd SJ, Kovats RS, Chalabi Z (2011) Climate change, crop yields, and undernutrition: development of a model to quantify the impact of climate scenarios on child undernutrition. Environ Health Perspect 119(12):1817–1823

Markandya A, Sampedro J, Smith SJ, Van Dingenen R, Pizarro-Irizar C, Arto I, González-Eguino M (2018) Health co-benefits from air pollution and mitigation costs of the Paris Agreement: a modelling study. Lancet Planetary Health 2(3):e126–e133

McCollum DL, Echeverri LG, Busch S, Pachauri S, Parkinson S, Rogelj J, Krey V, Minx JC, Nilsson M, Stevance A, Riahi K (2018a) Connecting the sustainable development goals by their energy inter-linkages. Environ Res Lett 13:589–599

McCollum DL, Zhou W, Bertram C, de Boer H, Bosetti V, Busch S, Després J, Drouet L, Emmerling J, Fay M, Fricko O, Fujimori S, Gidden M, Harmsen M, Huppmann D, Iyer G, Krey V, Kriegler E, Nicolas C, Pachauri S, Parkinson S, Poblete-Cazenave M, Rafaj P, Rao N, Rozenberg J, Schmitz A, Schoepp W, van Vuuren D, Riahi K (2018b) Energy investment needs for fulfilling the Paris Agreement and achieving the Sustainable Development Goals. Nat Energy 3:589–599

Metcalf GE (1999) A distributional analysis of green tax reforms. Natl Tax J 52:655–682

Narayanan G, Aguiar BA, McDougall R (2012) Center for Global Trade Analysis: Purdue University

Nelson GC, Rosegrant MW, Palazzo A, Gray I, Ingersoll C, Robertson R (2010) Food security, farming, and climate change to 2050. International Food Policy Research Institute, Washington

OECD/NEA/IEA (2005) Projected costs of generating electricity 2005. OECD Publishing, Paris. https://doi.org/10.1787/9789264008274-en

O'Neill BC, Kriegler E, Ebi KL, Kemp-Benedict E, Riahi K, Rothman DS, van Ruijven BJ, van Vuuren DP, Birkmann J, Kok K, Levy M, Solecki W (2017) The roads ahead: narratives for shared socioeconomic pathways describing world futures in the 21st century. Glob Environ Change 42:169–180

Paltsev S, Capros P (2013) Cost concepts for climate change mitigation. Clim Change Econ 4:1340003

Paltsev S, Reilly JM, Jacoby HD, Eckaus RS, McFarland J, Sarofim M, Asadoorian M, Babiker M (2005) The MIT Emissions Prediction and Policy Analysis (EPPA) Model: Version 4. Report No. 125

Ragwitz M, Held A, Klein A, Resch G, Faber T, Haas R (2006) Report (D9) of the IEE project, OPTRES: assessment and optimisation of renewable energy support schemes in the European electricity market, Best practice support schemes for RES-E in a dynamic European electricity market, Karlsruhe (Germany)

Rao S, Klimont Z, Leitao J, Riahi K, Van Dingenen R, Reis LA, Calvin K, Dentener F, Drouet L, Fujimori S, Harmsen M, Luderer G, Heyes C, Strefler J, Tavoni M, Van Vuuren DP (2016) A multi-model assessment of the co-benefits of climate mitigation for global air quality. Environ Res Lett 11:124013

Ravallion M (1997) Can high-inequality developing countries escape absolute poverty? Econ Lett 56:51–57

Ravallion M (2001) Growth, inequality and poverty: looking beyond averages. World Dev 29:1803–1815

Ravallion M, Chen S (1997) What can new survey data tell us about recent changes in distribution and poverty? World Bank Econ Rev 11:357–382

Reis LA, Drouet L, Van Dingenen R, Emmerling J (2018) Future global air quality indices under different socioeconomic and climate assumptions. Sustainability 10:1–27

Riahi K, van Vuuren DP, Kriegler E, et al (2017) The Shared Socioeconomic Pathways and their energy, land use, and greenhouse gas emissions implications: an overview. Glob Environ Change 42:153–168

Rogelj J, Shindell D, Jiang K, Fifita S, Forster P, Ginzburg V, Handa C, Kheshgi H, Kobayashi S, Kriegler E, Mundaca L, Séffian R, Vilariño MV (2019) Mitigation pathways compatible with 1.5°C in the context of sustainable development. In: Masson-Delmotte V, Zhai P, Pörtner H-O, Roberts D, Skea J, Shukla PR, Pirani A, Moufouma-Okia W, Péan C, Pidcock R, Connors S, Matthews JBR, Chen Y, Zhou X, Gomis MI, Lonnoy E, Maycock T, Tignor M, Waterfield T (eds) Global warming of 1.5°C. An IPCC Special Report on the impacts of global warming of 1.5°C above pre-industrial levels and related global greenhouse gas emission pathways, in the context of strengthening the global response to the threat of climate change, sustainable development, and efforts to eradicate poverty

Roy J, Tschakert P, Waisman H, Abdul Halim S, Antwi-Agyei P, Dasgupta P, Hayward B, Kanninen M, Liverman D, Okereke C, Pinho PF, Riahi K, Suarez Rodriguez AG (2019) Sustainable development, poverty eradication and reducing inequalities. In: Masson-Delmotte V, Zhai P, Pörtner H-O, Roberts D, Skea J, Shukla PR, Pirani A, Moufouma-Okia W, Péan C, Pidcock R, Connors S, Matthews JBR, Chen Y, Zhou X, Gomis MI, Lonnoy E, Maycock T, Tignor M, Waterfield T (eds) Global warming of 1.5°C. An IPCC Special Report on the impacts of global warming of 1.5°C above pre-industrial levels and related global greenhouse gas emission pathways, in the context of strengthening the global response to the threat of climate change, sustainable development, and efforts to eradicate poverty

United Nations (UN) (2015) Resolution adopted by the General Assembly on 25 September 2015. A/RES/70/1. United Nations, New York

United Nations (UN) (2017a) Resolution adopted by the General Assembly on Work of the Statistical Commission pertaining to the 2030 Agenda for Sustainable Development. A/RES/71/313. United Nations, New York

United Nations (UN) (2017b) World Income Inequality Database (WIID3.4). UNU-WIDER. United Nations, New York

United Nations (UN) (2018) UN data

United Nations Environment Programme (UNEP) (2015) The Emissions Gap Report 2015. Nairobi

van Vuuren DP, Kriegler E, O'Neill BC, Ebi KL, Riahi K, Carter TR, Edmonds J, Hallegatte S, Kram T, Mathur R, Winkler H (2014) A new scenario framework for Climate Change Research: scenario matrix architecture. Clim Change 122(3):373–386 von Stechow C, McCollum D, Riahi K, Minx JC, Kriegler E, van Vuuren DP, Jewell J, Robledo-Abad C, Hertwich E, Tavoni M, Mirasgedis S, Lah O, Roy J, Mulugetta Y, Dubash NK, Bollen J, Ürge-Vorsatz D, Edenhofer O (2015) Integrating global climate change mitigation goals with other sustainability objectives: a synthesis. Ann Rev Environ Resour 40(1):363–394

von Stechow C, Minx JC, Riahi K, Jewell J, McCollum DL, Callaghan MW, Bertram C, Luderer G, Baiocchi G (2016) 2 °C and SDGs: united they stand, divided they fall? Environ Res Lett 11(3)

Vona F, Marin G, Consoli D, Popp D (2018a) Environmental regulation and green skills: an empirical exploration. J Assoc Environ Resour Econ 5(4):713–753

Vona F, Marin G, Consoli D (2018b) Green employment: what, where and how much? Blog OFCE. SciencesPo

World Bank (WB), The (2018) World Development Indicators (WDI)

Youssef A, Lannes L, Rault C, Soucat A (2015) Energy consumption and health outcomes in Africa. J Energy Dev 41:175–200

Co-benefits Under the Market Mechanisms of the Paris Agreement

Axel Michaelowa, Aglaja Espelage, and Stephan Hoch

1 Introduction

Greenhouse gas (GHG) mitigation action in developing countries has always been voluntary. Under the Kyoto Protocol (KP), only industrialised countries had quantified emission reduction targets. While under the Paris Agreement (PA) all countries have to submit nationally determined contributions (NDCs), each country is free to decide how to define such contributions. At the same time, the PA explicitly 'welcomes' the Agenda 2030 for Sustainable Development and prominently mentions that the long-term objectives of the PA should take place 'in the context of sustainable development' (UNFCCC 2015, Article 2). Given this situation, any mitigation in developing countries needs to show other benefits that justify the outlays. Such (co-)benefits (Mayrhofer and Gupta 2016) relate to sustainable development (SD), and have traditionally been differentiated into economic, social, and environmental benefits.

International carbon market mechanisms allow developing countries to generate revenues by the sale of emission reduction credits. These can accrue from a wide range of technologies, which will have different SD impacts. We now have 20 years of experience with international carbon markets and the way how they treated SD benefits (and negative impacts). Given that the market mechanism under Article 6.4 of the PA establishes path dependency by continuing to assign the dual objective of promoting both mitigation and SD in the establishment of the new multilateral market mechanism, it is important to draw lessons from past experience. This

---

A. Michaelowa · A. Espelage · S. Hoch
Perspectives Climate Research, Freiburg, Germany

A. Michaelowa (✉)
Department of Political Science, University of Zurich, Zürich, Switzerland
e-mail: michaelowa@perspectives.cc; axel.michaelowa@pw.uzh.ch

© Springer Nature Switzerland AG 2020
W. Buchholz et al. (eds.), *Ancillary Benefits of Climate Policy*, Springer Climate,
https://doi.org/10.1007/978-3-030-30978-7_3

chapter presents the different phases of this history and provides an outlook on how SD can be brought into the new market mechanisms under the PA.

1.1 The Kyoto Mechanisms and SD Benefits

The Clean Development Mechanism (CDM) allows emission reduction projects in developing countries to generate certified emission reductions (CERs). As per Article 12 of the KP, the CDM has a dual objective: helping industrialised countries to reach their mitigation targets in a cost-effective manner, and to assist developing countries in achieving SD while voluntarily contributing to the ultimate objective of the UNFCCC. The CDM operates under the guidance and authority of the conference of the parties to the Kyoto Protocol (CMP), and is overseen by the CDM Executive Board (CDM EB) which is supported by staff of the Secretariat of the UN Framework Convention on Climate Change (UNFCCC). The other Kyoto mechanisms Joint Implementation (JI) and International Emissions Trading (IET) do not have an SD target.

1.2 The Official Rules Regarding SD Under the CDM

Each CDM project needs to be formally registered by the CDM EB before it can start to generate CERs. In order to file a registration request, a project developer needs an official letter of approval by the Designated National Authority (DNA) of the country where the project is located. This letter needs to state explicitly 'that the project activity contributes to sustainable development in the country'.

The official template for the documentation of a CDM project (Project Design Document, PDD) contains a section on SD benefits. In contrast to the GHG emissions which are monitored and verified according to approved methodologies, SD benefits do not require monitoring. The documentation needs to be published on the internet for a period of a month, and any stakeholder can submit a comment to the UNFCCC-accredited auditor undertaking the check (validation) of the project documentation.

Any CDM project needs to undertake local stakeholder consultations. However, the format of this consultation was never specified. While good practice has been to hold a public meeting where stakeholders can ask questions and comment on the project design, in some important CDM host countries simple written questionnaires have been used.

With regard to the check of SD benefits by DNAs in the phase of CDM emergence after the Marrakech Accords of 2001, Michaelowa (2003) provides examples of SD criteria lists from India, Indonesia and Morocco. In all cases, development assistance supported the engagement of stakeholders with the DNAs leading to a set of criteria. But it took several years to achieve an operationalisation of SD benefit testing. However, a systematic alignment of the contribution of mitigation activities to multilateral SD goals (at that time the Millenium Development Goals) has never been attempted either at conceptual or country level.

1.3 Stakeholder Views Regarding SD Benefits Under International Carbon Markets

According to Dransfeld et al. (2017), lenient approaches to SD benefits under the CDM are supported by governments that have a limited budget for acquisition of CERs. These fear that overly stringent rules limit credit supply and increase CER prices. Many host countries put a high value on their sovereign right of defining SD within their territory. This is often linked to the view that SD needs to be specified at the local level and cannot be imposed from the outside. Michaelowa (2005) stressed that in the early CDM phase only few host countries put an emphasis on defining SD criteria and explained it by the lacking clout of stakeholders' interested in SD with regards to government decision-making.

More stringent approaches are supported by project developers in some voluntary carbon standards who see a comparative advantage in transparent SD assessments that showcase high SD benefits. Likewise, buyers in the voluntary markets require access to projects with high SD contributions in order to convince customers that their efforts to become climate neutral are credible and not 'tainted' by negative SD impacts. The strongest SD supporters are naturally environment and development NGO representatives who want to achieve a 'holistic' outcome not just focusing on GHG mitigation.

Matsuo (1998) and Thorne and La Rovere (1999) were the first ones to propose a criteria and indicator set for assessment of SD benefits of CDM projects. These criteria sets informed many of the early host country attempts to operationalise SD assessment, as well as early research (Sutter 2003). Compared to the development of methodologies and parameters for defining baselines for mitigation, SD criteria and indicators took a back seat. This is also due to the fact that monitoring SD benefits involves high transaction costs. In particular indirect SD benefits, such as job benefits, require elaborate methodologies such as randomised control trials, which can be extremely expensive. Dransfeld et al. (2017) stress that striking an acceptable balance between costs and depth of SD assessment is a key challenge.

2 The CDM 'Gold Rush' and SD Benefits

From 2005 onwards, the CDM expanded in a remarkable fashion as CER demand from the EU triggered a rapid increase in CER prices. Consultants raced around the world to identify the most attractive CDM opportunities. Thousands of projects generating over one billion CERs were developed in a few years in over 90 developing countries. CDM project developers became quoted on stock exchanges valued at hundreds of million Euros. In such a booming market, the question of SD benefits was initially sidelined, but emerged in a powerful way when the initial shortage of CERs gave way to a 'supply glut' and enabled buyers to differentiate between CERs of different characteristics.

2.1 National Race to the Bottom Regarding SD Contributions

The national policies with regard to SD benefits differed significantly in the CDM gold rush phase, but overwhelmingly displayed 'laissez faire' characteristics. Generally, DNAs of the countries with the largest CDM project pipelines did not apply stringent SD benefit criteria. TERI (2012, p. 118ff) provides a summary of SD criteria and indicators published by DNAs, finding that only 16 out of 28 assessed DNAs published their criteria on the internet. Adding results from a survey, TERI found three DNAs applying numerical scores (Bhutan, Georgia and Thailand), while nine DNAs provided criteria with differentiated indicator lists and 14 DNAs only provided general criteria (TERI 2012, p. 20ff).

In Brazil, Malaysia, Peru, Rwanda and Uruguay, a more stringent assessment of SD benefits takes place. The Rwandan DNA requires project developers to update SD checklist at each verification, being the only case of a mandate of ex-post SD monitoring (TERI 2012, p. 36). Uruguay applied a complex process of weighting SD benefits proposed by Sutter (2003) and Olsen and Fenhann (2008). Peru's DNA visited project sites and asked local communities about their needs and their potential contribution to the project (TERI 2012). Brazil made a thorough check of the SD as well as additionality argumentation in the PDD (Hultman et al. 2012). Indonesia required project developers to provide explanation and justification for a set of 17 indicators, to be checked by the DNA's Technical Team. Failure to fulfil one indicator would lead to rejection. Malaysia applied strict technology transfer criteria which led to rejection of about 20% of proposals (Michaelowa 2006).

In other countries, SD contributions play a marginal to negligible role for host country approval of projects. India and South Africa applied a set of highly generic criteria for social, environmental, economic and technological well-being, and officially checked PDDs regarding those (Boyd et al. 2009, p. 827). The South African case is surprising given it had an early, intense stakeholder consultation process on SD criteria (Kim 2004). NGOs criticised India and South Africa heavily for not safeguarding SD (Tiwari 2014 for India, see TERI (2012, p. 92ff), Johnson (2018) and Sangam (2017) for a good summary of the contentious discussions on a landfill gas CDM project in Durban, and South African DNA (2009) for a widely uncritical assessment of the SD benefits of South Africa's CDM portfolio). Chile was criticised by Rindefjall et al. (2011) for seeing CDM credits just as another export commodity and thus not checking SD benefits at all. China, the largest CDM host globally did never publish any SD criteria or indicators (TERI 2012, p. 20); indirectly it stated its preferences by introducing a domestic tax on CER revenues differentiated according to project types.

2.2 Potential Policy Options for Ensuring SD Under International Market Mechanisms

Specific policy suggestions in order to promote SD benefits were first made by Boyd et al. (2009, p. 828). They suggest different approaches: international minimum SD standards, global SD checklists where countries could add or delete criteria, an SD benefit point allocation where projects would have to reach a minimum score or emission credit multipliers for projects with high SD benefits. With regards to a point-based allocation, Parnphumeesup and Kerr (2011) found that experts and the local population weight SD criteria differently in the context of biopower projects in Thailand.

Torvanger et al. (2013) proposed a special CDM track focusing on SD benefits, with a presumably higher credit price. While officially, such policies have not been introduced, some of them have 'de facto' been applied by CER buyers since the end of the gold rush period.

2.3 Do CDM Projects Generally Have a Low SD Contribution?

A seminal review of academic literature on SD benefits of CDM projects undertaken during the gold rush period (Olsen 2007) concluded that there is a general trade-off between the economic attractiveness of a CDM project and its SD contribution. If one wanted to minimise CER generation costs, spending on SD benefits would always be an obstacle. Olsen thus saw little incentive for project developers to actually promote SD. Olsen (2007) as well as a similarly critical report by Schneider (2007) triggered a wealth of contributions trying to assess whether the situation was actually so problematic.

Some researchers tried to do large-scale text analysis of PDDs. Watson and Fankhauser (2009) did this for 409 documents and concluded that end-of-pipe projects tend to generate lower SD benefits than renewable energy or forestry projects in particular. They did not find a difference between small- and large-scale projects. TERI (2012, p. 57ff) applied a similar approach to 202 PDDs using stratified random sampling. They find that 95% claimed economic, 86% social and 74% environmental benefits. Small-scale projects claim more SD benefits than large ones.

Multiple 'small n sample assessments' of SD benefits following Olsen (2007) concluded mostly with negative results. All these were based on an assessment of

PDDs, so probably still led to an optimistic outlook given that project developers are unlikely to provide a negative outlook in their documentation (Horstmann and Hein 2017). These assessments can be differentiated into global, country and project-type specific ones. Sutter and Parreño (2007) found none of 16 projects contributing to SD. Ellis et al. (2007) gave a high-level view on SD benefits of 18 projects, arguing that there was a trade-off between project size and SD benefits. Boyd et al. (2009) assessed 10 randomly chosen CDM projects in five countries whether they provided direct or indirect SD benefits, finding mixed results. Dirix et al. (2016) doing a meta-analysis sweepingly conclude that 'the CDM has failed to deliver poverty alleviation'.

Many analyses focused on specific host countries. Alexeew et al. (2010) assessed 40 projects in India and found a trade-off between project additionality and SD benefits. Subbarao and Lloyd (2011) found that five small renewable energy projects in India failed to provide SD benefits. Karakosta et al. (2013) looked at 14 projects in Kenya and found that they score well regarding economic benefits, whereas environmental and social benefits are limited. Uddin et al. (2015) focused on 30 coal mine methane reduction projects in China finding benefits regarding mine safety, employment and technology transfer. Yan (2016) assessed 69 projects in China and found a trade-off between SD contribution and CER generation.

With regard to project type-specific assessment, Gupta et al. (2008) assessed five small renewable energy projects and found only one to have high SD benefits. Fernández et al. (2014) found positive social benefits of 46 hydro projects in Brazil. Aggarwal (2014) found negative impacts of three out of four forestry CDM projects in India.

More recently, the methods for SD assessment have been broadened. Tiwari and Goga d'Souza (2009) made field visits to seven typical CDM projects in tribal areas in India, covering energy efficiency in industry, forestry, hydro and biomass power. Their assessment is scathing, stating 'most projects violate promises made for sustainable development' (ibid., p. 44), and listing land grabbing without or only limited compensation, local environment pollution, lack of long-term employment benefits and limitation of access to local common resources as key negative impacts. Rousseau's (2017) detailed field study of a CDM hydropower project in Southern China finds that the CDM status did not change the SD outcome of the project to the better compared to non-CDM hydro projects.

Wang et al. (2013) did an input–output analysis to assess employment impacts of power sector CDM projects in China, resulting in insignificant direct job losses, but much larger indirect job gains.

He et al. (2014) applied a panel regression for the period 2005–2010 to 58 CDM host countries with the Human Development Index as dependent variable and found 'convincing evidence' that CDM projects contribute to SD in these countries. Du and Takeuchi (2018) econometrically assessed SD benefits of renewable energy CDM projects in China and found that biomass energy projects had significant income and employment effects in rural communities while wind and PV projects had employment benefits only.

Mori-Clement (2019) applied a novel empirical framework for SD assessment of 338 CDM projects in Brazil. She compared data for economic and social development in 425 municipalities with at least one CDM project for the period before 2000 (pre-CDM) and 2010 (post-CDM) for hydro, biomass energy, landfill gas and methane avoidance projects. She found positive local income and job effects for all project types, but can attribute poverty reduction only to hydro projects. However, Mori-Clement and Bednar-Friedl (2019) in another paper qualified the employment effects as 'small and transitory'. Her results run counter the earlier literature which general sees very low SD benefits of hydropower projects.

A rich body of literature has developed regarding the technology transfer benefits of CDM projects which relatively early applied empirical approaches (Seres et al. 2010; Lema and Lema 2013; Murphy et al. 2015; TERI 2012, p. 68ff). These analyses are consistently finding a relatively high incidence of technology transfer, but a heterogeneity across project types and a decline over time in the emerging economies which have hosted the majority of CDM projects.

2.4 How Media and NGO Campaigns Triggered CER Buyer Differentiation to Safeguard SD Benefits

Already in 2003 the EU had introduced specific checks to prevent that hydro CDM projects generate negative impacts on local populations. Any hydro project above 20 MW capacity wanting to export CERs into the EU had to prove that the World Commission on Dams standards were respected.

As discussed by Schade and Obergassel (2014), during the gold rush period a number of projects emerged that were seen as negatively impacting on SD and human rights. These included hydropower dams that led to resettlement as well as biopower plants in palm oil plantations that clamped down on local protesters asking for decent work conditions. NGOs focused on specific cases, especially the Aguan biogas power palm oil project in Honduras and the Barro Blanco hydro plant in Panama. In the former case, the project owners were accused of intimidating and even killing members of the local community who were protesting against land grabbing. In the latter case, project owners were reproached to override the interests of a local indigenous community. TERI (2012, p. 103) stresses that the abuses, if existing, did not relate to the CDM project as such but the underlying economic activity. In contrast to TERI (2012), Schade and Obergassel (2014) argue that CER revenues from Aguan helped to finance a human rights violator. Similarly, Obergassel et al. (2017) found a 'number of human rights infringements for Barro Blanco, the Ugandan hydropower project Bujagali and the Kenyan geothermal project Olkaria'. In response to the criticism, Panama's DNA revoked the CDM approval letter for the Barro Blanco project, an unprecedented step (Chatziantoniou and Alford-Jones 2016).

Other project types such as industrial gas destruction plants and landfill gas collection were seen as providing strong CER revenues while not benefitting the local population. For certain large projects reducing the industrial gas HFC-23 perverse incentives to increase production of the refrigerant HCFC-22 with negative climate impacts were feared (Wara 2007). A small NGO called 'CDM Watch' singlehandedly organised a campaign against industrial gas projects (see Bryant 2019, p. 131ff for an account of the campaign, and CDM Watch 2010 for its argumentation), achieving a prohibition of import of those credits by the EU Commission from 2013 onwards. This essentially made CERs from industrial gas projects a non-sellable commodity anywhere.

The campaigns of NGOs also led to changes in host country DNA rules for SD. For example, in India, from 2012 onwards, the DNA required project developers to spend at least 2% of CER revenues on SD activities with local communities and to include a monitorable action plan in the project documentation. However, this provision was not really enforced (Joshi 2013) and came at a time when the CER prices crashed. It thus essentially remained 'dead letter'.

2.5 The Emergence of Voluntary Private Standards Fostering SD

On the voluntary carbon markets, demand is driven mostly by corporate and consumer social and environmental responsibility considerations. Therefore, buyers seek not only to offset their emissions in the most cost-effective way as it is the case for demand in compliance schemes, but want to support activities with overall societal and environmental benefits. Already in 2003, a private organization called Gold Standard (GS) was set up by NGOs and government institutions to provide an enhanced CDM procedure in order to ensure strong SD benefits (Headon 2009). Over the years, the GS has developed its own, elaborate approach to assess SD benefits, and has eventually fully embraced indicators that measure contributions of GS projects to Sustainable Development Goals (SDGs). This approach includes multiple stakeholder consultations. Subsequently, the Climate, Community & Biodiversity Standard (CCBS) (Melo et al. 2014) and the Social Carbon Standard (SCS) followed. In 2015, almost 40% of transactions on the voluntary markets used one of these three standards, and price premia reached 2.7 $/credit for CCBS credits and 0.6 $ for GS credits (Hamrick 2015). In 2017, three million GS CERs were issued (GS 2018a, p. 12).

All three standards require ex ante SD benefit assessments as well as monitoring the identified SD indicators. The monitored values are verified by an independent party, which 'ensures compliance and therefore adds to the reliability and credibility of the SD assessment' (Arens et al. 2015b, p. 50). Moreover, meetings with local stakeholders are mandatory and grievance mechanisms exist (for more details see Arens et al. 2015b). While the GS has a stronger focus on safeguards ('do no harm' approach), Social Carbon is geared towards a continuous improvement ('do good' approach) and the CCBS includes both ('do no harm' and 'do good') approaches (Dransfeld et al. 2017). The Fair Trade Climate Standard, operated in cooperation with GS, focuses specifically on social SD impacts (Fair Trade 2019).

Using a sample of 39 projects, Nussbaumer (2009) found that GS projects slightly outperform 'normal' CDM projects with respect to SD benefits. Drupp (2011) found a similar result for a sample of 18 GS projects compared with 30 other projects. Crowe (2013) found a clear overperformance of CCBS projects regarding poverty benefits, while GS projects only perform somewhat better than normal CDM projects.

2.6 The SD Tool

In 2011, the CDM EB invited input on how SD benefits from CDM projects could be enhanced, in the context of a campaign to improve the reputation of the CDM. This led to UNEP DTU developing a voluntary SD tool under the guidance of the EB, which can be used by project developers. The tool which is based on earlier work by Olsen and Fenhann (2008) contains standardised evaluation criteria for the key categories of SD impacts. All project developers who use the tool prepare a Sustainable Development Co-benefit (SDC) report. By March 2019, 63 reports were available on the website,[1] i.e. 20.4% of CDM projects registered since the publication of the tool in April 2014, but only 0.8% of all registered CDM projects. Olsen et al. (2018a) interviewed eight users of the SD tool who criticised that the tool had no safeguards against negative SD impacts and did not provide monitoring and verification approaches. Similar statements had already been collected by Olsen et al. (2015). Arens et al. (2015a, c) provide recommendations on how to reform the SD tool.

3 SD in the Context of the Paris Mechanisms and the Agenda 2030

3.1 The Crisis of International Market Mechanisms After 2013

Between 2011 and 2013, the secondary market CER price fell by 95% as the import ceiling for CERs under the EU ETS was reached, and other industrialised countries like Japan stopped the acquisition of CERs due to the uncertainty about the international climate policy regime in the aftermath of the Copenhagen failure.

---

[1] Accessible via: https://www4.unfccc.int/sites/sdcmicrosite/Pages/SD-Reports.aspx (last accessed on April 1, 2019).

This led to a sudden halt in new CDM activities and frantic effort of the existing ones to secure buyers. Only a small subset of buyers including some Nordic countries and the World Bank continued to acquire CERs, some at substantial premia compared to the secondary market price. These buyers focused on project types perceived as inherently having high SD benefits, such as distributed household energy efficiency technologies. This led to the emergence of many efficient cookstove CDM programmes of activities (PoA) in Africa, which became the largest category under the programmatic CDM (see Dransfeld et al. 2015 for a discussion of such PoAs). Lately, also CER demand from Korea has led to further demand for such project types. With regard to project types commonly seen as providing high SD benefits, lately a discussion has come up whether the benefits are actually achieved (see Pickering et al. (2016) for a scathing criticism of water filtre activities in Kenya, and Aung et al. (2016) for a similar criticism of cookstove activities in India).

3.2 The SDGs: New Support for SD Contributions

The development of the 2030 Agenda for Sustainable Development, under whose umbrella the Sustainable Development Goals (SDGs) have been formulated, coincided with the negotiation of the PA. The PA welcomes in its preamble the adopted Agenda and is seen as first international agreement falling under this new universal approach to SD. Both agreements mark the culmination of decades of negotiations on governing climate and development policy together, influenced by a growing body of research literature in this regard (Gomez-Echeverri 2019). The Agenda 2030 establishes, as does the PA, a bottom-up system allowing government to prioritise their own needs and approaches in achieving a common goal (on the governance approach, see for instance Biermann et al. 2017). The 17 SDGs are further substantiated through 169 targets on sustainable development.

SDG 13 directly calls the international community to take action to combat climate change and its impacts and specifically mentions actions in mitigation, adaptation and resilience. Related targets and indicators of the specific goal are more procedural in nature: implementing disaster risk reduction strategies, integrating climate change measures in policy planning, building capacities and institutions, strengthen education and mobilise funding for the UNFCCC (UNGA 2015). Mitigation, adaptation and resilience actions are then linked to several SDGs that have an impact on climate change (such as SDG7—clean energy, SDG 9—industry and infrastructure, SDG 11—sustainable cities and SDG 12—production and consumption) or are impacted by climate change (SDG 1—poverty, SDG 2—food security, SDG3—health) (Gomez-Echeverri 2019). Other SDGs relate to areas that both contribute to mitigation and adaptation goals of the PA such as SDG 14—life below water and SDG 15—life on land.

The unanimous adoption of the 2030 Agenda gave it significant political weight at the international level, supported also by NGOs active in the environment and social spheres of society. The SDGs are also influencing the debate on measuring

SD benefits in international carbon markets. The usefulness of monitoring carbon market activities against different SDGs is however not very straightforward. Translating the 169 targets under the SDGs in measurable and simultaneously fitting indicators to be monitored at a national level already poses an 'unprecedented statistical challenge' to the international community (MacFeely 2018). Their translation into monitoring, reporting and verification systems at the level of economic sectors, local communities or even of singular projects or programmes, for instance in the context of carbon market activities, is highly challenging and requires the development of specific methodologies for the relevant context.

Some efforts are being undertaken by voluntary standards in this regard. The organisation Verra active on the voluntary market is currently developing the Sustainable Development Verified Impact Standard (SD VISta) as framework to assess and report upon the SD co-benefits of project-based activities. To be certified, project developers will have to demonstrate positive impact of their project on at least one SDG covering either the dimension 'people and prosperity' or 'planet' (Verra 2018). Also, the GS demands SD reporting from its project developers. One option is to select and monitor against the SDG targets and indicators as approved by the United Nations. Another option is to follow an already approved SDG tool in line with an SDG methodology. The GS is rebranding its standard as 'Global Standard for Global Goals' aiming to certify climate and SD benefits in a holistic approach and not considering SD as a 'co-benefit' anymore. Under this standard, specific tools for monitoring for some SDGs have been developed for SDGs 13 (climate action), 7 (affordable and clean energy), 6 (clean water and sanitation), 5 (gender equality), and 3 (good health and well-being) (GS 2018b).

3.3 Article 6 Market Mechanisms Under the PA and Negotiations About the Characteristics of Their SD Regulation

The decision to introduce new market mechanisms through Article 6 of the PA had not been expected by many, given that negotiations on new market mechanisms had been sluggish and rather contentious (see Hoch et al. 2015). The market mechanism under Article 6.4, often seen as the successor to the CDM, explicitly refers to SD as key objective besides achieving mitigation. The Paris Mechanisms are likely to be more dominated by governments than the Kyoto Mechanisms, and their scope is likely to extend to policy instruments and entire sectors.

The negotiations on rules for the Article 6 mechanisms progressed slowly after Paris. With regard to SD, Olsen et al. (2018a, b) summarise a convergence that SD remains to be decided on the national level, while a tendency to use a tool with comparable indicators and standards was seen. Olsen et al. (2019) thus propose a voluntary SD labelling tool, building on the experiences achieved with the CDM SD tool.

At COP 24 in Katowice in December 2018, the so-called 'Paris Rulebook' was to be agreed. While the COP indeed took such a decision, it could not agree on the detailed rules for the market mechanisms and the decision was deferred to COP 25.

While the rules have not been formally agreed, the discussions of SD had led to an unbracketed text (UNFCCC 2018, para 36) that stated that SD contributions remain to be defined in the host country approval process. A push by some parties to include a specific reference to the SDGs was unsuccessful, even if given the nature of SDGs, it would not infringe upon the prerogative of sovereign governments to set their own SD priorities. However, under the Article 6.4 mechanism now a grievance procedure is foreseen which will allow stakeholders to raise questions of SD impacts (UNFCCC 2018, para 54).

3.4 Will Demand Make a Difference?

Under the bottom-up architecture of the PA demand for credits from the Article 6 mechanisms is likely to be rather fragmented. Especially in the context of activities undertaken without UNFCCC oversight under Article 6.2, the buyer is likely to 'call the shots'. Greiner et al. (2019) assessed the currently known pilot activities for Article 6 and found that the majority of them put an emphasis on ensuring SD benefits. The concrete operationalisation, however, remains to be seen. Article 6.2 establishes an accounting framework for transfers of mitigation outcomes. Underlying safeguards and principles for the activities as such are likely to be introduced only as reporting requirements with some sort of international review on the reported information. It is likely that a 'no harm' principle like the one agreed under the airline offset scheme CORSIA could play an important role in these reporting and review processes.

4 Conclusions

SD benefit accrual from international carbon markets has been made difficult from the beginning by the fact that defining SD is universally seen as issue of national sovereignty. The relevance of this fundamental principle of international relations has not changed between the Kyoto Protocol and the Paris Agreement.

While the CDM gold rush phase showed a clear tendency for a race to the bottom with regards to SD benefit testing by DNAs when CERs were a scarce commodity, the end of the phase also saw the emergence of a bottom-up approach to ensure SD benefits by CER buyers. This was aided by a supply surplus of CERs, which allowed buyers to become 'picky'. A significant price premium for high SD project types emerged, but volumes remained small. For the Paris Mechanisms, the role of SD benefit checks remains uncertain. Yet, the elevated role of achieving the SDGs may offer potential to work towards a greater alignment between UNFCCC policy instruments and Agenda 2030. First indications about the political will of governments should become clearer over the next 2 years as Article 6 pilot activities emerge.

The shift from project-specific case studies to more aggregated checks of SD contributions led to an erosion of the earlier perception that CDM projects generally fail to contribute to SD. However, whenever specific operational CDM projects are assessed on the ground the outcomes are more negative than on an aggregated level. This is particularly striking for project types like hydropower, which perform well in aggregated studies, but are heavily criticised in specific case studies.

Voluntary standards tools for SD assessment exist and have been refined over the years but are still only used by a minority of project developers. SD performance of projects developed under these standards seems to be better than of the 'normal' projects, but only slightly. The adoption of the Agenda 2030 by the UN boosted the discussion on integrating SD benefits in climate action and therefore in carbon market activities. Voluntary standards are actively promoting the development of SDG-inspired methodologies to credit mitigation activities and their SD benefits. This development had so far no impact on the development of the rules for the Paris mechanisms, but could influence the design of bilateral cooperative approaches and the demand for international credits with high SD benefits in the future.

References

Aggarwal A (2014) How sustainable are forestry clean development mechanism projects?—a review of the selected projects from India. Mitig Adapt Strateg Glob Chang 19:73–91

Alexeew J, Bergset L, Meyer K, Petersen J, Schneider L, Unger C (2010) An analysis of the relationship between the additionality of CDM projects and their contribution to sustainable development. Int Environ Agreements 10:233–248

Arens C, Mersmann F, Beuermann C, Rudolf F (2015a) Reforming the CDM SD tool: recommendations for improvement. DEHSt Discussion Paper, Berlin

Arens C, Mersmann F, Beuermann C, Rudolf F, Olsen KH, Fenhann, J (2015b) Mapping the indicators: an analysis of sustainable development requirements of selected market mechanisms and multi-lateral institutions. DEHSt discussion paper, Berlin

Arens C, Beuermann C, Mersmann F, Rudolf F, Holm Olsen K, Fenhann J, Hinostroza M, Bakhtiari F (2015c) Final report of the Project "Evaluation and development of recommendations on the CDM EB's Sustainable Development tool including the sustainability requirements of other flexible mechanisms". Climate Change 23/2015, Federal Environmental Office, Dessau

Aung T, Jain G, Sethuraman K, Baumgartner J, Reynolds C, Grieshop A, Marshall J, Brauer M (2016) Health and climate-relevant pollutant concentrations from a carbon-finance approved cookstove intervention in rural India. Environ Sci Technol 50:7228–7238

Biermann F, Kanie N, Kim RE (2017) Global governance by goal-setting: the novel approach of the UN sustainable development goals. Curr Opin Environ Sustain 26–27:26–31

Boyd E, Hultman N, Timmons Roberts J, Corbera E, Cole J, Bozmoski A, Ebeling J, Tippman R, Mann P, Brown K, Liverman D (2009) Reforming the CDM for sustainable development: lessons learned and policy futures. Environ Sci Policy 12:820–831

Bryant G (2019) Carbon markets in a climate-changing capitalism. Cambridge University Press, Cambridge

CDM Watch (2010) CDM Watch submission to the European Commission on design aspects of quality restrictions on the use of credits from industrial gas projects, Brussels

Chatziantoniou A, Alford-Jones K (2016) Panama withdraws problematic Barro Blanco Dam project from CDM registry. https://www.ciel.org/panama-withdraws-problematic-barro-blanco-dam-project-cdm-registry/. Accessed 8 March 2019

Crowe T (2013) The potential of the CDM to deliver pro-poor benefits. Clim Pol 13:58–79

Dirix J, Peeters W, Sterckx S (2016) Is the clean development mechanism delivering benefits to the poorest communities in the developing world? A critical evaluation and proposals for reform. Environ Dev Sustain 18:839–855

Dransfeld B, Hoch S, Honegger M, Michaelowa A (2015) Developing sectoral mechanisms in the transition period towards a new climate treaty, Climate Change 01/2015. Federal Office of the Environment, Dessau

Dransfeld B, Wehner S, Bagh T, Bürgi P, Puhl I, Zegg M, Friedmann V, Hoch S, Honegger M, Michaelowa A, Warland L (2017) SD-benefits in future market mechanisms under the UNFCCC, Climate Change 04/2017. Federal Environmental Office, Dessau

Drupp M (2011) Does the Gold standard label hold its promise in delivering higher sustainable development benefits? A multi-criteria comparison of CDM projects. Energy Policy 39:1213–1227

Du Y, Takeuchi K (2018) Can climate mitigation help the poor? Measuring impacts of the CDM in rural China, discussion paper no.1808. Graduate School of Economics, Kobe University, Kobe

Ellis J, Winkler H, Corfee-Morlot J, Gagnon-Lebrun F (2007) CDM: taking stock and looking forward. Energy Policy 35:15–28

Fair Trade (2019) Carbon credits. https://info.fairtrade.net/product/carbon-credits. Accessed 23 Mar 2019

Fernández L, de la Soto C, Silveira Andrade J, Lumbreras J, Mazorra J (2014) Social development benefits of hydroelectricity CDM projects in Brazil. Int J Sust Dev World Ecol 21:246–258

Gold Standard (2018a) Annual report 2017, Geneva

Gold Standard (2018b) Certified SDG impacts. https://www.goldstandard.org/our-work/new-certification-solutions. Accessed 21 Mar 2019

Gomez-Echeverri L (2019) Climate and development: enhancing impact through stronger linkages in the implementation of the Paris Agreement and the Sustainable Development Goals (SDGs). Philos Trans A Math Phys Eng Sci 376:20160444

Greiner S, Chagas T, Krämer N, Michaelowa A, Brescia D, Hoch S (2019) Moving towards next generation carbon markets. Observations from art. 6 pilots. Climate Finance Innovators, Freiburg

Gupta J, van Beukering P, van Asselt H, Brander L, Hess S, van der Leeuw K (2008) Flexibility mechanisms and sustainable development: lessons from five AIJ projects. Clim Pol 8:261–276

Hamrick K (2015) Ahead of the curve. State of the voluntary carbon markets. Forest Trends Ecosystems Marketplace, Washington, DC

He J, Huang Y, Tarp F (2014) Has the clean development mechanism assisted sustainable development? Nat Res Forum 38:248–260

Headon S (2009) Whose sustainable development? Sustainable development under the Kyoto Protocol, the Coldplay effect, and the CDM Gold Standard. Colo J Int Environ Law Policy 20:127–156

Hoch S, Horstmann B, Michaelowa A, Hein J (2015) New climate investments must strengthen sustainable development and minimize trade-offs. DIE Briefing Paper, Bonn

Horstmann B, Hein J (2017) Aligning climate change mitigation and sustainable development under the UNFCCC: a critical assessment of the clean development mechanism, the green climate fund and REDD+. German Development Institute, Bonn

Hultman N, Pulver S, Guimarães L, Deshmukh R, Kane J (2012) Carbon market risks and rewards: firm perceptions of CDM investment decisions in Brazil and India. Energy Policy 40:90–102

Johnson J (2018) The clean development mechanism's contribution towards sustainable development in South Africa: the Bisasar landfill case study. Masters thesis, North West University, Potchefstroom

Joshi F (2013) Pressure on India's unique CDM sustainable development fund, 25 July 2013. https://carbonmarketwatch.org/2013/07/25/pressure-on-indias-unique-cdm-sustainable-development-fund-watch-this-6/. Accessed 8 Mar 2019

Karakosta C, Marinakis V, Letsou P, Psarras J (2013) Does the CDM offer sustainable development benefits or not? Int J Sustain Dev World Ecol 20:1–8

Kim J (2004) Sustainable development and the CDM: a South African case study. J Environ Dev 13:201–219

Lema A, Lema R (2013) Technology transfer in the clean development mechanism: insights from wind power. Glob Environ Chang 23:301–313

MacFeely S (2018) The 2030 agenda: an unprecedented statistical challenge, international policy analysis. Friedrich Ebert Stiftung, New York

Matsuo N (1998) How is the CDM compatible with sustainable development? A view from project guidelines and adaptation measures. IGES, Kanagawa

Mayrhofer J, Gupta J (2016) The science and politics of co-benefits in climate policy. Environ Sci Pol 57:22–30

Melo I, Turnhout E, Arts B (2014) Integrating multiple benefits in market-based climate mitigation schemes: the case of the climate, community and biodiversity certification scheme. Environ Sci Pol 35:49–56

Michaelowa A (2003) CDM host country institution building. Mitig Adapt Strateg Glob Chang 8:201–220

Michaelowa A (2005) Creating the foundations for host country participation in the CDM. In: Yamin F (ed) Climate change and carbon markets. A handbook of emission reduction mechanisms. Earthscan, London, pp 305–320

Michaelowa A (2006) Examples of operational sustainable development criteria, presentation at workshop European CDM linking directive: how to strategize CDM opportunities in Thailand? Bangkok, December 14, 2006. https://www.powershow.com/view/9a353-NzZiN/Examples_of_operational_sustainable_development_criteria_Axel_Michaelowa_powerpoint_ppt_presentation. Accessed 8 Mar 2019

Mori-Clement Y (2019) Impacts of CDM projects on sustainable development: improving living standards across Brazilian municipalities? World Dev 113:222–236

Mori-Clement Y, Bednar-Friedl B (2019) Do clean development mechanism projects generate local employment? Testing for sectoral effects across Brazilian municipalities. Ecol Econ 157:47–60

Murphy K, Kirkman G, Seres S, Haites E (2015) Technology transfer in the CDM: an updated analysis. Clim Pol 15:127–145

Nussbaumer P (2009) On the contribution of labelled certified emission reductions to sustainable development: a multi-criteria evaluation of CDM projects. Energy Policy 37:91–101

Obergassel W, Peterson L, Mersmann F, Schade J, Hofbauer J, Mayrhofer M (2017) Human rights and the clean development mechanism: lessons learned from three case studies. J Hum Rights Environ 8:51–71

Olsen KH (2007) The clean development mechanisms contribution to sustainable development: a review of the literature. Clim Chang 84:59–73

Olsen KH, Fenhann J (2008) Sustainable development benefits of the clean development mechanism projects: a new methodology for sustainable assessment based on text analysis of the project design documents submitted for validation. Energy Policy 36:2819–2830

Olsen KH, Fenhann J, Hinostroza M, Arens C, Mersmann F, Rudolf F, Beuermann C (2015) Assessing usefulness – do stakeholders regard the CDM's SD tool as practical? DEHSt discussion paper, Berlin

Olsen KH, Arens C, Mersmann F (2018a) Learning from CDM SD tool experience for article 6.4 of the Paris Agreement. Clim Pol 18:383–395

Olsen KH, Taibi F-Z, Braden S, Verles M (2018b) Criteria for sustainable development and how to use the sustainable development goal (SDG) framework: defining criteria for sustainable development nationally and using the global SDG framework for implementation of article 6 of the Paris Agreement. UNEP DTU Partnership, Copenhagen

Olsen KH, Bakhtiari F, Duggal VK, Fenhann J (2019) Sustainability labelling as a tool for reporting the sustainable development impacts of climate actions relevant to article 6 of the Paris Agreement. Int Environ Agreements https://doi.org/10.1007/s10784-018-09428-1

Parnphumeesup P, Kerr S (2011) Classifying carbon credit buyers according to their attitudes towards and involvement in CDM sustainability labels. Energy Policy 39:6271–6279

Pickering A, Arnold B, Dentz H, Colford J, Null C (2016) Climate and health co-benefits in low-income countries: a case study of carbon financed water filters in Kenya and a call for independent monitoring. Environ Health Perspect 125:278–283

Rindefjall T, Lund E, Stripple J (2011) Wine, fruit, and emission reductions: the CDM as development strategy in Chile. Int Environ Agreements 11:7–22

Rousseau J-F (2017) Does carbon finance make a sustainable difference? Hydropower expansion and livelihood trade-offs in the Red River valley, Yunnan Province, China. Singap J Trop Geogr 38:90–107

Sangam A (2017) Barriers to and determinants of funding sustainable development projects in developing countries: a case study of the EThekwini municipality. PhD thesis, Durban University of Technology, Durban

Schade J, Obergassel W (2014) Human rights and the clean development mechanism. Camb Rev Int Aff 27:717–735

Schneider L (2007) Is the CDM fulfilling its environmental and sustainable development objectives? An evaluation of the CDM and options for improvement. Report for WWF, Oeko Institut, Berlin

Seres S, Haites E, Murphy K (2010) The contribution of the clean development mechanism under the Kyoto Protocol to technology transfer. United Nations Framework Convention on Climate Change, Bonn

South Africa, Designated National Authority for the CDM (2009) CDM status review, Energy Department of the Republic of South Africa, Cape Town

Subbarao S, Lloyd B (2011) Can the clean development mechanism deliver? Energy Policy 39:1600–1611

Sutter C (2003) Sustainability check-up for CDM projects. ETH Zurich, Zurich

Sutter C, Parreño J (2007) Does the current clean development mechanism (CDM) deliver its sustainable development claim? An analysis of officially registered CDM projects. Clim Chang 84:75–90

TERI (2012) Assessing the impact of the clean development mechanism on sustainable development and technology transfer. Paper for the CDM Dialogue, New Delhi

Thorne S, La Rovere E (1999) Criteria and indicators for appraising clean development mechanism projects. HELIO International, Paris

Tiwari A (2014) Mapping & Analysis of CDM projects in India from a sustainable development perspective. https://carbonmarketwatch.org/wp-content/uploads/2014/09/A.-Tiwari-Mapping-Analysis-of-CDM-Projects-in-India-from-a-Sustainable-Development-Perspective.pdf. Accessed 8 Mar 2019

Tiwari A, Goga d'Souza N (2009) CDM for sustainable development? Money for Nothing!!! A people's perspective. Laya Resource Center, Visakhapatnam

Torvanger A, Shrivastava M, Pandey N, Tornblad S (2013) A two-track CDM: improved incentives for sustainable development and offset production. Clim Pol 13:471–489

Uddin N, Blommerde M, Taplin R, Laurence D (2015) Sustainable development outcomes of coal mine methane clean development mechanism projects in China. Renew Sust Energ Rev 45:1–9

UN General Assembly (2015) Transforming our world: the 2030 agenda for sustainable development, A/RES/70/1

UNFCCC (2015) Paris Agreement, FCCC/CP/2015/10/Add.1, Paris

UNFCCC (2018) SBSTA 49 agenda item 11(b). Matters relating to Article 6 of the Paris Agreement: Rules, modalities and procedures for the mechanism established by Article 6, paragraph 4, of the Paris Agreement. Version 2 of 8 December 10:00 hrs, Katowice

Verra (2018) Sustainable development verified impact standard. https://verra.org/project/sustainable-development-verified-impact-standard/. Accessed 10 Nov 2018

Wang C, Zhang W, Cai W, Xi X (2013) Employment impacts of CDM projects in China's power sector. Energy Policy 59:481–491

Wara M (2007) Is the global carbon market working? Nature 445:595–596

Watson C, Fankhauser S (2009) The clean development mechanism: too flexible to produce sustainable development benefits? Grantham Research Institute on Climate Change and the Environment working paper no. 2. LSE, London

Yan D (2016) Sustainable development benefits of clean development mechanism projects in China: evidence from Yangtze River Delta region. Manag Eng 25:1838–5745

Technological Transition and Carbon Constraints Under Uncertainty

Alexander Golub

1 Introduction

The economics of ancillary benefits of greenhouse gas mitigation is an important branch of environmental economics (see, for example OECD 2000). Several insightful theoretical and applied studies have been published over the last several decades. The analytical methodology is now significantly better thanks to improvements in epidemiology and economic and air quality modeling. Most of these ancillary benefits can be attributed to the simultaneous reduction of conventional air pollution and the corresponding reduction of human health risk. Lelieveld et al. (2019) found that global fossil-fuel emissions account for about two-thirds of the additional mortality attributable to air pollution. About 3.6 million cases of mortality could be avoided each year by reductions in fossil fuel combustion. Air quality improvements prevent premature mortality, reduce morbidity, and reduce losses of labor productivity. All of this leads to a positive effect on economic growth. Keramidas et al. (2017) show that achieving the $2\,°C$ temperature target yields co-benefits that largely offsets cost of climate policy. This paper is an attempt to examine the positive effects of climate policy on long-term economic growth. Under some conditions a climate policy that targets carbon emissions may induce the transition of a low- or middle-income economy to a higher steady state, while in the absence of climate policy an economy may end up in a development trap. According to neoclassical economic growth theory, globalization and increased international competition in the long

---

A. Golub (✉)
American University, Washington, DC, USA
e-mail: agolub@american.edu

© Springer Nature Switzerland AG 2020
W. Buchholz et al. (eds.), *Ancillary Benefits of Climate Policy*, Springer Climate,
https://doi.org/10.1007/978-3-030-30978-7_4

run should lead to a steady productivity increase, faster capital accumulation, and therefore faster economic growth in countries with a relatively low initial per capita capital accumulation. Then in theory lesser developed countries should eventually catch up with the developed world as represented by OECD countries. Nevertheless, empirical evidences suggest that some countries end up in a so-called trap. Detailed discussions on poverty traps are found in Azariadis and Stachurski (2005), and the phenomenon of club convergence is well described in Barro and Martin (2003). Most recent empirical evidence can be found in Galvao et al. (2013) and in Vollmer et al. (2013). Multiple equilibrium is the main reason for divergence to different well distinct steady states that corresponds to different stage of development (see for example, Graham and Temple 2006) instead of convergence to a unique one. Fodha and Seegmuller (2014) connect environmental quality with poverty traps and introduced the term "environmental poverty trap." However, Bassetti et al. (2013) analyzed the joint distribution of per capita income and carbon dioxide emissions, and concluded that evidence does not support theoretical models predicting the existence of a poverty-environment trap. Obviously more research in this field is needed. The optimal growth theory explains development traps as a result of multiple equilibriums (multiple steady states). Non-concavity of the aggregated production function creates necessary conditions for multiple equilibrium. Local increasing returns (triggered by accelerated productivity growth) result in convex-concave or concave-convex-concave shape of the aggregate production function. In this paper we explain how the transition of an economy from an "old" to a "new" technological structure may generate non-concavity in the aggregated production function and prevent convergence to the highest steady state, thereby holding the economy in a development trap. Furthermore, given that the "old" technological structure is as a rule carbon intensive, we are arguing that timely climate policy may facilitate transition to a higher steady state. And climate policy uncertainty will increase the cost of capital (Golub et al. 2018, 2019) and may increase the probability of the economy being trapped in a lower steady state. An increased probability of convergence to a higher steady state could be treated as an important ancillary benefit of climate policy previously not described in the economic literature.

2 The Model Under Perfect Foresight

Suppose that a society has access to two production technologies F_1 and F_2 to produce a single, undifferentiated output Y. Given L units of labor and K units of capital, what is this society's *aggregate production function*, $F(K, L)$? That is, what level of output will this society attain by dividing available resources between the two technologies?

2.1 The Central Planner Problem

The problem facing the central planner is given by

$$\max_{L_1,L_2,K_1,K_2} F_1(K_1, L_1) + F_2(K_2, L_2) \tag{1}$$

subject to resource constraints

$$L_1 + L_2 \leq L \tag{2}$$

$$K_1 + K_2 \leq K \tag{3}$$

and non-negativity constraints on L_1, L_2, K_1, and K_2. The aggregate production function $F(K, L)$ is simply the value function associated with the optimization problem.

Under the natural assumption that both F_1 and F_2 are strictly increasing in both capital and labor, the resource constraints must bind at the optimum. If they did not, it would be possible to increase output by using the leftover labor and capital in *either* F_1 or F_2. With this simplification, an interior optimum—one in which a positive quantity of capital and labor is devoted to each production technology—must satisfy the first order conditions

$$\frac{\partial F_1(K_1^*, L_1^*)}{\partial K} = \frac{\partial F_2(K_2^*, L_2^*)}{\partial K} \tag{4}$$

$$\frac{\partial F_1(K_1^*, L_1^*)}{\partial L} = \frac{\partial F_2(K_2^*, L_2^*)}{\partial L} \tag{5}$$

where $K_1^* + K_2^* = K$ and $L_1^* + L_2^* = L$ give the optimal allocation of capital and labor between the two production technologies. The aggregate production function F evaluated at K and L is simply

$$F(K, L) = F_1(K_1^*, L_1^*) + F_2(K_2^*, L_2^*) \tag{6}$$

where K_1^*, K_2^*, L_1^*, and L_2^* are defined implicitly by Eqs. 4 and 5.

In this stylized model, "labor" represents an aggregated labor force without specification of skilled and unskilled labor, education degree, etc. The "capital" includes capital investment and material resources (including energy, raw materials, etc.). The central planner is looking for a socially optimal distribution of labor and capital resources between two competing technologies.

2.2 The Decentralized Solution

Now suppose that output is directed by the decentralized decisions of perfectly competitive firms, some of which operate F_1 and some of which operate F_2. Since both technologies produce the same output, we may treat it as the numeraire. Let w_1 be the wage paid to labor in the sector operating F_1 and r_1 be the rental rate of capital in this sector. Define w_2 and r_2 analogously. Under perfect competition, labor and capital are paid their marginal value products. Since Y is the numeraire, in the present example this is equivalent to labor and capital being paid their marginal products. Thus,

$$r_1 = \frac{\partial F_1(K_1^*, L_1^*)}{\partial K} \tag{7}$$

$$w_1 = \frac{\partial F_1(K_1^*, L_1^*)}{\partial L} \tag{8}$$

and

$$r_2 = \frac{\partial F_2(K_2^*, L_2^*)}{\partial K} \tag{9}$$

$$w_2 = \frac{\partial F_2(K_2^*, L_2^*)}{\partial L} \tag{10}$$

In equilibrium, we must have $w_1 = w_2$ and $r_1 = r_2$. If this were not the case, labor or capital would flow from one sector to the other. Imposing this condition gives us exactly the same first order conditions as we derived in the central planner problem.

2.3 Adding Constant Returns to Scale

We have established that both the central planner and decentralized solutions give rise to the value function in Eq. 6 characterized by the first order conditions from Eqs. 4 and 5. Yet this tells us nothing about the *form* of the aggregate production function. To say anything more, we need an additional assumption. If both F_1 and F_2 exhibit constant returns to scale, we can rewrite the problem in per labor terms according to

$$f_1(k_1) = F_1(K_1, L_1)/L_1 \tag{11}$$

$$f_2(k_2) = F_2(K_2, L_2)/L_2 \tag{12}$$

where $k_1 = K_1/L_1$ and $k_2 = K_2/L_2$ are the respective capital-labor ratios devoted to each technological process. The functions f_1 and f_2 give output per unit labor as a function of the capital-labor ratio used in each production process. The overall capital-labor ratio $k = K/L$ is related to the technology-specific capital-labor ratios according to

$$k = \left(\frac{L_1}{L}\right) k_1 + \left(1 - \frac{L_1}{L}\right) k_2 \tag{13}$$

provided the labor resource constraint binds, as it will at an optimum. Similarly, aggregate output per unit labor, $f(k) = F(K, L)/L$, can be expressed as

$$f(k) = \left(\frac{L_1}{L}\right) f_1(k_1) + \left(1 - \frac{L_1}{L}\right) f_2(k_2) \tag{14}$$

when the labor resource constraint binds.

Now, since L is a constant, maximizing total output is equivalent to maximizing output per unit labor. Therefore, the first order conditions given in Eqs. 11 and 12 continue to apply. After a little algebra, we can rewrite these as

$$f_1'(k_1^*) = f_2'(k_2^*) \tag{15}$$

$$f_1(k_1^*) - k_1^* f_1'(k_1^*) = f_2(k_2^*) - k_2^* f_2'(k_2^*) \tag{16}$$

where $k_1^* = K_1^*/L_1^*$ and $k_2^* = K_2^*/L_2^*$.

The first condition, Eq. 15, is straightforward. It says that, at the optimum, the marginal products of the two technologies must be equal. If this were not the case, it would be possible to increase output per labor, and hence total output, by decreasing the capital-labor ratio devoted to one technology to increase the other. The second condition, Eq. 16 is less obvious, but by substituting Eq. 15 and rearranging, we can see that it is equivalent to

$$f_1'(k_1^*) = \frac{f_2(k_2^*) - f_1(k_1^*)}{k_2^* - k_1^*} \tag{17}$$

The right hand side is simply the slope of the line through the points $\left(k_1^*, f_1(k_1^*)\right)$ and $\left(k_2^*, f_2(k_2^*)\right)$. The left hand side is the slope of f_1 evaluated at k_1^* which, as we know from Eq. 15, is equal to the slope of f_2 evaluated at k_2^*. The first order conditions, then, amount to a tangency condition, as depicted in Fig. 1.

Now we can characterize $f(k)$, and hence $F(K, L)$. Solving Eq. 13 for (L_1/L) gives

$$\left(\frac{L_1}{L}\right) = \frac{k - k_2}{k_1 - k_2} \tag{18}$$

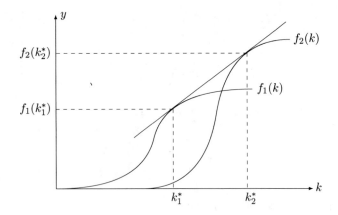

Fig. 1 The tangency condition

This holds whenever the labor resource constraint binds, so, in particular, it holds at an interior optimum, with k_1^* and k_2^* replacing k_1 and k_2. Substituting this into Eq. 14,

$$f(k) = \left(\frac{k - k_2^*}{k_1^* - k_2^*}\right) f_1(k_1^*) + \left(\frac{k - k_1^*}{k_2^* - k_1^*}\right) f_2(k_2^*) \tag{19}$$

Rearranging and simplifying,

$$f(k) = \left(\frac{f_2(k_2^*) - f_1(k_1^*)}{k_2^* - k_1^*}\right) k + \left(\frac{k_2^* f_1(k_1^*) - k_1^* f_2(k_2^*)}{k_2^* - k_1^*}\right) \tag{20}$$

and thus, multiplying both sides by L,

$$F(K, L) = \left(\frac{f_2(k_2^*) - f_1(k_1^*)}{k_2^* - k_1^*}\right) K + \left(\frac{k_2^* f_1(k_1^*) - k_1^* f_2(k_2^*)}{k_2^* - k_1^*}\right) L \tag{21}$$

Therefore, whenever it is optimal to use both technologies at once, if the underlying production functions satisfy constant returns to scale then the aggregate production function will be of the form $F(K, L) = aK + bL$.

But the question remains: when will both technologies be in use? If the problem is well-behaved, the first order conditions identify the optimal capital-labor ratio for each technology. Regardless of the *particular* mix of f_1 and f_2 employed, as long as both technologies are present it is always best to operate the first at k_1^* and the second at k_2^*. Yet it is not always possible to do so without violating the most basic necessary condition for an optimum.

From Eq. 13, we know that the overall capital-labor ratio k is a convex combination of the technology-specific capital-labor ratios k_1 and k_2 whenever the

Technological Transition and Carbon Constraints Under Uncertainty 75

labor resource constraint binds. This is true regardless of whether the first order conditions are satisfied. But we also know that the labor resource constraint *must* bind at the optimum, so Eq. 13 holds. For both technologies to be in use, the convex combination must be *strict*. That is, L_1^* must be strictly less than L. Geometrically, a strict convex combination of two points lies strictly between those points. It follows that unless k lies strictly between k_1^* and k_2^*, it is impossible to operate at these capital-labor ratios without wasting labor. And if any labor is wasted, the allocation cannot be optimal.

Thus, although the first order conditions characterize only an interior optimum, we can nonetheless describe the full solution as follows. Whenever the overall capital-labor ratio k lies strictly between k_1^* and k_2^*, the firm should operate both technologies in the linear combination indicated by Eq. 19. Aggregate output will be linear in K and L according to Eq. 21. If, on the other hand, k does *not* lie between k_1^* and k_2^*, society should operate only one technology: whichever produces more output at k. In this case, aggregate output will be given by either F_1 or F_2, depending on which technology is in use. In the situation depicted in Fig. 1, for example, society will first operate f_1 exclusively, until $k > k_1^*$, at which point it will begin a steady transition to f_2 along the tangent line. Eventually, at very high levels of capital per labor, $k > k_2^*$, society will operate only f_2.

3 Adding Uncertainty

The results of the previous section characterize society's optimal use of two technologies under perfect foresight. As we have seen, it is not optimal to switch suddenly from one technology to the other: the transition should be smooth and linear. Both the central planner and decentralized solutions yield this result. Real firms and real planners, however, lack perfect foresight. In this section we explore the consequences of adding uncertainty to the model.

Following Batra (1974) suppose that production uncertainty takes the form of an unknown multiplier α on f_2 with $\mathbb{E}[\alpha] = \mu$. This represents the idea that the second technology, the "new technology," is uncertain, while the "old technology" is fully specified.

3.1 Risk Neutrality

Under risk neutrality, the central planner's problem is

$$\max_{L_1, L_2, K_1, K_2} F_1(K_1, L_1) + \mathbb{E}[\alpha] F_2(K_2, L_2) \tag{22}$$

with first order conditions

$$\frac{\partial F_1(K_1^*, L_1^*)}{\partial K} = \mu \frac{\partial F_2(K_2^*, L_2^*)}{\partial K} \tag{23}$$

$$\frac{\partial F_1(K_1^*, L_1^*)}{\partial L} = \mu \frac{\partial F_2(K_2^*, L_2^*)}{\partial L} \tag{24}$$

In the decentralized version, risk-neutral firms equate the wage and rental rate with the expected marginal products of labor and capital, yielding the same first order conditions. The centralized and decentralized solutions are identical. Because this problem also exhibits certainty equivalence, assuming constant returns to scale, the analysis given above in the case of perfect foresight holds with only slight modification. Expressed in per labor terms, the first order conditions become

$$f_1'(k_1^*) = \mu \cdot f_2'(k_2^*) \tag{25}$$

$$f_1(k_1^*) - k_1^* f_1'(k_1^*) = \mu \cdot \left[f_2(k_2^*) - k_2^* f_2'(k_2^*) \right] \tag{26}$$

Now, by analogy with the case of perfect foresight,

$$\mathbb{E}[f(k)] = \left(\frac{\mu f_2(k_2^*) - f_1(k_1^*)}{k_2^* - k_1^*} \right) k + \left(\frac{k_2^* f_1(k_1^*) - k_1^* \mu f_2(k_2^*)}{k_2^* - k_1^*} \right) \tag{27}$$

and thus, multiplying both sides by L,

$$\mathbb{E}[F(K, L)] = aK + bL \tag{28}$$

where

$$a = \frac{\mu f_2(k_2^*) - f_1(k_1^*)}{k_2^* - k_1^*} \tag{29}$$

and

$$b = \frac{k_2^* f_1(k_1^*) - k_1^* f_2(k_2^*)}{k_2^* - k_1^*} \tag{30}$$

Therefore, under risk neutrality and multiplicative uncertainty, a smooth linear transition between technologies remains optimal. The only difference is that optimality is characterized in terms of expected rather than certain output. Again, under these conditions, both the central planner and profit-maximizing firms will arrive at the optimum.

3.2 Expected Utility

More generally, firms or the central planner may be risk averse. Here we assume that their preferences for risk can be represented in the von Neumann–Morgenstern expected utility framework. The central planner maximizes the expected utility of *output* while firms maximize the expected utility of *profits*.

The central planner's problem is given by

$$\max_{L_1, L_2, K_1, K_2} \mathbb{E}\{u[F_1(K_1, L_1) + \alpha F_2(K_2, L_2)]\} \tag{31}$$

As long as u is an increasing function, the resource constraints must bind at the optimum. If this were not the case, unused labor and capital could be applied to the first technology which, because its output is certain, would increase expected utility. Thus, assuming the regularity conditions necessary to exchange differentiation and expectation, the first order conditions are

$$\frac{\partial F_1(K_1^*, L_1^*)}{\partial K} = \frac{\mathbb{E}[\alpha \cdot u'(Y^*)]}{\mathbb{E}[u'(Y^*)]} \cdot \frac{\partial F_2(K_2^*, L_2^*)}{\partial K} \tag{32}$$

$$\frac{\partial F_1(K_1^*, L_1^*)}{\partial L} = \frac{\mathbb{E}[\alpha \cdot u'(Y^*)]}{\mathbb{E}[u'(Y^*)]} \cdot \frac{\partial F_2(K_2^*, L_2^*)}{\partial L} \tag{33}$$

where

$$Y^* = F_1(K_1^*, L_1^*) + \alpha F_2(K_2^*, L_2^*) \tag{34}$$

In contrast, firms maximized the expected utility of *profits*. For simplicity, suppose that firms and the central planner share the same utility function. In the first sector, firms continue to set

$$r_1 = \frac{\partial F_1(K_1^*, L_1^*)}{\partial K} \tag{35}$$

$$w_1 = \frac{\partial F_1(K_1^*, L_1^*)}{\partial L} \tag{36}$$

because there is no uncertainty. But in the second sector, firms solve

$$\max_{L_2, K_2} \mathbb{E}\{u[\alpha F_2(K_2, L_2) - w_2 L_2 - r_2 K_2]\} \tag{37}$$

Again, assuming the necessary regularity conditions, the first order conditions are given by

$$r_2 = \frac{\mathbb{E}[\alpha \cdot u'(\pi_2^*)]}{\mathbb{E}[u'(\pi_2^*)]} \cdot \frac{\partial F_2(K_2^*, L_2^*)}{\partial K} \tag{38}$$

$$w_2 = \frac{\mathbb{E}\left[\alpha \cdot u'(\pi_2^*)\right]}{\mathbb{E}\left[u'(\pi_2^*)\right]} \cdot \frac{\partial F_2(K_2^*, L_2^*)}{\partial L} \tag{39}$$

where

$$\pi_2^* = \alpha F_2(K_2^*, L_2^*) - w_2 L_2^* - r_2 K_2^* \tag{40}$$

Once again, in general equilibrium we must have $w_1 = w_2$ and $r_1 = r_2$ so

$$\frac{\partial F_1(K_1^*, L_1^*)}{\partial K} = \frac{\mathbb{E}\left[\alpha \cdot u'(\hat{\pi}_2^*)\right]}{\mathbb{E}\left[u'(\hat{\pi}_2^*)\right]} \cdot \frac{\partial F_2(K_2^*, L_2^*)}{\partial K} \tag{41}$$

$$\frac{\partial F_1(K_1^*, L_1^*)}{\partial L} = \frac{\mathbb{E}\left[\alpha \cdot u'(\hat{\pi}_2^*)\right]}{\mathbb{E}\left[u'(\hat{\pi}_2^*)\right]} \cdot \frac{\partial F_2(K_2^*, L_2^*)}{\partial L} \tag{42}$$

where

$$\hat{\pi}_2^* = \alpha F_2(K_2^*, L_2^*) - \frac{\partial F_1(K_1^*, L_1^*)}{\partial L} L_2^* - \frac{\partial F_1(K_1^*, L_1^*)}{\partial K} K_2^* \tag{43}$$

Thus, in the case of more general risk preferences than risk neutrality, the centralized and decentralized solutions may differ. Further, neither solution, in general, exhibits certainty equivalence.

3.3 Another Form of Uncertainty

Risk neutrality no longer implies certainty equivalence if a more general form of production uncertainty is assumed. Feldstein (1971) suggested a model of production under uncertainty using Cobb–Douglas technology $\tilde{A} K^{\tilde{\alpha}} L^{1-\tilde{\alpha}} \tilde{V}$ with uncertain parameters $(\tilde{\alpha}, \tilde{A})$ and a multiplicative random error term $\tilde{V}$. Here we follow a similar approach. This is in contrast to our earlier assumption that uncertainty could be represented by a multiplicative term on the production function alone.

To begin, we assume that $f_1(k_1) = A_1 k_1^{\alpha_1}$, where both A_1 and α_1 are known constants: the firm or central planner has full knowledge of the old technology. In contrast, we suppose that $f_2(k_2) = \tilde{A}_2 k_2^{\tilde{\alpha}_2}$ with both $\tilde{A}_2$ and $\tilde{\alpha}_2$ unknown: the second technology is known to be Cobb-Douglas, but its parameter values are uncertain.[1] We require only that $\tilde{A}_2 > 0$, so that the second technology produces positive output, and $\tilde{\alpha}_2 > \alpha_1$ so that the old technology dominates for low levels of k and the new technology dominates for high levels. This means that society knows it should

---

[1] We place a tilde over unknown quantities.

eventually transition completely from f_1 to f_2, but the optimum trajectory of the transition is uncertain.

We suppose uncertainty about $(\tilde{A}_2, \tilde{\alpha}_2)$ can be summarized in terms of a prior probability distribution in which $\tilde{A}_2$ and $\tilde{\alpha}_2$ are independent. Because we have assumed that $\tilde{\alpha}_2 > \alpha_1$, the distribution of $\tilde{\alpha}_2$ must have as its support $(\alpha_1, 1)$. Although we could choose any number of distributions on this interval, a particularly attractive choice is the uniform, indicating complete prior ignorance. That is, we suppose that $\tilde{\alpha}_2$ distributes according to the probability density function $g(x) = \frac{1}{1-\alpha_1}$. We leave the prior distribution for $\tilde{A}_2$ unspecified, requiring only that its mean $\mathbb{E}[\tilde{A}_2] = \mu$ exists and is finite. Since $\tilde{A}_2 > 0$, $\mu > 0$.

Under risk neutrality, the central planner's problem becomes

$$\max_{L_1, L_2, K_1, K_2} \mathbb{E}\left[A_1 K_1^{\alpha_1} L_1^{1-\alpha_1} + \tilde{A}_2 K_2^{\tilde{\alpha}_2} L_2^{\tilde{\alpha}_2}\right] \tag{44}$$

subject to the same non-negativity and resource constraints as given above. Now, using our assumptions about the prior distribution of $\tilde{A}_2$ and $\tilde{\alpha}_2$ we can rewrite the objective function as

$$
\begin{aligned}
\mathbb{E}[Y] &= \mathbb{E}\left[A_1 K_1^{\alpha_1} L_1^{1-\alpha_1} + \tilde{A}_2 K_2^{\tilde{\alpha}_2} L_2^{1-\tilde{\alpha}_2}\right] \\
&= A_1 K_1^{\alpha_1} L_1^{1-\alpha_1} + \mathbb{E}\left[\tilde{A}_2 K_2^{\tilde{\alpha}_2} L_2^{1-\tilde{\alpha}_2}\right] \\
&= A_1 K_1^{\alpha_1} L_1^{1-\alpha_1} + \mathbb{E}\left[\tilde{A}_2\right] \mathbb{E}\left[K_2^{\tilde{\alpha}_2} L_2^{1-\tilde{\alpha}_2}\right] \\
&= A_1 K_1^{\alpha_1} L_1^{1-\alpha_1} + \mu \cdot \mathbb{E}\left[K_2^{\tilde{\alpha}_2} L_2^{1-\tilde{\alpha}_2}\right]
\end{aligned}
$$

where the second to last equality follows because functions of independent random variables are themselves independent. Now, using the assumption that $\tilde{\alpha}_2$ follows a Uniform $(\alpha_1, 1)$ distribution, we have

$$
\begin{aligned}
\mathbb{E}\left[K_2^{\tilde{\alpha}_2} L_2^{1-\tilde{\alpha}_2}\right] &= \frac{1}{1-\alpha_1} \int_{\alpha_1}^{1} K_2^{\tilde{\alpha}_2} L_2^{1-\tilde{\alpha}_2} \, d\tilde{\alpha}_2 \\
&= \frac{L_2}{1-\alpha_1} \int_{\alpha_1}^{1} k_2^{\tilde{\alpha}_2} \, d\tilde{\alpha}_2 \\
&= \frac{L_2(k_2 - k_2^{\alpha_1})}{(1-\alpha_1) \log k_2} \\
&= \frac{K_2 - K_2^{\alpha_1} L_2^{1-\alpha_1}}{(1-\alpha_1)(\log K_2 - \log L_2)}
\end{aligned}
$$

Thus, the optimization problem under uncertainty given in Eq. 44 is equivalent to the following optimization problem *under certainty*

$$\max_{L_1, L_2, K_1, K_2} A_1 K_1^{\alpha_1} L_1^{1-\alpha_1} + \mu \left(\frac{K_2 - K_2^{\alpha_1} L_2^{1-\alpha_1}}{(1 - \alpha_1)(\log K_2 - \log L_2)} \right) \tag{45}$$

subject to the resource and non-negativity constraints. Because this problem cannot be solved analytically, we will consider a numerical solution below.

Now consider the non-stochastic problem in which all parameters of both technologies are known. Calculating the first order conditions from Eqs. 15 and 16 under Cobb–Douglas technologies, we have

$$A_1 \alpha_1 k^{*(\alpha_1 - 1)} = A_2 \alpha_2 k^{*(\alpha_2 - 1)} \tag{46}$$

$$A_1 (1 - \alpha_1) k^{*\alpha_1} = A_2 (1 - \alpha_2) k^{*\alpha_2} \tag{47}$$

Taking logarithms, solving the resulting 2×2 linear system and exponentiating the result, we have

$$k_1^* = \left[\frac{A_1}{A_2} \left(\frac{\alpha_1}{\alpha_2} \right)^{\alpha_2} \left(\frac{1 - \alpha_1}{1 - \alpha_2} \right)^{1-\alpha_2} \right]^{\frac{1}{\alpha_2 - \alpha_1}} \tag{48}$$

and

$$k_2^* = \left[\frac{A_1}{A_2} \left(\frac{\alpha_1}{\alpha_2} \right)^{\alpha_1} \left(\frac{1 - \alpha_1}{1 - \alpha_2} \right)^{1-\alpha_1} \right]^{\frac{1}{\alpha_2 - \alpha_1}} \tag{49}$$

To show that the problem presented in Eq. 45 does not exhibit certainty equivalence, consider the following numerical example. Let $K = 200$, $L = 100$, $A_1 = 1$, and $\alpha_1 = 0.2$. Suppose further that $\mathbb{E}[\tilde{A}_2] = 1$. Since $\tilde{\alpha}_2 > \alpha_1$ and $\alpha_1 = 0.2$, it follows that $\tilde{\alpha}_2 \sim$ Uniform(0.2, 1). Table 1 contrasts the solution of Eq. 45, "Optimum," with that obtained by substituting $\mathbb{E}[\tilde{A}_2]$ and $\mathbb{E}[\tilde{\alpha}_2]$ for A_2 and α_2 in Eqs. 48 and 49, "Substituting Means." If this problem exhibited certainty equivalence, the two would be identical. Clearly, this is not the case.

Table 1 A numerical example

	K_1^*	L_1^*	K_2^*	L_2^*	Y^*
Optimum	11.53	31.57	188.47	68.43	154.95
Substituting Means	6.19	16.08	193.81	83.92	151.95

This table presents a numerical example comparing the solution to the optimization equation 45, "Optimum," to that obtained by substituting $\mathbb{E}[\tilde{A}_2]$ and $\mathbb{E}[\tilde{\alpha}_2]$ into Eqs. 49 and 49, "Substituting Means." The calculations were carried out with $K = 200$, $L = 100$, $A_1 = 1$, $\alpha_1 = 0.2$ and $\mathbb{E}[\tilde{A}_2] = 1$

In this example, the solution obtained by substituting means overuses the second technology, resulting in lower total output.

Because this problem does not exhibit certainty equivalence, our result concerning the optimality of a smooth linear transition between technologies no longer holds. Thus, risk neutrality by itself is not enough to guarantee that a linear transition is optimal. The result depends sensitively on the nature of production uncertainty.

4 Implications for Economic Growth

Consider a simple Solow growth model, with no technological progress and exogenous population growth rate n, so that

$$\dot{k} = sf(k) - (n + \delta)k \tag{50}$$

where $\dot{k}$ is the time derivative of the capital-labor ratio, δ is the rate of depreciation, and s the savings rate.

4.1 Optimal Use of Technology

Now consider an aggregate per labor production function resulting from the optimal combination of two convex-concave technologies, as depicted in Fig. 1. Figure 2 shows the implications for the Solow model. For ease, we consider the case of

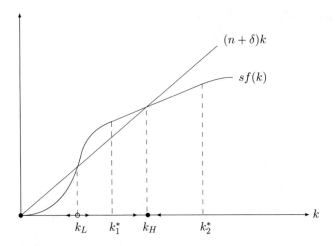

Fig. 2 Solow model with optimal use of technology

perfect foresight, but as we proved above, the analysis is identical in the case of multiplicative uncertainty and risk neutrality.

There are two possibilities. First, if the rate of effective depreciation, $n + \delta$, is very high, the savings rate is very low, or some combination of these, the only steady state is $k = 0$. In the more interesting second case, there are three steady states: two stable and one unstable. For sufficiently low initial capital-labor ratios, $k < k_L$, the economy converges zero capital—a poverty trap. The boundary, k_L, is itself a steady state but an unstable one. Unless the economy begins with a capital-labor ratio *precisely* equal to k_L, its steady state will lie elsewhere. For any initial capital-labor ratio above k_L, the economy will converge to k_H.

The precise location of k_L and k_H depends on the parameter values, and the underlying technologies. It could be the case that k_H lies between k_1^* and k_2^*, as depicted in the figure, so that both technologies are present in the steady state, or it could be that only one is present. Unless the slope of the linear segment of f is precisely equal to $n + \delta$, a very strange situation in which there could potentially be an infinity of steady states, only the first technology will be present at k_L.

Using the results derived above, we can characterize the transitional dynamics of the economy between k_1^* and k_2^*, that is during the transition from one technology to the other. From Eq. 20, we see that along the linear segment $f(k) = ak + b$ where

$$a = \frac{f_2(k_2^*) - f_1(k_1^*)}{k_2^* - k_1^*} \tag{51}$$

and

$$b = \frac{k_2^* f_1(k_1^*) - k_1^* f_2(k_2^*)}{k_2^* - k_1^*} \tag{52}$$

Substituting into Eq. 50 and rearranging,

$$\dot{k} = (sa - n - \delta)k + sb \tag{53}$$

along the transition between technologies. This is a first order linear differential equation with constant coefficients, whose solution is

$$k(t) = Ce^{(sa-n-\delta)t} - \frac{sb}{sa - n - \delta} \tag{54}$$

If we impose the initial condition $k(0) = k_1^*$,

$$C = k_1^* + \frac{sb}{sa - n - \delta} \tag{55}$$

Technological Transition and Carbon Constraints Under Uncertainty

and thus

$$k(t) = \left(k_1^* + \frac{sb}{sa - n - \delta}\right) e^{(sa-n-\delta)t} - \frac{sb}{sa - n - \delta} \tag{56}$$

Now that we have derived the dynamics of k along the transition, we can characterize how the proportion of resources devoted to each technology changes over time. From Eq. 19, we see that the weight given to the first technology is

$$\omega_1 = \frac{L_1}{L} = \frac{k - k_2^*}{k_1^* - k_2^*} \tag{57}$$

and the weight given to the second technology is

$$\omega_2 = \frac{L_2}{L} = \frac{k - k_1^*}{k_2^* - k_1^*} \tag{58}$$

Therefore, along the transition

$$\omega_1(t) = \frac{1}{k_1^* - k_2^*} \left[\left(k_1^* + \frac{sb}{sa - n - \delta}\right) e^{(sa-n-\delta)t} - \frac{sb}{sa - n - \delta} - k_2^*\right] \tag{59}$$

and

$$\omega_2(t) = \frac{1}{k_2^* - k_1^*} \left[\left(k_1^* + \frac{sb}{sa - n - \delta}\right) e^{(sa-n-\delta)t} - \frac{sb}{sa - n - \delta} - k_1^*\right] \tag{60}$$

4.2 Sub-optimal Use of Technology

As we have seen, it is not optimal for society to switch suddenly from one technology to another: the transition should be smooth and linear. This is true both under perfect foresight and risk neutrality with multiplicative uncertainty. Consequently, if the firm is behaving optimally, when both technologies are present, there should be *no* increasing returns. The presence of either increasing returns, or a sudden "leap" from one technology to another is an indication of sub-optimal resource usage. We can, in fact, go further. Along the transition from one technology to the other, any sub-optimal aggregate production function must by definition lie below the tangent line. Whatever specific shape it happens to take, this implies a convex region, and hence a region of increasing returns.

Figure 3 illustrates a possible outcome of sub-optimal resource usage, in which society faces more than one stable, positive output steady state. With an initial level of capital per labor $k_a < k < k_b$, the economy will converge to k_b. Only if the initial level of capital per labor is between k_c and k_d will the economy converge to the high output steady state k_d. The question remains: when will this occur?

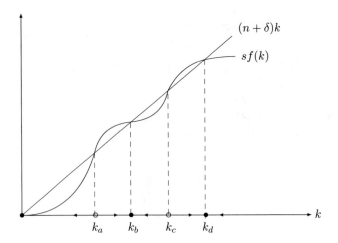

Fig. 3 Solow model with sub-optimal use of technology

Assuming perfect foresight, if the second technology generates a positive research and development externality, then the central planner and competitive solutions differ. If the second technology generates a positive research and development externality, then the central planner and decentralized solutions will differ. Specifically, profit-maximizing firms will underutilize F_2 as they fail to internalize all of the benefits of production. This is one way in which a region of increasing returns could arise.

Alternatively, consider the case in which the second technology faces multiplicative uncertainty of the kind described above. If firms are risk averse, the decentralized solution will fail to maximize expected output, leading to a convex region in the expected per labor production function. If firms have mean-variance preferences, they are willing to accept lower expected profits in exchange for lower variance of profits. Under multiplicative uncertainty, the only way to reduce the variance of profits is to produce less with the second technology. Under these conditions, it also appears that the decentralized solution will lead to the systematic underutilization of F_2.

These scenarios suggest the possibility that a policy intervention leading to greater use of the second technology could send the economy to a higher-output steady state.

5 Discussion

Transition of an economy from one technological structure that perhaps exhausted major sources of economic growth (and consistently exhibits decreasing returns to scale) to a new technological structure that exhibits significantly higher productivity is the major pathway for developing countries to catch up with developed world. An "ideal" convergence to a new technological stricture produces a linear AK segment of the aggregated production function. In this paper we demonstrated how uncertain returns on new technologies may prevent accumulation of capital in new and less carbon intensive sectors. Presence of increasing returns on a macro level results in multiple equilibrium and in multiple steady states.In this stylized model the new technological structure represents an aggregated cluster of new technologies. Complementarity between renewable energy and power storage capacities (including the fleet of electric cars), new building materials that provide storage for captured carbon (for example wood products) etc., makes the new technological structure much less carbon intensive than the old one. Climate policy uncertainty increases the risk of investing into the new technological structure and reduces chances of convergence of the economy to higher steady state. However, timely introduction of climate policy will reduce the risk of investing into the new low carbon technological structure and therefore facilitates conversion to the higher stated state. Then conversion to a higher steady state could be considered as one of the most important of the ancillary benefits of climate policy. While the diffusion of the new technologies is the engine of economic growth, climate policy could be a catalyst this growth. According to the new economic growth theory (Romer 1990) accumulation of human capital is the major engine for economic growth. Convergence to higher steady states is associated with knowledge and human capital accumulation and adoption of new technologies. An increasing returns phenomenon as a by-product of learning and human capital accumulation plays an important role in understanding processes of technological innovations while explicitly introducing it into a growth model. There have been several publications during the last few years that have been dedicated to the increasing returns phenomenon (Romer 1990; Ros 2001; Perälä 2008 etc.). Complementarity between new technological structures and qualified labor is another reason to pay attention to ancillary benefits. High concentrations of PM2.5 in the atmosphere is the leading cause of premature mortality and excessive morbidity attributed to environmental factors in developing and middle-income countries. Insufficient human capital is one of the most important reasons for development traps and club convergence. Another reason could be a heterogeneous quality of labor due to income inequality. Low income families may suffer from malnutrition in addition to low educational and skill levels and as a result cannot supply high quality labor. Gong et al. (2010) treats health as a simple function of consumption, which enables the study of health and growth in an aggregate macroeconomic model with increasing returns. They conclude that rich countries may end up with higher capital, better health, and higher consumption than poor countries. Semmler and Ofori (2007), Mayer-Foulkes (2008) demonstrated existence of twin-peak income distribution in developing countries. They argue that twin-peak of income distribution exists due to the local increasing returns. Promotion of health and education is an important factor in escaping development traps (Berthélemy 2011). High local pollution may stimulate migration of highly qualified labor from countries like China, India, Russia, etc. to Europe and the USA where air quality is much better. These "environmental constraints" on exogenous supply of highly qualified labor could be another barrier for economic growth in countries with poor air quality. The proposed analytical approach in this paper could be further extended to study the entire range of intended and unintended consequences of transition to the new technological structure. The ability of an economy to converge to the highest steady state is the most important collateral benefit, but there are many other benefits we did not consider in this paper. For example, the new technological structure requires more educated labor and more educated labor as a rule is better paid. Increase of income leads to shifts in individual preferences. Increase in income stimulates an increasing demand for environmental quality (well known in the economic literature "Environmental Kuznets Curve"). Thus deployment of the new technologies may trigger a positive feedback loop and result in further tightening of environmental policy. Study of this feedback may be of academic interest.

Acknowledgement The author wishes to thank Francis DiTraglia for his research assistance.

References

Azariadis C, Stachurski J (2005) Poverty traps. In: Handbook of economic growth, Chapter 5. North Holland, Amsterdam

Barro RJ, i Martin XS (2003) Economic growth. MIT Press Books, Cambridge

Bassetti T, Benos N, Karagiannis S (2013) CO2 emissions and income dynamics: What does the global evidence tell us? Environ Resour Econ 54(1):101–125

Batra RN (1974) Resource allocation in a general equilibrium model of production under uncertainty. J Econ Theory 8(1):50–63

Berthélemy J (2011) Health, education and emergence from the development trap. Afr Dev Rev 23(3):300–312

Feldstein MS (1971) Production with uncertain technology: some economic and econometric implications. Int Econ Rev 12(1):27–36

Fodha M, Seegmuller T (2014) Environmental quality, public debt and economic development. Environ Resour Econ 57:487–504

Galvao AF Jr, Montes-Rojas G, Olmo J (2013) A panel data test for poverty traps. Appl Econ 45(14):1943–1952

Golub AA, Fuss S, Lubowski R, Hiller J, Khabarov N, Koch N, Krasovskii A, Kraxner F, Laing T, Obersteiner M, Palmer C (2018) Escaping the climate policy uncertainty trap: options contracts for REDD+. Clim Pol 18(10):1227–1234

Golub A, Lugovoy O, Potashnikov V (2019) Quantifying barriers to decarbonization of the Russian economy: real options analysis of investment risks in low-carbon technologies. Clim Pol 19:716–724

Gong L, Li H, Wang D, Zou H-f (2010) Health, taxes, and growth. Ann Econ Finance 11(1):73–94

Graham BS, Temple JRW (2006) Rich nations, poor nations: How much can multiple equilibria explain? J Econ Growth 11(1):5–41

Keramidas K et al (2017) Global energy and climate outlook 2017 how climate policies improve air quality: global energy trends and ancillary benefits of the Paris agreement. Publications Office, Luxembourg. Web

Lelieveld J, Klingmüller K, Pozzer A, Burnett RT, Haines A, Ramanathan V (2019) Effects of fossil fuel and total anthropogenic emission removal on public health and climate. Proc Natl Acad Sci 116:7192–7197

Mayer-Foulkes D (2008) The human development trap in Mexico. World Dev 36(5):775–796

OECD (2000) Ancillary benefits and costs of greenhouse gas mitigation: proceedings of an IPCC Co-Sponsored Workshop, Held on 27–29 March 2000, Washington, DC/Paris

Perälä MJ (2008) Increasing returns in the aggregate: fact or fiction? J Econ Stud 35(2):112–153

Romer P (1990) Endogenous technological change. J Polit Econ 98(October, Part 2):S71–S102

Ros J (2001) Development theory and the economics of growth. The University of Michigan Press, Ann Arbor

Semmler W, Ofori M (2007) On poverty traps, thresholds and take-offs. Struct Change Econ Dyn 18(1):1–26

Vollmer S, Holzmann H, Ketterer F, Klasen S, Canning D (2013) The emergence of three human development clubs. PLoS One 8:e57624

Part II
Conceptual and Theoretical Approaches for the Analysis of Ancillary Benefits

Sustainable International Cooperation with Ancillary Benefits of Climate Policy

Nobuyuki Takashima

1 Introduction

Climate change is currently causing significant environmental damage worldwide and it is therefore imperative that international environmental public goods, such as the reduction of greenhouse gas (GHG) emissions, are provided. However, because there is no governing supranational authority to control such trans-boundary pollutants, international environmental agreements (IEAs) must be implemented to coordinate actions among countries. For example, in the first commitment period of the Kyoto Protocol, from 2008 to 2012, Annex I countries of the United Nations Framework Convention on Climate Change (UNFCCC) were required to limit or abate GHG emissions. After the first commitment period, the Paris Agreement (adopted in 2015) aimed to strengthen the global response to the threat of climate change.[1] To ensure that the global temperature increase in this century is well below 2 °C above pre-industrial levels and to encourage countries to limit a temperature increase even further to 1.5 °C, broad participation is required. Furthermore, such participation must include developing countries with no obligation for emission reduction under the Kyoto Protocol.

---

[1]For more details, see the website of UNFCCC. URL: https://unfccc.int/process-and-meetings/the-paris-agreement/the-paris-agreement

---

This research did not receive any specific grants from public, commercial, or not-for-profit funding agencies.

N. Takashima (✉)
Kyushu University Platform of Inter/Transdisciplinary Energy Research (Q-PIT), Nishi-ku, Fukuoka, Japan
e-mail: ntakashima@econ.kyushu-u.ac.jp

© Springer Nature Switzerland AG 2020
W. Buchholz et al. (eds.), *Ancillary Benefits of Climate Policy*, Springer Climate,
https://doi.org/10.1007/978-3-030-30978-7_5

Some developed countries, such as Japan, the United States, and Canada, have not participated in the second commitment period of the Kyoto Protocol (from 2013 to 2020). Moreover, in 2017, the United States announced its withdrawal from the Paris Agreement. Therefore, in a world with both developed and developing countries, it is now necessary to find a way to achieve international environmental cooperation, not only to reach a far-reaching agreement but also how to achieve long-term cooperation.

From a theoretical perspective, a repeated game model can be used to sustain long-term cooperation for agreed emission abatement. A repeated game model typically assumes an infinitely repeated game in which the game is repeated infinitely. It also assumes that countries agree on a contract in the first stage, and this has to be enforced in subsequent stages using credible threats (see Hovi et al. 2015; Asheim et al. 2006). Assuming that each participating country in the agreement has an incentive to free ride on the abatement of others, a repeated game model focuses on compliance. That is, generally, the analysis of an IEA using a repeated game model aims to analyze the conditions under which participating countries will meet their commitments. Compliance is ensured by the threat of the future decreased total abatement; that is, "punishments" in the case that a country deviates from the agreement. If the punishment is a credible one, an equilibrium in which every participant cooperates can be sustained. More precisely, an IEA within a repeated game model analyzes the condition under which participants cooperate in accordance with their commitments under a "strategy" that specifies the participants' action plans.

The equilibrium concept in the repeated game is a weakly renegotiation-proof (WRP) equilibrium (Farrell and Maskin 1989, pp. 330–331). A WRP equilibrium requires two conditions: the first is subgame perfection where each player cannot gain a payoff by a solo defection from the strategy and the second is renegotiation-proofness where not all players will be strictly worse off by carrying out the punishment rather than by renegotiating.

A considerable body of literature has addressed the provision of global international pollution reduction, considering the public good nature of the reduction of such pollutants. For example, climate change mitigation generates public benefits on a global scale that all countries enjoy on an equal basis (called "primary benefits"). However, in the real world, such climate protection can be considered as impure public goods. That is, it provides not only primary benefits, but also private benefits that only abating countries receive (called "ancillary benefits").[2] For example, climate protection behavior not only reduces GHG emissions but also secondary environmental pollutants (which are emitted with the combustion of fossil fuels) on a local scale. Such pollutants include sulfur dioxide (SO_2), nitrogen oxides (NO_x), and particulate matter (PM) emissions, and these can be reduced simultaneously.

---

[2]Rive and Rübbelke (2010) and Rübbelke (2003) elucidate the difference between primary and ancillary (secondary) benefits of climate change mitigation.

The ancillary benefits from the reduction of secondary pollutants resulting from GHG emission reduction have been highlighted as a potential catalyst for countries to engage in climate policy (Aunan et al. 2007; Ekins 1996a, b; Finus and Rübbelke 2013; Rive 2010; Rübbelke 2003; Takashima 2017a). Ekins (1996a, b) suggests that the secondary benefits of reducing CO_2 emissions are of the same magnitude as the abatement costs of significant levels of CO_2 abatement and can be larger than the estimates of the primary benefits of CO_2 abatement. Using a new computable general equilibrium (CGE) model of the Chinese economy, considering the costs and benefits of climate protection to China, Aunan et al. (2007) argue that climate protection gives significant ancillary benefits to China because secondary benefits such as the reduction of PM and NO_x improve public health and increase agricultural yields. Rive (2010) describes that the abatement behavior of CO_2 emissions reduces air pollutants such as PM, SO_2, and NO_x, and assesses the co-benefit of the reduction of these secondary emissions (associated with regional climate policy in Western Europe) using an augmented CGE model, and that the recognition of the ancillary benefits from the reduction of SO_2, NO_x, and PM when designing policies increases the attainability of the abatement goals and enhances the political feasibility of climate policies.

Rübbelke (2003) theoretically analyzes countries' decisions on climate policy in an impure public goods model that considers the effect of ancillary benefits and shows that the various environmental programs require cautious coordination because environmental programs that do not directly aim to mitigate climate change can affect the design efficiency of climate change programs.

In the context of the Clean Development Mechanism (CDM), Rive and Rübbelke (2010) consider that GHG abatement in developing countries is subsidized by developed countries. They assume that developing countries' payoffs do not depend on total climate protection, but on private consumption and domestic environmental protection including ancillary benefits such as the reduction of SO_2 emissions. Considering this assumption, they assess the effect of the CDM subsidy rate under different national environmental policy regimes to determine the conditions which are most favorable for generating positive effects on the reduction of developing countries' poverty interpreted as a lack of private consumption options or as a low welfare level.[3]

Finus and Rübbelke (2013) present a pioneering study investigating the theoretical effect of ancillary benefits on IEA participation using a one-shot game model. The result of Finus and Rübbelke (2013) shows that ancillary benefits have negative or neutral effects on agreement size, retaining the pessimistic view that an agreement can be sustained if entered into by a few countries, as shown in Barrett (1994) and Carraro and Siniscalco (1993) in one-shot game model. Finus and Rübbelke (2013) also explain that those countries that consider private ancillary benefits to a greater

---

[3]They also perform simulations to assess the effects of CDM subsidy rates on welfare, poverty, and air quality in China.

extent will abate more emissions irrespective of the IEA. In that case, the relative importance of an IEA for climate protection is reduced.

In the context of a repeated game model, Takashima (2017a) assumes that ancillary benefits are generated by environmental protection. Using the *Penance-m* strategy presented in Froyn and Hovi (2008) and Takashima (2017a) examines the effect of those ancillary benefits on full IEA participation within two types of abatement cost functions: linear and quadratic. As a result, looking at the situation where a game is repeated and compliance is ensured by future credible punishment threats, then the negative effect of ancillary benefits on a stable IEA with linear costs, as shown by Finus and Rübbelke (2013), disappears. In Takashima (2017a), however, this number remains unchanged with convex costs. Takashima (2018) presents the *Flexible Penance* strategy, whereby compliance is enforced by specifying the punishment to a deviation, and shows that this condition depends on the benefit–cost ratios of the two types of countries.

The present study and Takashima (2018) assume asymmetry among the countries: the abatement benefit and cost parameters are high and low, respectively. Moreover, this study considers the effect of ancillary benefits and assumes that the two types of countries can have different levels. On this assumption, we investigate the condition in which an agreement consisting of both types of countries is sustained as a WRP equilibrium using the *Flexible Penance* strategy.[4]

Our results lead to several conclusions. First, ancillary benefits affect the number of punishing countries to deter a country's deviation from the strategy, which is the condition for WRP equilibrium. Second, we obtain the minimum number of participating countries for the IEA to always be satisfied as the WRP equilibrium. Finally, ancillary benefits can have positive effects on the minimum participation number, which must be taken into consideration as a condition before the agreement starts, depending on the levels of ancillary benefits of type 1 and type 2 countries.

The remainder of the chapter is structured as follows: Sect. 2 introduces the models, Sect. 3 presents the *Flexible Penance* strategy and the WRP equilibrium outcomes, Sect. 4 discusses the conditions regarding *Flexible Penance* for the success of agreement, and Sect. 5 presents our conclusions.

2 Model

There are two types of countries: type 1 and type 2. The number of type i countries is n_i ($i = 1, 2$). In every period, type i countries can choose to *cooperate* (i.e., reduce emissions) or *defect* (i.e., not reduce emissions). Given that n_i type i countries

---

[4]As described in Asheim et al. (2006) and Hovi et al. (2015), to analyze an IEA's formation, two theoretical models are generally employed: one is called as a one-shot game model and the other is a repeated game model.

contribute to the public good ($i = 1, 2$), the payoff of a contributing type i country is

$$b_i (n_1 + n_2) + \alpha_i b_i - c_i,$$

where the marginal benefit from a unit of the public good (b_i) is smaller than the marginal cost of cooperating (c_i) ($b_i < c_i$; $i = 1, 2$), and α_i denotes the parameter of the ancillary benefits where $\alpha_i b_i$ is the private good aspect of the benefits. Based on Pavlova and de Zeeuw (2008), the present study assumes a two-sided asymmetry for all countries: the public and ancillary benefits and cost parameters are high and low, respectively. That is, this study also assumes asymmetry in the ancillary benefits.[5]

In a real world, the public benefits denote the reduction of GHGs, which is derived from the abatement efforts of participating countries. In contrast, ancillary benefits of abatement denote an improvement in domestic air pollution. It is natural for the degree of improvement in domestic air pollution to differ between developed and developing countries when both types of countries address the abatement of global emissions.

Given n_1 and n_2, if contributing type i country changes its action to *defect*, its payoff becomes

$$b_i (n_1 + n_2 - 1).$$

In this case, the defector receives benefits from other countries' abatement without owing abatement costs. However, the defector cannot receive ancillary benefits because it made no effort to engage in environmental protection. We assume that $b_i(n_1 + n_2 - 1) > b_i(n_1 + n_2) + \alpha_i b_i - c_i > 0$, which means that the full participation state is not a Nash equilibrium at this stage of the game and full participation Pareto dominates zero participation. Rearranging the left part of above inequalities, we obtain $c_i > b_i + \alpha_i b_i$, which means that each country cannot gain its payoff by individual abatement efforts. If this inequality is not satisfied, every country individually engages in abatement irrespective of the agreement.

Regarding the public benefit of emission abatement, several studies assume that the benefits of countries choosing cooperative behavior are greater than or equal to the benefits when they choose to defection (e.g., Asheim et al. 2006; Barrett 1994, 1999; Froyn and Hovi 2008).[6] In this study, we simplify the varying beneficial effects of *cooperate* and *defect* based on previous studies (Barrett 2001; Finus and Rübbelke 2013). Moreover, we add the ancillary benefits as a private benefit because the benefits of *cooperate* and *defect* may differ if an abatement action also has private good characteristics. Specifically, a private benefit is a spillover effect

---

[5]Rive and Rübbelke (2010) introduce the difference in the "co-benefit rate" of GHG abatement between developed and developing countries. This can be considered the asymmetry in ancillary benefits in our model.

[6]In other words, the slope of the benefit function when a country cooperates is steeper than or equal to the slope of the benefit function when the country defects.

that accompanies pollution mitigation. Therefore, we assume that this difference occurs because cooperating countries receiving some additional benefits and regard these additional benefits as ancillary benefits. In the present study and in Takashima (2017a), the effects of the difference in the benefits of *cooperate* and *defect* are expressed simply as a parameter of ancillary benefits in the payoff functions.

Froyn and Hovi (2008) show that the difference in benefits between *cooperate* and *defect* decreases the number of punishing countries for a WRP equilibrium, while Takashima (2017a) shows that ancillary benefits decrease the number of punishing countries.[7] Therefore, we consider that the difference between these benefit parameters has a similar effect to ancillary benefits. Essentially, a larger benefit difference between *cooperate* and *defect* corresponds to a smaller stipulated number of punishing countries.

For simplicity, here we use $\gamma_i \equiv c_i/b_i$. Furthermore, as in Barrett (2002, 2003), Asheim et al. (2006), and Takashima (2017a, b), each country discounts its future payoffs using a common discount factor, $\delta \in (0, 1)$, which is close to 1.

3 The Strategy and the WRP Equilibrium

3.1 The Strategy

A repeated game model assumes that countries agree, before the game begins (or in the first period), on the content of the strategy (i.e., contract) that must be enforced in subsequent periods for fear of credible threats (i.e., future punishments).[8]

This study uses *Flexible Penance* strategy presented by Takashima (2018) which specifies that (1) a signatory plays *cooperate* unless another signatory has been the sole deviator from *Flexible Penance* in the previous period and that (2) if a unilateral deviation by type i ($i = 1, 2$) country occurs, m ($<n$) countries excluding the deviator are selected from among the type 1 and type 2 countries and m countries play *defect* in the next period. Conversely, $n - m$ countries play *cooperate*. If (2) is satisfied, the cooperative relationship will restart in the next period.

Regarding renegotiation possibilities, the *Flexible Penance* strategy assumes that a potential renegotiation by punishing countries can occur when the punisher group consists of type 1 and type 2 countries and when the sub-punisher groups consist of either type 1 or type 2 countries. This is because this study considers that all

---

[7]In Froyn and Hovi (2008), when δ is close (but not equal) to 1, the number of punishing countries for a WRP equilibrium is decided as $[c - d - (d - b)(n - 1)]/b < m \leq [c - (d - b)n]/b$. Moreover, when $d = b$ in this condition, we have $(c - b)/b < m \leq c/b$.

[8]For this implicit assumption, see Hovi et al. (2015). Similarly, Asheim et al. (2006) assume that countries agree in the first period of the contract. Generally, the strategies in a repeated game model implicitly make this kind of assumption. For example, see Asheim et al. (2006), Asheim and Holtsmark (2009), Barrett (1999, 2002, 2003), Froyn and Hovi (2008), and Takashima (2017a, b, 2018).

countries owe the same abatement levels and therefore the countries' payoffs depend on the number of cooperators.[9]

This study assumes a single-period punishment, as seen in previous studies such as Asheim et al. (2006), Asheim and Holtsmark (2009), Froyn and Hovi (2008), and Takashima (2017a, b, 2018). For example, the *Penance (Regional Penance)*, *Penance-m*, *Regional Cooperative*, and *Flexible Penance* strategies prescribe that a deviator must receive a punishment in a single period before the restarting of the cooperative relationship among all participants. Additionally, from the assumption that countries agree on a strategy before the game begins, a repeated game model assumes that the content of the punishments under which participants play *cooperate* in accordance with the strategy for fear of credible threats is common knowledge among participants before cooperation starts.

The total number of punishing countries m to deter a country's deviation is determined as follows:

$$m = \theta m + (1 - \theta) m,$$

where $\theta(0 \leq \theta \leq 1)$ is the ratio of type 1 countries selected as punishing countries. For example, integer θm is the number of punishing countries selected from the type 1 countries, and integer $(1 - \theta)m$ is the number of punishing countries selected from the type 2 countries, to deter a country's deviation. First, in Sect. 3.2, we assume that the punishing countries are randomly selected from type 1 and 2 countries.[10] After that, in Sect. 4, we consider that the ratio of type 1 and 2 punishing countries is decided by choice.

3.2 WRP Equilibrium

3.2.1 The WRP Equilibrium Concept

According to Asheim et al. (2006), Froyn and Hovi (2008), Takashima (2017a, b, 2018), the strategy must satisfy two requirements for IEAs to be sustained as a WRP equilibrium. The first requirement is subgame perfection: In the context of a repeated game in which countries discount their future payoffs, no player can gain by a single-period deviation after any history.[11] The second requirement is that the

---

[9]As mentioned in Takashima (2018), it is sufficient to consider a potential renegotiation by the whole group or sub-groups of punishing countries for the reason that in our model the incentives for renegotiation depend on the number of punishing countries. This incentive can be different between type 1 and type 2 punishing countries. For additional detail, see Lemma 2.

[10]Takashima (2018) implicitly assumes that the punishing countries are randomly selected.

[11]In the theory of repeated games with a discount factor, because a player cannot gain by multiple period deviations if he/she cannot gain by a one-period deviation (Abreu 1988, p. 390), we need only check that no player can gain by a one-period deviation from the strategy after any history.

strategy profile must be renegotiation-proof: When a unilateral deviation occurred in the previous period, this requirement is satisfied when not all players strictly gain by immediately collectively restarting cooperation rather than by carrying out the punishment, because we assume that the punishments last only one period.

If subgame perfection is satisfied, then in the agreement, no player ever changes its actions, as specified by the strategy. Additionally, an agreement is renegotiation-proof if threats deter a deviation. This makes not only the deviator but also all non-punishing countries worse off in the punishment scenario. Therefore, it must be in at least the punishing countries' best interests to punish a deviator without accepting an invitation to renegotiate. In summary, a punishment in a repeated game implies that all punishing countries play *defect* after the deviation.

Considering the effect of ancillary benefits, we examine the condition that all participants in the agreement agree that *Flexible Penance* is sustained as a WRP equilibrium: subgame perfection and renegotiation-proofness in the next section.

3.2.2 Subgame Perfection and Renegotiation-Proof Requirements

Lemma 1 shows the conditions for the subgame perfection requirement.

Lemma 1
Considering the effect of ancillary benefits, the Flexible Penance strategy with m punishing countries satisfies subgame perfection if

$$m \geq \frac{max\{\gamma_1 - \alpha_1, \gamma_2 - \alpha_2\} - 1}{\delta}.$$

Proof See Appendix 1. ∎

Lemma 1 shows that the number of punishing countries must be an integer larger than or equal to the values of $(\gamma_1 - \alpha_1 - 1)/\delta$ and $(\gamma_2 - \alpha_2 - 1)/\delta$ for subgame perfection, and that the condition to deter both types of countries from deviating. If this condition is not satisfied, each country tends to deviate from the agreement to increase its payoff.

Let us compare Lemma 1 in this study with Lemma 1 found in Takashima (2018). The condition of subgame perfection in Takashima (2018), $m \geq (\gamma_1 - 1)/\delta$ or $m \geq (\gamma_2 - 1)/\delta$, shows that benefit–cost ratios γ_1 and γ_2 decide the subgame perfection requirement. However, in this study, from $m \geq (\gamma_1 - \alpha_1 - 1)/\delta$ or $m \geq (\gamma_2 - \alpha_2 - 1)/\delta$, it is not only the public benefit and costs but also the ancillary benefits that can affect subgame perfection. That is, the ancillary benefits can decrease the lower bound of the condition of subgame perfection.

Next, Lemma 2 shows the conditions for the renegotiation-proof requirement.

Lemma 2

Considering the effect of ancillary benefits, the Flexible Penance strategy with m punishing countries is renegotiation-proof if

$$m \leq \min \{\gamma_1 - \alpha_1, \gamma_2 - \alpha_2\}.$$

Proof See Appendix 2. ■

For renegotiation-proofness to be satisfied, from Lemma 2, the number of punishing countries must be an integer that is smaller than or equal to the values of $\gamma_1 - \alpha_1$ and $\gamma_2 - \alpha_2$. The intuition behind this result is as follows. The incentive for renegotiation increases as the level of the benefit–cost ratio decreases. Additionally, the ancillary benefits increase the punishing countries' incentive for abatement. Therefore, to decrease the punishing countries' incentive to renegotiate, their payoffs for punishment must be increased by decreasing the number of punishing countries.

The renegotiation-proof requirement in Takashima (2018) is that $m \leq \min \{\gamma_1, \gamma_2\}$, while the requirement in this study is that $m \leq \min \{\gamma_1 - \alpha_1, \gamma_2 - \alpha_2\}$. The renegotiation-proof requirement is the lowest value because of the assumption that the punishing countries are randomly selected among two types of countries. From these conditions, we know that the ancillary benefits can affect the upper bound of m, which depends on the magnitude relationship between $\gamma_1 - \alpha_1$ and $\gamma_2 - \alpha_2$, and not on the magnitude relationship between γ_1 and γ_2.

Proposition 1 is directly obtained from Lemmas 1 and 2.

Proposition 1

An agreement where all n_1 and n_2 countries cooperate in accordance with the Flexible Penance strategy as a weakly renegotiation-proof equilibrium if m exists such that

$$\frac{\max \{\gamma_1 - \alpha_1, \gamma_2 - \alpha_2\} - 1}{\delta} \leq m \leq \min \{\gamma_1 - \alpha_1, \gamma_2 - \alpha_2\}.$$

We obtain the following implications from this proposition. First, the number of punishing countries depends not only on the magnitude relationship of the benefit–cost ratios between both types of countries but also on the value of ancillary benefits.

If m is determined as an integer, then from the inequalities in Proposition 1, all participating countries play *cooperate* in accordance with *Flexible Penance*. In contrast, a cooperative relationship cannot be achieved if the exogenous parameters do not satisfy the condition in Proposition 1. In Sect. 4, we consider the condition in which *Flexible Penance* always works by presenting some rules on the selection of punishing countries.

4 Additional Rules on Flexible Penance with Ancillary Benefits

This section focuses on the condition in which *Flexible Penance* is effective, considering that the ratio of type 1 and 2 punishing countries is decided by choice.

For simplicity, we assume that $\delta = 1$ in Proposition 1. For *Flexible Penance* to be effective, m must be always selected as an integer in Proposition 1. If $\gamma_2 - \alpha_2 < \gamma_1 - \alpha_1 < (\gamma_2 - \alpha_2)/(1 - \theta)$ (or if $\gamma_1 - \alpha_1 < \gamma_2 - \alpha_2 < (\gamma_1 - \alpha_1)/\theta$), the upper bound of m can be decided as $\gamma_1 - \alpha_1(\gamma_2 - \alpha_2)$. Therefore, we obtain conditions that $\gamma_1 - \alpha_1 - 1 < m \le \gamma_1 - \alpha_1$ $(\gamma_2 - \alpha_2 - 1 < m \le \gamma_2 - \alpha_2)$. Under these conditions, m is always decided as an integer. Therefore, the gap between upper and lower bounds of m can be increased compared with Proposition 1. However, if $\gamma_1 - \alpha_1 > (\gamma_2 - \alpha_2)/(1 - \theta)$ (or if $\gamma_2 - \alpha_2 > (\gamma_1 - \alpha_1)/\theta$), a minimum number of participants is required for m to be selected as an integer.

Then, let m^* be an integer where $\gamma_1 - \alpha_1 - 1 < m^* \le \gamma_1 - \alpha_1$ when $\gamma_2 - \alpha_2 < \gamma_1 - \alpha_1$, and let m^{**} be an integer where $\gamma_2 - \alpha_2 - 1 < m^{**} \le \gamma_2 - \alpha_2$ when $\gamma_1 - \alpha_1 < \gamma_2 - \alpha_2$. Proposition 2 shows the minimum number of participants required for *Flexible Penance* always to work.

Proposition 2
On the condition that $n_1 + n_2 > m$, Flexible Penance works if

$$n_1 > m^* - (\gamma_2 - \alpha_2)$$

when $\gamma_2 - \alpha_2 < \gamma_1 - \alpha_1$, and

$$n_2 > m^{**} - (\gamma_1 - \alpha_1)$$

when $\gamma_1 - \alpha_1 < \gamma_2 - \alpha_2$.
Proof See Appendix 3. ∎

In Proposition 2, the right-hand side of each inequality shows that the condition of minimum participation of type i $(=1, 2)$ countries must exist for the number of punishing countries to always be decided as an integer. When the punishing countries are randomly selected from type 1 and 2 countries, the feasibility of IEAs depends on whether the exogenous parameters satisfy the condition in Proposition 1. If the ratio of type 1 and 2 punishing countries is decided by choice and Proposition 2 is satisfied, the number of punishing countries is always decided as an integer.

On the assumption that $n_1 + n_2 > m$, if the upper condition in Proposition 2 and $n_2 > m^* - n_1$ are satisfied, the number of punishing countries, m^*, is sustained as the equilibrium; if the bottom condition and $n_1 > m^{**} - n_2$ are satisfied, then the number of punishing countries, m^{**}, is sustained as the equilibrium.

In other words, on the assumption that $n_1 + n_2 > m$, it requires a clause that the agreement would not enter into force until at least type 1 (type 2) countries more than or equal to $m^* - (\gamma_2 - \alpha_2)$ $(m^{**} - (\gamma_1 - \alpha_1))$ have acceded to it.

This requirement is similar to the concept of the *minimum participation clause* in Barrett's (1997) IEA model with asymmetric countries using a one-shot game. If this requirement is satisfied, *Flexible Penance* is always effective.

Additionally, Proposition 2 shows that α_1 and α_2 can affect the levels of $m^* - (\gamma_2 - \alpha_2)$ and $m^{**} - (\gamma_1 - \alpha_1)$. The lower bound of n_1 (n_2), $m^* - (\gamma_2 - \alpha_2)$ $(m^{**} - (\gamma_1 - \alpha_1))$ weakly decreases with α_1 and α_2 because n_1 (n_2) is an integer.[12] There is a possibility that the lowest number of type i participating countries decreases by α_1 and α_2 compared with the case of no ancillary benefits, depending on the values of α_1 and α_2. Therefore, inversely, the lowest number of type i participating countries can increase. That is, there is a possibility that ancillary benefits decrease the tipping point at which *Flexible Penance* with additional rules on selecting the punishing countries is effective.

5 Summary and Discussion

During the past several decades, the effect of secondary benefits that accompany pollution mitigation due to climate policy against global warming has received much attention. The Paris Agreement aims to uphold and promote international cooperation by developed and developing countries as a key challenge for environmental protection. Not only domestically, but also internationally, the recognition of secondary benefits and analysis of their effects is a central theme in enhancing the success of climate change mitigation.

Developed and developing countries coexist in reality, and they have different levels of ancillary benefit. Therefore, it is necessary to assume a world with asymmetric countries to reflect this case. That is, this study introduces the effect of ancillary benefits such as the improvement in domestic environmental problems (e.g., the reduction of secondary pollutants such as SO_2, NO_x, and PM) that accompany climate policy against global warming (e.g., reduced air pollution or improved biodiversity) into the IEA formation in a world with asymmetric countries.

We assume two-sided asymmetry for all countries: the public and ancillary benefits and cost parameters are high and low, respectively. This study presents new theoretical findings on the effect of ancillary benefits on IEAs with two types of asymmetric countries using the *Flexible Penance* strategy in the repeated game model adopted from Takashima (2018). We show that the conditions under which participants cooperate in accordance with the *Flexible Penance* can be relaxed by adding the selection rule of the type of punishing countries in *Flexible Penance*. Additionally, under *Flexible Penance* with our selection rule, we reveal the minimum participation clause under which the *Flexible Penance* always works.

---

[12] When $\gamma_2 - \alpha_2 < \gamma_1 - \alpha_1$ $(\gamma_1 - \alpha_1 < \gamma_2 - \alpha_2)$, the number of punishing countries, m^* (m^{**}), decreases discontinuously while α_1 (α_2) increases continuously, leading to a jump in equilibrium.

For the effects of ancillary benefits on IEA, we obtain the following results. First, ancillary benefits affect the number of punishing countries, which is the condition for the WRP equilibrium in which all countries cooperate in accordance with the *Flexible Penance* strategy. More precisely, ancillary benefits can reduce the punishment levels. Second, the ancillary benefits can relax the condition of the minimum participation clause, depending on the ancillary benefit parameters.

Our results show that the ancillary benefits are a significant key factor in enhancing the feasibility of sustained, full participation in an IEA with asymmetric countries for the long term. Our approach thus suggests considering the ancillary benefits as a solution to climate change.

Acknowledgment I would like to thank the editor, Dirk T.G. Rübbelke, for his very helpful comments and suggestions. I am also grateful to Toshiyuki Fujita, Makoto Okamura, Akira Maeda, Tamotsu Nakamura, Yasunori Ouchida, for their helpful comments.

Appendix 1: Proof of Lemma 1

In a similar manner to Takashima (2018), we obtain the conditions for subgame perfection when the ancillary benefits are considered. We consider deviation by type $i \, (=1, 2)$ countries.

First, let us consider a deviation by a type 1 country.

(A) The incentive constraint for each country to play *cooperate* when no deviation occurs in any period. A participating country, j, receives $b_1(n_1 + n_2) + \alpha_1 b_1 - c_1$ in each period if no deviation occurs in the previous period. If country j deviates in period t and reverts to the strategy in period $t + 1$, it receives $b_1(n_1 - 1) + b_1 n_2$ in period t and $b_1(n_1 - \theta m) + b_1(n_2 - (1 - \theta)m) + \alpha_1 b_1 - c_1$ in period $t + 1$. Thereafter, each country receives $b_1 n_1 + b_1 n_2 + \alpha_1 b_1 - c_1$ from period $t + 2$ onward. Each country plays *cooperate* if

$$
\begin{aligned}
(1 + \delta) \, (b_1 \, (n_1 + n_2) + \alpha_1 b_1 - c_1) &\geq b_1 \, (n_1 - 1) + b_1 n_2 \\
&+ \delta \, (b_1 \, (n_1 - \theta m) + b_1 \, (n_2 - (1 - \theta) \, m) + \alpha_1 b_1 - c_1) .
\end{aligned}
\tag{1}
$$

By rearranging inequality (1), we obtain

$$
m \geq (c_1 - \alpha_1 b_1 - b_1) \, / \delta b_1 .
$$

Given that $\gamma_1 = c_1/b_1$, we have

$$
m \geq (\gamma_1 - \alpha_1 - 1) \, / \delta .
\tag{2}
$$

Similarly, if a type 2 country deviates, we obtain the lower bound for the number of punishing countries needed to maintain cooperation:

$$m \geq (\gamma_2 - \alpha_2 - 1)/\delta. \tag{3}$$

Inequalities (2) and (3) represent those conditions under which each signatory plays *cooperate* in every period, provided that the other signatories also *cooperate*. If the number of punishing countries m is less than the right-hand side of inequalities (2) and (3), a deviation occurs in period t to increase that country's payoff.

(B) The incentive constraint for $n_1 + n_2 - m$ countries to play *cooperate* after a unilateral deviation in period $t - 1$. First, we consider deviation by a type 1 country. If countries play *cooperate* in period t, they first receive $b_1(n_1 - \theta m) + b_1(n_2 - (1 - \theta)m) + \alpha_1 b_1 - c_1$ and then $b_1(n_1 + n_2) + \alpha_1 b_1 - c_1$ from period $t + 1$ onward. If one country deviates in period t but cooperates in period $t + 1$, that country first receives $b_1(n_1 - \theta m - 1) + b_1(n_2 - (1 - \theta)m)$ and then $b_1(n_1 - \theta m) + b_1(n_2 - (1 - \theta)m) + \alpha_1 b_1 - c_1$ in period $t + 1$ as a result of punishment by m countries. Thereafter, each country receives $b_1(n_1 + n_2) + \alpha_1 b_1 - c_1$ from period $t + 2$ onward. Therefore, type 1 countries play *cooperate* after a unilateral deviation if

$$b_1 (n_1 - \theta m) + b_1 (n_2 - (1 - \theta) m) + \alpha_1 b_1 - c_1 + \delta (b_1 (n_1 + n_2) + \alpha_1 b_1 - c_1)$$
$$\geq b_1 (n_1 - \theta m - 1) + b_1 (n_2 - (1 - \theta) m) + \delta (b_1 (n_1 - \theta m)$$
$$+ b_1 (n_2 - (1 - \theta) m) + \alpha_1 b_1 - c_1) . \tag{4}$$

By rearranging inequality (4), we obtain

$$m \geq (c_1 - \alpha_1 b_1 - b_1)/\delta b_1.$$

Given that $\gamma_1 = c_1/b_1$, we have

$$m \geq (\gamma_1 - \alpha_1 - 1)/\delta. \tag{5}$$

Condition (5) is obtained when a type 2 country deviates in period t.

Second, we consider deviation by a type 2 country after a type 1 country's unilateral deviation. Type 2 countries play *cooperate* after a type 1 country's deviation if

$$b_2 (n_1 - \theta m) + b_2 (n_2 - (1 - \theta) m) + \alpha_2 b_2 - c_2 + \delta (b_2 (n_1 + n_2) + \alpha_2 b_2 - c_2)$$
$$\geq b_2 (n_1 - \theta m) + b_2 (n_2 - (1 - \theta) m - 1) + \delta (b_2 (n_1 - \theta m)$$
$$+ b_2 (n_2 - (1 - \theta) m) + \alpha_2 b_2 - c_2) . \tag{6}$$

By rearranging inequality (6), we obtain

$$m \geq (c_2 - \alpha_2 b_2 - b_2) / \delta b_2.$$

Given that $\gamma_2 = c_2/b_2$, we have

$$m \geq (\gamma_2 - \alpha_2 - 1) / \delta. \tag{7}$$

Inequalities (5) and (7) represent those conditions under which each signatory plays *cooperate* in every period, provided that the other signatories also play *cooperate*.

If the number of punishing countries is less than or equal to the right-hand side of inequality (5) and (7), a deviator in period $t - 1$ increases its payoff by deviating in period t. That is, the deviator in period $t - 1$ deviates again in the next period.

(C) The incentive constraint for m punishing countries to punish a deviation. First, we consider the payoff for a punishing country that fails to punish—that is, when it plays *cooperate* in period t after a deviation in $t - 1$. As the country defecting in period t will be punished in period $t + 1$, this defection leads to a loss in period $t + 1$.

Type 1 punishing countries implement the punishment if

$$b_1 (n_1 - \theta m) + b_1 (n_2 - (1 - \theta) m) \geq b_1 (n_1 - \theta m + 1) + b_1 (n_2 - (1 - \theta) m) + \alpha_1 b_1 - c_1,$$

or

$$c_1 \geq b_1 + \alpha_1 b_1.$$

Type 2 punishing countries implement the punishment if

$$b_2 (n_1 - \theta m) + b_2 (n_2 - (1 - \theta) m) \geq b_2 (n_1 - \theta m) + b_2 (n_2 - (1 - \theta) m + 1) + \alpha_2 b_2 - c_2,$$

or

$$c_2 \geq b_2 + \alpha_2 b_2.$$

From the assumption that $c_i > b_i + \alpha_i b_i$ ($i = 1, 2$), which denotes a solo cooperation is not profitable, the above inequalities always hold.

We know that the definition of subgame perfection requires both types of countries not to deviate. Therefore, from inequalities (2), (3), (5), and (7), the condition for subgame perfection is

$$m \geq (\gamma_1 - \alpha_1 - 1) / \delta \text{ and } m \geq (\gamma_2 - \alpha_2 - 1) / \delta.$$

This condition can be rewritten as

$$m \geq (\max \{\gamma_1 - \alpha_1, \gamma_2 - \alpha_2\} - 1) / \delta.$$

Appendix 2: Proof of Lemma 2

Because of the assumption that a potential renegotiation can occur when the entire group consists of type 1 and type 2 punishing countries and when sub-groups consist solely of type 1 or type 2 punishing countries, the following potential renegotiations under these three scenarios must be considered.

We consider the type i punishing countries' incentive for renegotiation after deviation by a type i ($i = 1, 2$) country.

(A) Consider the case where type 1 and type 2 punishing countries (θm and $(1 - \theta)m$) renegotiate in period t. Type 1 punishing countries receive $b_1(n_1 - \theta m) + b_1(n_2 - (1 - \theta)m)$ if they adopt the strategy, and $b_1(n_1 + n_2) + \alpha_1 b_1 - c_1$ if they do not punish by renegotiation. They receive $b_1 n_1 + b_1 n_2 + \alpha_1 b_1 - c_1$ in each period irrespective of their action from period $t + 1$ onward. Therefore, renegotiation is deterred if

$$b_1 (n_1 - \theta m) + b_1 (n_2 - (1 - \theta) m) \geq b_1 (n_1 + n_2) + \alpha_1 b_1 - c_1.$$

Assuming that $\gamma_1 = c_1/b_1$, we have

$$m \leq \gamma_1 - \alpha_1. \tag{8}$$

For type 2 countries, renegotiation is avoided if

$$b_2 (n_1 - \theta m) + b_2 (n_2 - (1 - \theta) m) \geq b_2 (n_1 + n_2) + \alpha_2 b_2 - c_2.$$

Assuming that $\gamma_2 = c_2/b_2$, we have

$$m \leq \gamma_2 - \alpha_2. \tag{9}$$

(B) Consider that only type 1 punishing countries (θm) renegotiate in period t. Type 1 punishing countries will not renegotiate if

$$b_1 (n_1 - \theta m) + b_1 (n_2 - (1 - \theta) m) \geq b_1 n_1 + b_1 (n_2 - (1 - \theta) m) + \alpha_1 b_1 - c_1.$$

Assuming that $\gamma_1 = c_1/b_1$, we have:

$$m \leq (\gamma_1 - \alpha_1) / \theta. \tag{10}$$

(C) Consider that only type 2 punishing countries ($(1 - \theta)m$) renegotiate in period t. Type 2 punishing countries will not renegotiate if

$$b_2 (n_1 - \theta m) + b_2 (n_2 - (1 - \theta) m) \geq b_2 (n_1 - \theta m) + b_2 n_2 + \alpha_2 b_2 - c_2,$$

or

$$m \leq (\gamma_2 - \alpha_2) / (1 - \theta).$$ (11)

To prevent a punishing country's renegotiation, (8), (9), (10), and (11) are necessary and sufficient. These four conditions are summarized as

$$m \leq \min \{\gamma_1 - \alpha_1, \gamma_2 - \alpha_2, (\gamma_1 - \alpha_1)/\theta, (\gamma_2 - \alpha_2)/(1 - \theta)\}.$$

This is equivalent with

$$m \leq \min \{\gamma_1 - \alpha_1, \gamma_2 - \alpha_2\},$$

since $\theta \in [0, 1]$.

Appendix 3: Proof of Proposition 2

We show the requirements for the minimum number of participants required for agreements to be sustained as a WRP equilibrium using the *Flexible Penance* strategy.

(a) *Case $\gamma_2 - \alpha_2 < \gamma_1 - \alpha_1$*

Let m^* be an integer where $\gamma_1 - \alpha_1 - 1 < m^* \leq \gamma_1 - \alpha_1$. If $m^* \leq (\gamma_2 - \alpha_2)/(1 - \theta)$, renegotiation by type 2 punishing countries is always prevented. Rearranging this equation, we obtain the lowest number of type 1 punishing countries θm^* required to deter renegotiations by type 2 punishing countries:

$$\theta m^* \geq m^* - (\gamma_2 - \alpha_2).$$ (12)

Condition (12) represents the lowest number of type 1 punishing countries required to deter renegotiation by type 2 punishing countries. Therefore, the number of type 1 participants must be larger than θm^*. The minimum number of type 1 participants, n_1, can be written as:

$$n_1 > m^* - (\gamma_2 - \alpha_2).$$ (13)

On the assumption that $n_1 + n_2 > m$, the total number of participants must be greater than m^*. Therefore, (13) and $n_2 > m^* - n_1$ need to be satisfied.

(b) *Case $\gamma_1 - \alpha_1 < \gamma_2 - \alpha_2$*

Let m^{**} be an integer where $\gamma_2 - \alpha_2 - 1 < m^{**} \leq \gamma_2 - \alpha_2$. If $m^{**} \leq (\gamma_1 - \alpha_1)/\theta$, then only deviation of type 1 punishing countries is prevented. Rearranging this, we obtain the lowest number of type 1 punishing countries $(1 - \theta)m^{**}$ required to deter their renegotiation:

$$(1 - \theta)\, m^{**} \geq m^{**} - (\gamma_1 - \alpha_1)\,. \tag{14}$$

Similar to case (a), (14) allows us to obtain the lowest number of type 2 participants n_2:

$$n_2 > m^{**} - (\gamma_1 - \alpha_1)\,. \tag{15}$$

On the assumption that $n_1 + n_2 > m$, the total number of participants must be greater than m^{**}. Therefore, (15) and $n_1 > m^{**} - n_2$ need to be satisfied.

References

Abreu D (1988) On the theory of infinitely repeated games with discounting. Econometrica 56:383–396

Asheim GB, Holtsmark B (2009) Renegotiation-proof climate agreements with full participation: conditions for pareto-efficiency. Environ Resour Econ 43(4):519–533

Asheim GB, Froyn CB, Hovi J, Menz FC (2006) Regional versus global cooperation for climate control. J Environ Econ Manage 51(1):93–109

Aunan K, Berntsen T, O'Connor D, Persson T, Vennemo H, Zhai F (2007) Benefits and costs to China of a climate policy. Environ Dev Econ 12:471–497

Barrett S (1994) Self-enforcing international environmental agreements. Oxf Econ Pap 46(special issue):878–894

Barrett S (1997) The strategy of trade sanctions in international environmental agreements. Resour Energy Econ 19:345–361

Barrett S (1999) A theory of full international cooperation. J Theor Polit 11(4):519–541

Barrett S (2001) International cooperation for sale. Eur Econ Rev 45(10):1835–1850

Barrett S (2002) Consensus treaties. J Inst Theor Econ 15(4):529–547

Barrett S (2003) Environment and Statecraft. Oxford University Press, New York

Carraro C, Siniscalco D (1993) Strategies for the international protection of the environment. J Public Econ 52(3):309–328

Ekins P (1996a) How large a carbon tax is justified by the secondary benefits of CO_2 abatement? Resour Energy Econ 18:161–187

Ekins P (1996b) The secondary benefits of CO_2 abatement: how much emission reduction do they justify? Ecol Econ 16:13–24

Farrell J, Maskin E (1989) Renegotiation in repeated games. Games Econ Behav 1(4):327–360

Finus M, Rübbelke DTG (2013) Public good provision and ancillary benefits: the case of climate agreements. Environ Resour Econ 56(2):211–226

Froyn CB, Hovi J (2008) A climate agreement with full participation. Econ Lett 99(2):317–319

Hovi J, Ward H, Grundig F (2015) Hope or despair? Formal models of climate cooperation. Environ Resour Econ 62(4):665–688

Pavlova Y, de Zeeuw A (2008) Asymmetries in international environmental agreements. Environ Dev Econ 18(1):51–68

Rive N (2010) Climate policy in Western Europe and avoided costs of air pollution control. Econ Model 27:103–115

Rive N, Rübbelke DTG (2010) International environmental policy and poverty alleviation. Rev World Econ 146:515–543

Rübbelke DTG (2003) An analysis of differing abatement incentives. Resour Energy Econ 25:269–294

Takashima N (2017a) International environmental agreements with ancillary benefits: repeated games analysis. Econ Model 61:312–320

Takashima N (2017b) The impact of accidental deviation by natural disaster-prone countries on renegotiation-proof climate change agreements. Environ Model Assess 22(4):345–361

Takashima N (2018) International environmental agreements between asymmetric countries: a repeated game analysis. Jpn World Econ 48:38–44

Impure Public Good Models as a Tool to Analyze the Provision of Ancillary and Primary Benefits

Anja Brumme, Wolfgang Buchholz, and Dirk Rübbelke

1 Introduction

Climate protection projects regularly involve a joint production of various effects with different degrees of publicness. While the mitigation of climate change represents a global public good, many ancillary effects are exerting only a local impact and thus are "private" from the point of view of a country hosting the project. For a long time, the theory of public goods has provided a powerful tool to analyze the provision of such so-called impure public goods (see, e.g., Cornes and Sandler 1996), which are not an exception but the normal case.

Take any example of an international public good and you may observe that the provision of these goods regularly also exhibits "private" components. In the context of international goods for health, Sandler and Arce (2002) give the example of immunizing populations where one's vaccination helps to prevent the (global) spread of infectious diseases, but at the same time it directly protects their own body. Another example of joint production of effects of different degrees of publicness are military alliances like the North Atlantic Treaty Organization (NATO) alliance; e.g., Sandler and Murdoch (2000) and Sandler and Hartley (2001) argue that participation in such alliances may yield both private and pure public defense outputs for its members. The public output could be seen in the provision of deterrence (e.g., by means of nuclear weapons) to forestall an enemy attack on members of the alliance and a private output could be that a member's defense spending may also address a domestic terrorist threat that does not threaten other allies.

A. Brumme (✉) · D. Rübbelke
Technische Universität Bergakademie Freiberg, Freiberg, Germany
e-mail: anja.brumme@vwl.tu-freiberg.de; dirk.ruebbelke@vwl.tu-freiberg.de

W. Buchholz
University of Regensburg, Regensburg, Germany
e-mail: wolfgang.buchholz@ur.de

© Springer Nature Switzerland AG 2020
W. Buchholz et al. (eds.), *Ancillary Benefits of Climate Policy*, Springer Climate,
https://doi.org/10.1007/978-3-030-30978-7_6

There are also examples where the public effect is perceived in a negative way. Atkinson et al. (1987, p. 5) argue that *"[t]errorists often have multiple demands,"* i.e., they may seek for the sense of purpose and identity (Maikovich 2005), which are private aspects. At the same time, the fear they spread via their terrorist actions could be disseminated over the whole world, so that we face a global impure public bad.[1] In the climate change context a few countries could benefit from a moderate temperature increase because of, e.g., reduced heating costs, increased agricultural output, and improved prospects for the tourism industry (see, e.g., Buchholz et al. 2018) so that for these countries, ambitious climate change mitigation would become a public bad.

However, also the ancillary private effects can be harmful, e.g., in the case of immunization against diseases there is an ancillary cost for the treated individual if there is some risk that the vaccination has exceptional adverse consequences for her health.

Beside these examples the scientific literature has dealt with a lot of other impure public goods as diverse as fund-raising charities (Posnett and Sandler 1986), agricultural research (Khanna et al. 1994), financing public radio stations (Kingma and McClelland 1995), provision of refugee protection (Betts 2003), performing arts (Pugliese and Wagner 2011), green goods (Chan and Kotchen 2014; Munro and Valente 2016), housing upkeep (Leonard 2016), peacekeeping (Sandler 2017), projects supported by the clean development mechanism (Rive and Rübbelke 2010), and multilateral financial mechanisms (Chan 2019).

In this chapter, we present the general features of the standard impure public good model where we focus on its specific application to climate policy. We describe how this theoretical approach evolved and outline its key concepts and techniques. Finally, we provide some scope for further research in the context of impure public goods.

2 Modeling Primary and Ancillary Effects of Climate Change Mitigation

2.1 Climate Change Mitigation: The Standard Pure Public Good Model

Economic analyses of international climate policy regularly depict climate change mitigation as a pure public good. Thus, "consumption" of the merits of mitigation is nonrival and nonexcludable. A standard approach is to use the summation technology as the technology of public supply aggregation (see, e.g., Sandler 2013 on technologies of public supply aggregation), i.e., the aggregate level G of global

---

[1]On joint production of effects exhibiting different degrees of publicness in the context of terrorism, see, e.g., Lee and Sandler (1989) and Rübbelke (2005).

climate change mitigation that is "consumed" in each country is the sum of all countries' individual contributions to climate protection. In a world of n countries (with $i = 1, \ldots, n$), the total level of climate protection is

$$G = g_i + \sum_{i \neq j} g_j = g_i + \widetilde{G}_i, \tag{1}$$

where $\widetilde{G}_i$ is the sum of climate change mitigation efforts by all countries except country i. Country i spends its monetary income w_i on its own mitigation effort g_i as well as on its consumption c_i of a private good, where the unit price of mitigation (e.g., of one ton of CO_2 abatement) is $p > 0$ and the unit price of the private good is one. Country i's utility function has standard properties and is continuous, strictly increasing, strictly quasi-concave, and everywhere twice differentiable. If the public good is provided noncooperatively, the utility maximization problem for country i is:

$$\max_{c_i, g_i} u_i \left(c_i, g_i + \widetilde{G}_i \right) = u_i \left(c_i, G \right) \tag{2}$$

s.t.

$$c_i + g_i p = w_i. \tag{3}$$

The result of this maximization problem is the standard first-order condition

$$\frac{\partial u_i / \partial G}{\partial u_i / \partial c_i} = p. \tag{4}$$

This condition states that an individual country, maximizing its domestic utility, will raise its climate change mitigation up to the level where the marginal rate of substitution between the pure public good "mitigation" and the private good (taken as absolute value) becomes equal to the price of the public good. This deviates from the Samuelson condition for efficient public good provision, which reads:

$$\sum_{i=1}^{n} \frac{\partial u_i / \partial G}{\partial u_i / \partial c_i} = p, \tag{5}$$

i.e., the sum of the marginal rates of substitution between the pure public good "mitigation" and the private good of all countries must become equal to the price of the public good in order to attain Pareto efficiency. It is straightforward from the comparison of conditions (4) and (5) that the Pareto efficient outcome is associated with a higher mitigation level than the outcome where countries selfishly

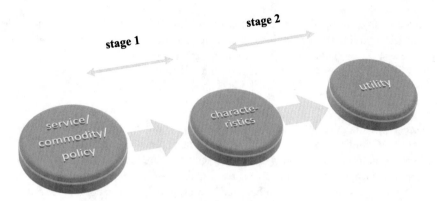

Fig. 1 A two-stage affair: Services, commodities, or policies generate characteristics (effects) which in turn provide utility to humans

maximize their domestic utility. Hence, global climate protection resulting from selfish uncoordinated behavior tends to be suboptimally low.[2]

This pure public good model of climate change mitigation, however, ignores the important category of ancillary benefits. For the integration of ancillary benefits in the analysis, we shall have recourse to Lancaster's (1966, 1971) characteristics approach.

2.2 Mitigation in the Standard Impure Public Good Model

The basic idea underlying the characteristics approach can easily be explained through an example from a different field of research, i.e., from performing arts where an audience enjoys a show in a theater. Beyond the pleasure of the audience, the promotion of theater culture may also raise local identity and international prestige (Baumol and Bowen 1966). Interpreted from the perspective of Lancaster's influential characteristics approach,[3] the performing arts then generate different characteristics (identity, prestige, an enjoyable show) at a first stage which in turn provide utility to humans at the second stage (see Fig. 1). Thus, it is not a good or service itself that provides utility, but the concomitant characteristics, i.e., relevant properties of the good or service, which affect humans' well-being.

---

[2] Only in some rather special cases it may occur that public good provision in an efficient allocation is lower than in a Nash equilibrium what has been coined as "overprovision anomaly" (Buchholz and Peters 2001).

[3] Lancaster's (1966) characteristics approach is a meaningful tool in several fields of economics. For example, it is involved in the methodology of discrete choice experiments (see, e.g., Hoyos 2010).

The characteristics approach can be traced back even to an earlier paper by Gorman (1980) from 1956,[4] in which he distinguishes classes of eggs in accordance with different measurable characteristics like the contents of vitamins. The characteristics approach also has some resemblance to Muth's household production approach (Muth 1966), where goods and services are considered as mere inputs to a household's production process whose outputs bring about utility to the household. In this regard, not the purchased food (the "good") but the nutrients enjoyed from the consumption of food (the "characteristics") provide utility to the household. Sandmo (1973) extends the characteristics approach by including some inputs that are collective goods.

Based on these characteristics approaches, Cornes and Sandler (1984) developed the standard impure public good model, where private and public characteristics are provided jointly by one good and where philanthropy served as the major field of application. In this context, Andreoni (1989) later coined the term "warm-glow giving," meaning that people who donate for a public-good charity do not exclusively provide a public characteristic, i.e., assistance for others, but also generate a private characteristic as they draw satisfaction from the act of giving itself.

In the context of climate policy,[5] a climate-change mitigation project also generates different characteristics that can be categorized as primary effects (i.e., climate change mitigation) and ancillary or secondary effects (i.e., improved local air quality, less noise from traffic etc.), respectively. Humans derive utility both from the primary effects, i.e., effects that were primarily intended by the climate policy, and from the ancillary effects (see Fig. 2).[6]

In our analysis of countries' behavior we take these two different categories of effects into account by assuming that mitigation provides two different characteristics, one of them representing the globally public primary effects and the other standing for ancillary effects that can only be enjoyed domestically in the considered country, i.e., the ancillary effects are private to the country pursuing the mitigation policy. We assume below that each unit of the impure public good "climate change mitigation policy" generates one unit of the public characteristic and ϕ (with $\phi > 0$) units of the private (ancillary) characteristic so that $z_i = \phi g_i$. The maximization problem for an individual country can now be expressed in characteristics space (c_i, z_i, G), which means that the goods no longer appear. The maximization problem of country i thus is described by

$$\max_{c_i, z_i, g_i} u_i \left(c_i, z_i, g_i + \widetilde{G}_i \right) = u_i \left(c_i, z_i, G \right) \tag{6}$$

---

[4]Even though it was written in 1956, this paper was only published as late as 1980.

[5]For impure public good modeling in the context of climate protection also see, e.g., Sandler (1996), Markandya and Rübbelke (2012), and Schwirplies and Ziegler (2016).

[6]For a distinction between primary and ancillary/secondary benefits also see, e.g., Rübbelke (2006).

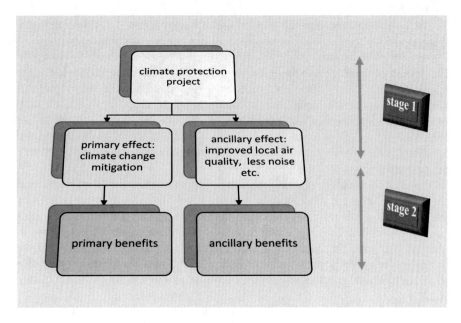

Fig. 2 Ancillary and primary benefits of a climate protection project and Lancaster's two stages

s.t.

$$c_i + \frac{p}{\phi} z_i = w_i \tag{7}$$

$$-z_i + \phi G = \phi \widetilde{G}_i, \tag{8}$$

where g_i is the level of the public characteristic provided by country i via its mitigation efforts (at the same level g_i). g_i contributes to the global public characteristic provision G, as $G = g_i + \widetilde{G}_i$. c_i represents the level of the private characteristic enjoyed from consuming the numéraire private good (at the same level c_i), i.e., one unit of the good c_i is assumed to generate one unit of its characteristic. By observing $g_i = \frac{z_i}{\phi}$ Eq. (7) follows from the budget constraint $c_i + pg_i = w_i$ stating that the extent of characteristics provision by country i is restricted by its monetary income w_i ($w_i > 0$). p (with $p > 0$) stands for the unit price of the impure public good.

Maximization of utility (in (6)) subject to restrictions (7) and (8) yields the first-order condition

$$\frac{\partial u_i / \partial G}{\partial u_i / \partial c_i} + \phi \frac{\partial u_i / \partial z_i}{\partial u_i / \partial c_i} = p. \tag{9}$$

The maximization problem stated in Eqs. (6)–(8) involves more than one linear constraint, which means that we face a so-called points-rationing problem.

The historic origin of economic concerns about such problems lies in rationing of goods during war times (and shortly thereafter).[7] During World War II, several states, e.g., England (see Booth 1985) and Germany (see Singer 1941), applied points rationing schemes (PRS) in order to influence the distribution of foods, clothes, and fuels and to guarantee a minimum supply for the poor. This meant that two prices were assigned in stores to the rationed goods—one price was listed in monetary units and one in "ration points" (or a "point currency," see also de Graaff 1948). In order to purchase rationed goods a consumer had to have both the cash and the necessary amount of ration points, i.e., he was facing two constraints (and not only the usual monetary one) restricting his utility maximization. Prominent alternative rationing systems are straight rationing and value rationing. Straight rationing is a scheme in which the ration currency is only applicable to one good (see, e.g., Tobin and Houthakker 1950), while value rationing "*sets the points prices of the goods covered by a particular ration currency in proportion to their money prices*" (Tobin 1952, p. 523).

After the 1940s, the interest of economists in points rationing problems declined,[8] since most rationing schemes introduced during World War II were abandoned in the 1950s. Yet, as Cornes (1996, pp. 2–3) points out, the diminishing interest in rationing "*... is a pity, because there are several problems of current interest that give rise to a formal structure which is precisely that of a points rationing problem.*"

In particular, Cornes and Sandler (1994, 1996) have taken advantage of the concept of virtual prices, which plays a major role in the rationing literature, for a further analysis of the points rationing problem stated in Eqs. (6)–(8). Rothbarth (1941) developed this concept of virtual prices when he explored by which amount a consumer's monetary income would have to be modified in a situation before a rationing scheme is applied in order to put him on the highest indifference curve that he can reach under rationing. The importance of a virtual price system for the welfare analysis of points rationing problems is stressed by Tobin (1952, p. 529) stating that a "*problem relevant to the estimation of the welfare consequences of rationing is to find the 'virtual' system of free market prices which would induce a consumer of a given money income to take precisely the same amount of each commodity as he consumes under rationing.*"

Consequently, through correctly set virtual prices, a consumer would get the incentive to choose the same bundle of goods as he would do under rationing (see also Neary and Roberts 1980).

---

[7] As Lloyd (1942, p. 49) puts it, "*[i]n war, when the free play of the price mechanism has to take a secondary role in the allocation of resources, rationing tends to become the rule rather than the exception.*"

[8] Among the notable exceptions is Baumol's (1982) analysis of fairness in the distribution of resources where he employs the issue of points rationing of commodities as an illustration.

Following this approach, a virtual world in which agents can purchase characteristics directly and do not have to take the detour of acquiring characteristics via the purchase of goods is created. We therefore have to move to the virtual world after having transferred the original problem from goods space to characteristics space [see Eqs. (6)–(8)]. Prices and income levels are now expressed as virtual magnitudes. The virtual price of characteristic j is then given by its inverse demand function ψ_j. It depends on the observable parameters p, w, $\widetilde{G}$, and ϕ and thus reads[9]

$$\psi_j = f_j\,(c, z, G) = \psi_j\left(p, w, \widetilde{G}, \phi\right) \tag{10}$$

(with $j = c, z, G$). Equivalently, the virtual income μ also depends on the observed parameters and therefore becomes

$$\mu = f_\mu\,(c, z, G) = \mu\left(p, w, \widetilde{G}, \phi\right). \tag{11}$$

Here, the concept of "virtual income" is equivalent to "full income"[10] or to Becker's idea of "social income," which he defines as "*the sum of a person's own income (his earnings, etc.) and the monetary value to him of the relevant characteristics of others*" (Becker 1974, p. 1063).

In the context of the impure public good approach, the virtual price system has allowed to conduct a comparative statics analysis and to investigate changes in consumption patterns due to changes in the exogenous parameters.[11] As one cannot observe market prices for the public characteristic, one has to establish virtual prices in order to determine responses of considered agents (countries) depending on the parameters associated with standard price taking functions.

In the pioneering comparative-statics joint-production model by Cornes and Sandler (1994), the connection between these virtual magnitudes and "real" observable magnitudes (like monetary income w) is based on the following three links:

1. The virtual income must be consistent with the value of the chosen allocation in the observable "real" world:

$$\mu = \psi_c\left(p, w, \widetilde{G}, \phi\right) c + \psi_z\left(p, w, \widetilde{G}, \phi\right) z + \psi_G\left(p, w, \widetilde{G}, \phi\right) G. \tag{12}$$

---

[9]The index i for the respective agent or country is omitted in the following expressions for simplicity.

[10]On the use of the "full income" conception see, e.g., Cornes and Sandler (1994) and Vicary and Sandler (2002).

[11]As Cornes and Sandler (1994, p. 405) put it: "*To forge the link between the less orthodox impure public good model and the parameters of orthodox price-taking behavior, we find it helpful to introduce virtual prices and income.*"

2. The price the considered agent is willing to pay for a unit of the impure public good, which is equal to the price p that is actually paid, must be equal to the valuation of the characteristics supplied by this unit of the impure public good:

$$p = \phi \psi_z \left(p, w, \widetilde{G}, \phi\right) + \psi_G \left(p, w, \widetilde{G}, \phi\right). \tag{13}$$

Here, $\phi \psi_z \left(p, w, \widetilde{G}, \phi\right)$ stands for the ancillary benefit per unit of the impure public good "climate change mitigation" and $\psi_G \left(p, w, \widetilde{G}, \phi\right)$ for the respective primary benefit.

3. The amount by which the ancillary private characteristic is augmented by an agent's provision of the impure public good must be proportional to the amount by which this country augments the provision of the public characteristic. Setting $\psi_c \left(p, w, \widetilde{G}, \phi\right) = 1$ we then obtain:

$$\begin{aligned} &\left\{D_z \left(\psi_z \left(p, w, \widetilde{G}, \phi\right), \psi_G \left(p, w, \widetilde{G}, \phi\right), \mu \left(p, w, \widetilde{G}, \phi\right)\right)\right\} \\ &= \phi \left\{D_G \left(\psi_z \left(p, w, \widetilde{G}, \phi\right), \psi_G \left(p, w, \widetilde{G}, \phi\right), \mu \left(p, w, \widetilde{G}, \phi\right)\right) - \widetilde{G}\right\}. \end{aligned} \tag{14}$$

Here, D_z and D_G stand for the demand of the considered agent for the characteristics z and G, respectively.

Figure 3 visualizes how the virtual magnitudes for a country are derived from constraints (7) and (8) in the characteristics space. Plane $ABCD$ represents the budget constraint originating from Eq. (7). If the considered country spends its income exclusively on the private good c and does not provide the impure public good, it will be situated somewhere along the upper edge AB, with its exact position depending on the amount of public-characteristic spill-ins it receives from the other countries (point E in the example shown in the figure). If, in contrast, the country spends its whole income on the impure public good, it will be situated along the line CD where $c = 0$. Its exact position again depends on the amount of the public characteristic $\widetilde{G}$ provided by the other countries, but now also includes the country's own contribution to the public characteristic (point F in the example shown in the figure). The slope of plane $ABCD$ in z-c space (along BC) is $-p/\phi$.

The second plane, TUV, depicts the technological constraints in the production of the three characteristics. To put it differently, it reflects the second constraint as given by Eq. (8) as well as the one-to-one relationships between the impure public good and its public characteristic and between the purely private good c and its characteristic, respectively. Point U is the country's initial public characteristic level where it enjoys the amount $\widetilde{G}$ of the public characteristic. Starting in this point, it can spend its income on the impure public and the private good and thus raise the amounts of all characteristics it can enjoy. If the country decides to consume only the private good, its position is shifted upwards along TU, up to point E where the budget constraint is met and the country's income has been completely spent on c. If it instead decides to spend its income only on the impure public good, it moves outwards along UV up to point F where, again, the budget constraint is met. The

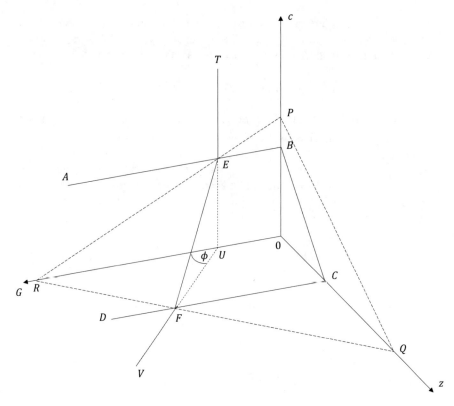

Fig. 3 Virtual magnitudes in characteristics space (based on a depiction in Cornes and Sandler 1994, p. 408)

slope of the ray UV in the G-z plane is ϕ since the impure public good generates characteristics z and G in the ratio ϕ.

As both constraints (7) and (8) have to hold simultaneously, the solution to our maximization problem must lie on the line EF where both planes intersect; the considered country's precise optimum position M on EF depends on its preferences and must be located in a point where an indifference surface (not shown in Fig. 3) is tangent to line EF.

The virtual budget constraint as expressed by Eq. (12) is represented by plane PQR in Fig. 3. It must incorporate the constraint EF so that the consumption bundle chosen in terms of virtual magnitudes is consistent with the allocation perceived in the "real" observable world, as stated above. Just as point B represents the country's observable monetary income w, i.e., the point where $c = w$, point P represents its virtual income $c = \mu$, where μ is the sum of monetary income w and the value of the public-characteristic spill-ins to the considered country. The other two intercepts, points Q and R, represent the quotient of virtual income and the virtual price of the respective characteristic, i.e., at Q, $z = \mu/\psi_z$ and at R, $G = \mu/\psi_G$ holds.

By means of Fig. 3, the effects of observable parameter changes on the country's optimal choice can be illustrated. If, for example, the country's income w increases, plane $ABCD$ is shifted in parallel outwards and its edges AB and CD both move along the c and z axes, respectively. Plane TUV is not affected by an increase in w. Hence, line EF moves outwards and with it the efficient point on it. Plane PQR will accommodate such that it incorporates the shifted line EF again and that it is compatible with the new optimal point M'.

The move of plane PQR, which represents the virtual budget constraint, is not necessarily parallel. This can be explained by observing that the adjustment of the system of virtual prices and income that is required to bring about the new optimum M' after the increase of income is affected by two factors:

1. Since the change in monetary income w may affect the demand for individual characteristics in different ways, demand changes may be different even if the characteristics of the impure public good are produced in a fixed relation. To give an example, let us assume that $\phi = 0.5$, i.e., that one unit of the impure public good produces 0.5 units of the ancillary private characteristic in addition to one unit of the public characteristic. Let us next suppose that the rise of income increases the demand for the ancillary benefit twice as much as for the public characteristic. Then, an increase of the country's impure public good contribution cannot produce the combination of private and public co-effects which ideally would be desired.
2. Virtual prices will have to adapt in order to bring about consistency between demand patterns and technical possibilities in order to attain M' also in the virtual system (see also Rübbelke 2002). Taking up the example above, this would imply that the ratio between virtual price of the public and the private (ancillary) characteristic has to decline, i.e., the private characteristic would have to become relatively more expensive.

Thus, the new optimum M' can only be attained if the plane PQR is not only shifted, but also if its slope changes in the relevant directions of the virtual budget plane. As Cornes and Sandler (1994) show, the extent of complementarity between the jointly supplied characteristics of the impure public good is of high importance for these changes and thus for the comparative static effects.

The graphical illustration in Fig. 3 indicates the complexity of the impure public good model, which demonstrates that the analysis is much more intricate than in the pure public good model. Nevertheless, the technique based on the points rationing approach has been employed for further comparative static analyses of co-benefits especially in the environmental context as, e.g., by Vicary (2000), Rübbelke (2003), Kotchen (2005, 2009) and Chan and Kotchen (2014). Vicary (1997, 2000) examines different available technologies for raising the level of the public characteristic. In his paper from 1997, he compares a model where the provision of the public characteristic is possible by means of both a pure public good and an impure public good with a model where only one of these options is available. In contrast to the case covered in Vicary (1997) where individuals' actions can only increase the level of the public characteristic, Vicary (2000) analyzes an offsetting framework, i.e., consumption of the impure public good is assumed to have an adverse effect on the environment and therefore contributes to a public bad (e.g., some kind of pollution) that can be voluntarily offset by the provision of donations as a pure public good. Rübbelke (2003) includes an alternative technology to produce the impure public good's private characteristic separately, while Kotchen (2005) allows for both a separate production of the private and of the public characteristic. His comparative static analysis suggests that, counterintuitively, a decrease in the price of the impure public ("green") good or an improvement of the technology for generating the joint characteristics can, due to crowding-out effects, cause a decline in demand for the public characteristic "environmental quality." Although consumption of the green good definitely rises in the new equilibrium, the improved possibilities of generating the private characteristic imply that this change might be achieved by a country's smaller direct public good contribution. Another interesting result is generated if a separate production of the joint characteristics is ruled out, i.e., the provision of the impure public good is the only way to obtain its private and public characteristics. In that case, demand for environmental quality can decrease when monetary income rises even if both characteristics have the properties of normal goods, provided that the joint characteristics are complements. Kotchen (2009) considers an offsetting model with two public goods, one of which is impure while the other is a pure public good. The joint public characteristic of the impure public good is an adverse effect on environmental quality that consumers, however, can voluntarily offset by contributing to environmental quality as a pure public good. He concludes that "*a technology improvement that makes a polluting good more environmentally friendly can actually diminish both environmental quality and social welfare*" (Kotchen 2009, p. 885). Moreover, enlarging the economy will not lead to individual public good contributions converging to zero, as is normally the case in a standard pure public good framework. Auld and Eden (1990) and Chan and Kotchen (2014) generalize the impure public good model to multiple impure public goods and jointly produced characteristics.

In environmental economics, the impure public goods approach has also been extended by considering such goods in dynamic frameworks, e.g., Corradini et al. (2015) investigate firms' investments in environmentally friendly innovation in a multi-period model in which the knowledge stock is treated as an impure public good.[12] Pittel and Rübbelke (2017) develop an impure public good model with local flow and global stock pollution in which there are two different types of abatement technologies, one of which is targeted exclusively on local pollution and the other one on both local and global pollution.

---

[12]Bahn and Leach (2008) use an overlapping generations model to investigate co-effects of climate policy.

3 Conclusions

As demonstrated above, the impure public good model—even in its static version—is more intricate than the pure public good model because one faces points rationing problems. To address such problems, a virtual system of prices and income is capable of facilitating the analysis, which is the conventional route taken by theoretical treatments of impure public goods. Yet, the transmission of effects from the "real" observable world to the virtual system is by far not obvious. Therefore, ways to ease the analysis are urgently needed, in particular because impure public goods are the rule and not the exception in many research areas, not only in the context of climate policy. One promising way is the application of the aggregative game approach (AGA), which was developed by Cornes and Hartley (2007). It somewhat allows to reduce complexity by referring only to the aggregate level of public characteristic provision, so there is no need to consider reaction functions for each country separately. Kotchen (2007) was among the first who used the AGA in the impure public good context. He employs it to prove that normality (w.r.t. full income) of all characteristics is sufficient for existence and uniqueness of a Nash equilibrium in the impure public good model. Cornes (2016) applies the AGA to a variety of environmental issues and suggests some extensions, e.g., with respect to technologies of public supply aggregation other than the summation technology of individual public good contributions. Brumme et al. (2019) also employ the AGA and use it to trace the impure public good model back to the pure public good framework in order to establish existence and uniqueness of a Nash equilibrium. As they in addition show, this procedure also eases the gaining of new insights into free riding behavior and the limitations of Warr neutrality.

References

Andreoni J (1989) Giving with impure altruism: applications to charity and Ricardian equivalence. J Polit Econ 97:1447–1458

Atkinson SE, Sandler T, Tschirhart J (1987) Terrorism in a bargaining framework. J Law Econ 30:1–21

Auld DAL, Eden L (1990) Public characteristics of non-public goods. Public Financ 3:378–391

Bahn O, Leach A (2008) The secondary benefits of climate change mitigation: an overlapping generations approach. Comput Manag Sci 5:233–257

Baumol WJ (1982) Applied fairness theory and rationing policy. Am Econ Rev 72:639–651

Baumol WJ, Bowen WG (1966) Performing arts: the economic dilemma. Twentieth Century Fund, New York

Becker GS (1974) A theory of social interactions. J Polit Econ 82:1063–1093

Betts A (2003) Public goods theory and the provision of refugee protection: the role of the joint-product model in burden-sharing theory. J Refug Stud 16:274–296

Booth A (1985) Economists and points rationing in the Second World War. J Eur Econ Hist 14:299

Brumme A, Buchholz W, Rübbelke D (2019) Impure public goods and the aggregative game approach. In: Buchholz W, Markandya A, Rübbelke D, Vögele S (eds) Ancillary benefits of climate policy—new theoretical developments and empirical findings. Springer, Berlin, pp 141–155

Buchholz W, Peters W (2001) The overprovision anomaly of private public good supply. J Econ 74:6378

Buchholz W, Cornes R, Rübbelke D (2018) Public goods and public bads. J Public Econ Theory 20:525–540

Chan NW (2019) Funding global environmental public goods through multilateral financial mechanisms. Environ Resour Econ 73:515–531

Chan NW, Kotchen MJ (2014) A generalized impure public good and linear characteristics model of green consumption. Resour Energy Econ 37:1–16

Cornes R (1996) Points rationing: applications and analysis, Keele University (unpublished manuscript)

Cornes R (2016) Aggregative environmental games. Environ Resour Econ 63:339–365

Cornes R, Hartley R (2007) Aggregative public good games. J Public Econ Theory 9:201–219

Cornes R, Sandler T (1984) Easy riders, joint production, and public goods. Econ J 94:580–598

Cornes R, Sandler T (1994) The comparative static properties of the impure public good model. J Public Econ 54:403–421

Cornes R, Sandler T (1996) The theory of externalities, public goods and club goods, 2nd edn. Cambridge University Press, New York

Corradini M, Costantini V, Mancinelli S, Mazzanti M (2015) Interacting innovation investments and environmental performances: a dynamic impure public good model. Environ Econ Policy Stud 17:109–129

de Graaff JV (1948) Towards an austerity theory of value 1. S Afr J Econ 16:35–50

Gorman WM (1980) A possible procedure for analysing quality differentials in the egg market. Rev Econ Stud 47:843–856

Hoyos D (2010) The state of the art of environmental valuation with discrete choice experiments. Ecol Econ 69:1595–1603

Khanna J, Huffman WE, Sandler T (1994) Agricultural research expenditures in the United States: a public goods perspective. Rev Econ Stat 76:267–277

Kingma BR, McClelland R (1995) Public radio stations are really, really not public goods: charitable contributions and impure altruism. Ann Public Coop Econ 66:65–76

Kotchen MJ (2005) Impure public goods and the comparative statics of environmentally friendly consumption. J Environ Econ Manag 49:281–300

Kotchen MJ (2007) Equilibrium existence and uniqueness in impure public good models. Econ Lett 97:91–96

Kotchen MJ (2009) Voluntary provision of public goods for bads: a theory of environmental offsets. Econ J 119:883–899

Lancaster K (1966) A new approach to consumer theory. J Polit Econ 74:132–157

Lancaster K (1971) Consumer demand: a new approach. Columbia studies in economics, vol 5. Columbia University Press, New York

Lee DR, Sandler T (1989) On the optimal retaliation against terrorists: the paid-rider option. Public Choice 61:141–152

Leonard T (2016) Housing upkeep and public good provision in residential neighborhoods. Hous Policy Debate 26:888–908

Lloyd EMH (1942) Some notes on point rationing. Rev Econ Stat 24:49–52

Maikovich AK (2005) A new understanding of terrorism using cognitive dissonance principles. J Theory Soc Behav 35:373–397

Markandya A, Rübbelke D (2012) Impure public technologies and environmental policy. J Econ Stud 39:128–143

Munro A, Valente M (2016) Green goods: are they good or bad news for the environment? Evidence from a laboratory experiment on impure public goods. Environ Resour Econ 65:317–335

Muth RF (1966) Household production and consumer demand functions. Econometrica 34:699–708

Neary JP, Roberts KWS (1980) The theory of household behaviour under rationing. Eur Econ Rev 13:25–42

Pittel K, Rübbelke D (2017) Thinking local but acting global? The interplay between local and global internalization of externalities. In: Buchholz W, Rübbelke D (eds) The theory of externalities and public goods: essays in memory of Richard C. Springer, Cham, pp 271–297

Posnett J, Sandler T (1986) Joint supply and the finance of charitable activity. Public Financ Q 14:209–222

Pugliese T, Wagner J (2011) Competing impure public goods and the sustainability of the theater arts. Econ Bull 31:1295–1303

Rive N, Rübbelke D (2010) International environmental policy and poverty alleviation. Rev World Econ 146:515–543

Rothbarth E (1941) The measurement of changes in real income under conditions of rationing. Rev Econ Stud 8:100–107

Rübbelke D (2002) International climate policy to combat global warming: an analysis of the ancillary benefits of reducing carbon emissions. Edward Elgar, Cheltenham

Rübbelke D (2003) An analysis of differing abatement incentives. Resour Energy Econ 25:269–294

Rübbelke D (2005) Differing motivations for terrorism. Def Peace Econ 16:19–27

Rübbelke D (2006) Analysis of an international environmental matching agreement. Environ Econ Policy Stud 8:1–31

Sandler T (1996) A game-theoretic analysis of carbon emissions. In: Congleton RD (ed) The political economy of environmental protection: analysis and evidence. University of Michigan Press, Ann Arbor, pp 251–272

Sandler T (2013) Public goods and regional cooperation for development: a new look. Int Trade J 36:13–24

Sandler T (2017) International peacekeeping operations: burden sharing and effectiveness. J Confl Resolut 61:1875–1897

Sandler T, Arce M DG (2002) A conceptual framework for understanding global and transnational public goods for health. Fisc Stud 23:195–222

Sandler T, Hartley K (2001) Economics of alliances: the lessons for collective action. J Econ Lit 39:869–896

Sandler T, Murdoch JC (2000) On sharing NATO defence burdens in the 1990s and beyond. Fisc Stud 21:297–327

Sandmo A (1973) Public goods and the technology of consumption. Rev Econ Stud 40:517–528

Schwirplies C, Ziegler A (2016) Offset carbon emissions or pay a price premium for avoiding them? A cross-country analysis of motives for climate protection activities. Appl Econ 48:746–758

Singer HW (1941) The German war economy in the light of economic periodicals. Econ J 51:192–215

Tobin J (1952) A survey of the theory of rationing. Econometrica 20:521–553

Tobin J, Houthakker HS (1950) The effects of rationing on demand elasticities. Rev Econ Stud 18:140–153

Vicary S (1997) Joint production and the private provision of public goods. J Public Econ 63:429–445

Vicary S (2000) Donations to a public good in a large economy. Eur Econ Rev 44:609–618

Vicary S, Sandler T (2002) Weakest-link public goods: giving in-kind or transferring money. Eur Econ Rev 46:1501–1520

Private Ancillary Benefits in a Joint Production Framework

Claudia Schwirplies

1 Introduction

Everyday decisions of individuals have an immense impact on the use of resources and on the climate. In industrialized countries like Germany, the consumption of private households causes approximately three quarters of total greenhouse gas emissions (BMUB 2016). Although climate-friendly activities and associated markets have expanded rapidly worldwide in recent years,[1] total direct carbon emissions produced by consumption even slightly increased (Mayer and Flachmann 2016). Clearly, the acceptance and participation of individuals is crucial for the success of climate change mitigation.

In general, individuals have two channels available for contributing to climate protection, i.e., changing their own behavior and supporting climate policies. However, the incentive to contribute is generally lower than necessary to reach the social optimum (Holländer 1990), since both options are contributions to a public good (IPCC 2001). No individual or country can be excluded from the benefits of climate protection and the enjoyment of these benefits by one individual or country does not reduce the benefits to others. This chapter discusses whether putting more emphasis on private ancillary benefits[2] from climate protection may attenuate the

---

[1] Examples include carbon-neutral certified goods, green energy programs, vehicles with alternative propulsion technologies, and energy-efficient appliances.

[2] By ancillary benefits we understand positive externalities from climate protection measures, i.e., secondary or co-benefits such as financial, economic, or health benefits.

C. Schwirplies (✉)
Department of Economics, University of Hamburg, Hamburg, Germany
e-mail: claudia.schwirplies@uni-hamburg.de

© Springer Nature Switzerland AG 2020
W. Buchholz et al. (eds.), *Ancillary Benefits of Climate Policy*, Springer Climate,
https://doi.org/10.1007/978-3-030-30978-7_7

incentive to free ride and provide an additional source of motivation for climate protection activities from the individual's point of view.

While it is increasingly recognized that climate change mitigation policies can have important ancillary benefits, e.g., positive impacts on the local environment, economy, and air quality (Pittel and Rübbelke 2008), little is known about private co-benefits and their role as motivating factors for more climate-friendly consumption decisions and the support of climate policies by individuals. We provide an overview of the existing evidence on the determinants of such choices and emphasize ancillary benefits that have the potential to motivate and support private and public climate protection activities.

Policy measures and behavioral changes can involve different channels through which they can contribute to climate protection. First is promoting and consuming green products g instead of conventional ones c (e.g., public transportation instead of the car). Second is making direct donations d to climate protection projects (e.g., funding climate research or offset carbon emissions via providers of voluntary carbon offsetting who invest in climate protection projects worldwide). In recent years, suppliers started creating a third alternative by linking their conventional products and services c with contributions to public goods or charitable causes d. Examples are green energy tariffs that combine the purchase of electricity with the provider's commitment to support the development of renewable energies or providers and retailers who plant trees for every product sold.[3]

Following the notation in Kotchen (2006), consumer i derives utility from a private characteristic X_i, a public characteristic Y_i, which is climate protection in our case, and the total amount of climate protection Y produced by i and all other actors in the economy, such that

$$U_i = U(X_i, Y, Y_i) \text{ subject to } X_i = c_i + \alpha g_i, Y_i = d_i + \beta g_i, \text{ and } w_i = c_i + g_i + d_i,$$

where w_i is the individual's income. In the traditional understanding, purely private goods and services c only generate the private characteristic, green products g as well as bundles of c and d produce both private and public characteristic. A typical example is conventional and organic food: both produce the private characteristic nutrition, while the latter also produces the public characteristic pollution or pesticide reduction. Alternatively, the producer or the consumer can also donate directly to a charity that promotes organic farming.

This framework, however, does not fully account for ancillary benefits so far. For instance, organic products can be good for another private characteristic, e.g., health, and donations for the promotion of organic farming can provide the private characteristic internal satisfaction. Therefore, in a broader definition of ancillary benefits, even d would generate X_i by enhancing internal satisfaction from doing

---

[3] The internet search engine Ecosia, the campaign "tree pate" by the online retailer OTTO or the Crombacher campaign to protect one square meter of the rainforest per case of beer sold are well-known examples in Germany.

something good or acting as an example (e.g., Rübbelke 2002). In this sense, green goods, donations, and bundles of conventional good and donation all potentially increase i's utility through different private characteristics.[4]

One possibility for enhancing both policy support and climate-friendly activities could be to shift the main justification or framing in public discussions and campaigns on greenhouse gas mitigation from the conventional argument of reducing climate change risks (i.e., laying emphasis on the public characteristic Y_i) to private co-benefits (i.e., laying more emphasis on the private characteristics X_i). While the single individual's contribution to the public characteristic is negligible, the amount of private characteristic produced is significant and personally more relevant to the individual (Bernauer and McGrath 2016). In contrast to the benefits for the climate that arise from public and private climate mitigation efforts, ancillary benefits are usually of short-term nature and associated with less uncertainty (Longo et al. 2012). These advantages should make mitigation efforts appear more attractive to individuals and serve as a motivator for behavioral changes and policy support.

The question what motivates individuals to voluntarily contribute to public goods has received great interest and much progress has been made in understanding the determinants of money or time donations (for summaries see e.g., List 2011; List and Price 2012). By now, substantial evidence suggests that economic, other extrinsic as well as intrinsic factors like altruism, feelings of warm glow and moral obligation, social norms and image motivation affect contributions to charities and other public goods (e.g., Andreoni 1995; Glazer and Konrad 1996; Harbaugh 1998; Crumpler and Grossman 2008; Ariely et al. 2009; Shang and Croson 2009). This literature also points at existing interactions of private activities with public efforts, incentives, or institutional settings (e.g., Frey and Oberholzer-Gee 1997; Bohnet et al. 2001; Brekke et al. 2003; Nyborg and Rege 2003).

Inspired by these findings, empirical evidence on green markets, the determinants of their growth, and the support for public climate change mitigation is constantly growing. Among others, this literature discusses economic incentives, internal satisfaction, and health benefits as potential drivers of the necessary support and behavioral changes to reduce greenhouse gas emissions. This chapter provides some insights on how beliefs about and experiences with these ancillary benefits are related to individuals' motivations to act on climate change. This knowledge can help researchers and decision makers from the public and private sectors develop effective local and global strategies for using ancillary benefits to foster the necessary policy acceptance and behavioral changes.

We start the chapter with a discussion of the findings regarding private ancillary benefits from climate-friendly activities in Sect. 2. Section 3 investigates whether private ancillary benefits also raise the support for public action. While we focus on money savings and economic benefits, internal satisfaction and health benefits

---

[4]This means that we extend the model with additional private characteristics and all three channels contribute to the generation of each private characteristic (e.g., health, internal satisfaction) depending on the respective production technology.

in Sect. 4, we discuss another internal factor that has rarely been linked to ancillary benefits so far, i.e., fairness and equity issues. Section 5 summarizes and draws some conclusions and points out further research needs.

2 Private Ancillary Benefits as Motivator for Climate-Friendly Activities

Carbon offsetting, green and bundled goods form different channels through which an individual may voluntarily contribute to climate protection. The channels, however, diverge along various dimensions. For example, contributions to climate protection may differ in terms of visibility, costs, and time spent on the activity such that the incentives and motives for using these channels may vary significantly. This section summarizes empirical evidence with a focus on incentives that can be considered ancillary benefits, i.e., monetary savings, internal satisfaction, and health benefits.

Nyborg et al. (2006), for instance, see potential in economic incentives for the initiation of changes in consumption norms or habits. Even without direct pricing interventions, such as taxes or subsidies, switching to green products or habits can come with economic incentives. One example is public transportation that might save time and nerves especially in big cities or for medium-distance trips. These time savings can be used for paid labor. Further examples involve changing travel habits and reducing long-distance flights, reducing water or energy consumption, using more energy-efficient appliances, car sharing, or buying fuel-efficient cars— all these measures not only reduce emissions but also save money.

In this respect, Miao and Wei (2013) use a survey among employees from a Midwestern university in the USA to elicit their main motives to reduce, reuse, recycle, and for green consumption (based on a 20 item activity scale) in hotels and at home. The authors find a significant increase in the willingness to reduce and to consume green if respondents believe in monetary savings or other economic benefits. In line with these findings, Cary and Wilkinson (1997) show that the perceived profitability of conservation practices is the most important factor influencing the use of these practices by the farmers in their sample. Still, there is also evidence that the effectiveness of monetary incentives may depend on visibility. Bolderdijk et al. (2012) compare an economic tyre-check appeal versus pro-environmental and neutral appeals and find significantly less compliance with an emphasis on economic benefits. The authors argue that individuals prefer conveying a responsible rather than a greedy picture of themselves.

Another channel of economic incentives is making the green consumption alternative more effective in providing the public characteristic. The donation literature provides some insights into this mechanism using rebate and matching schemes to exogenously vary the effectiveness of a gift to a charity. These studies conformingly find that rebate and matching schemes have the potential to increase donations (e.g., Karlan and List 2007; Meier 2007; Eckel and Grossman 2008; Rondeau and List 2008; Anik et al. 2014). In this respect, matching increases the effectiveness of the donation by multiplying the impact of the donor's gift.

The only field study, as we are aware of, that tests those mechanisms with carbon offsetting is the one by Kesternich et al. (2016) in cooperation with a long-distance bus provider in Germany. Carbon offsetting differs from pure charitable donations by compensating a previous polluting activity. Still, their results indicate that 1:1 matching is most promising scheme for inducing long-term participation. 1:1 matching increases not only the willingness to offset emissions produced by the present travel, but also the likelihood to offset emissions when customers return to book their next travels even if offsets are no longer matched by the provider. Schwirplies et al. (2019) add other contexts to the long-distance holiday trip by bus. They randomly vary the means of transportation (bus vs. plane) and the travel occasion (holiday vs. professional training) in a discrete choice setting. In all four choice situations, participants showed a significant willingness to pay for the 1:1 matching scheme, but not for lower amounts added by the provider.

Lange et al. (2017) and Schwirplies and Ziegler (2016) investigate the two aspects together, i.e., monetary savings and a higher effectiveness in providing climate protection. They compare seven climate-friendly measures[5] in two countries (Germany and the USA). In the USA, monetary savings clearly seem to be an important motivator for all seven activities. In contrast, the belief that the activity makes a high contribution to climate protection is only significantly correlated with the willingness to reduce meat and dairy products as well as flights. In Germany, monetary savings are also a highly significant determinant for all measures but using energy from renewable sources[6] as well as reducing flights. The effectiveness in providing the public characteristic plays a more important role than in the USA. Here, the authors find significant correlations for four measures (buying energy-efficient appliances, saving energy at home, reducing meat or dairy products, and reducing flights). Also for carbon offsetting, the belief in a high effectiveness in protecting the climate is a significant determinant in both countries.

The second block of ancillary benefits is emotions from acting in a climate-friendly way resulting in internal satisfaction. Schwirplies and Ziegler (2016) provide a comparison of these types of motives across Germany and the USA for carbon offsetting and the willingness to pay a price premium for climate-friendly products and services. Their findings indicate that a good feeling from contributing to climate protection plays an important role for both measures in both countries. In addition, the desire to act as an example provides an incentive for paying a price premium in Germany and for carbon offsetting in the USA.

---

[5]That is, buying energy-efficient appliances, saving energy at home, reducing meat or dairy products, using energy from renewable sources, buying a car with lower fuel consumption, reducing car use, and reducing flights.

[6]This makes sense, since consumers in Germany usually pay a price premium for green energy tariffs.

Further evidence on the effects of a warm glow feeling in the context of climate protection activities is ambiguous. In the study by Clark et al. (2003), for example, customers of a green power program rank warm glow as their least important motive, whereas Menges et al. (2005) find evidence for warm glow motives in their experiment on the willingness to pay for green electricity. Additionally, Blasch (2015) compares a donation frame ("With your donation for this climate protection project, you help ... ") and an offsetting frame ("Many of your everyday consumption activities cause CO_2 emissions and increase your carbon footprint. Supporting this project reduces CO_2 emissions ... "). The results reveal significant differences between the framings depending on the strength of experienced cold prickle of not contributing relative to the feeling of warm glow from contributing. Thus, the effect of internal satisfaction may depend on the design of the decision.

Another source of internal satisfaction is a feeling of responsibility for protecting the climate, which is comparable to the concept of cold prickle mentioned above. Individuals who feel responsible for climate protection might suffer a utility loss from consuming conventional products. Akter et al. (2009) and Lange and Ziegler (2015) show that such feelings are positively correlated with the probability to pay a carbon travel tax or to purchase carbon offsets and fuel-efficient vehicles. Further studies also support the hypothesis that a perceived moral obligation leads to a higher willingness to engage in carbon offsetting (e.g., Brouwer et al. 2008; Blasch and Farsi 2014). However, feelings of responsibility may be prone to external factors. Financial incentives may provide a signal and change the perception of responsibility (Nyborg et al. 2006). Specifically, a Pigouvian tax may be interpreted as a signal that the government is taking over the responsibility for climate protection. At the same time, the perception and acceptance of responsibility might depend on the beliefs about others' behavior, which has also been shown to be an important determinant for the take-up of climate-friendly activities (e.g., Welsch and Kühling 2009; Araghi et al. 2014; Blasch and Farsi 2014).

The final block of ancillary benefits are health benefits from climate-friendly consumption habits. Besides better local air quality, immediate health benefits may especially arise from changes in commuting habits, i.e., taking a walk or the bike more often, or dietary changes especially with more local or organic vegetables and less red meat. Still, economic studies on health benefits as ancillary benefits and motivator for climate-friendly behaviors are scarce. Bothner et al. (2019) see potential in highlighting health benefits for promoting climate-friendly activities in the sectors food and recycling, housing, other consumption, and mobility, if health information is combined with financial incentives. In contrast, in their experiment Perino and Schwirplies (2019) do not find any effect of emphasizing health benefits on the consumption of red meat. The authors let their participants fill in food diaries and asked them to answer some questions on newspaper articles. These articles randomly introduced climate protection, animal welfare and personal health as a reason to eat less meat. Participants in the health treatment neither showed higher stated intentions to reduce meat consumption after the treatment, nor did they reduce their consumption of red meat or loose satisfaction from eating red meat.

Altogether, monetary savings and the effectiveness of a climate protection activity seem to be of high relevance for private climate-friendly activities. But the finding also diverges across different countries and activities. The evidence on internal satisfaction and health benefits is mixed and at the same time still scarce. Clearly, more research is needed, ideally randomized controlled trials with real decisions of the relevant target group. Certainly, the papers discussed above also find numerous other factors that influence private contributions to climate protection, such as age, gender, income, education, environmental preferences, religiousness, and political views.

3 The Role of Ancillary Benefits for the Support of Climate Protection Policies

The second channel to lever climate protection is enhancing policy support. After great enthusiasm brought by the 2015 Paris Agreement to limit global warming to as much as 1.5 degrees, a countermovement has emerged in the world community. In the USA, for example, the climate change skeptic Donald Trump won the presidential elections and declared that the USA, one of the major players in this respect, is getting out of the Paris Agreement. However, without a broad acceptance and support of national and international climate policy by citizens all around the world, the comprehensive changes in economies and societies that are necessary to reach the targets of the Paris Agreement will not be realized. This section reviews the existing literature for evidence that public support of climate protection policies can be increased by reframing them with concerns that are more private.

So far, the evidence on the motivational power of the three blocks of co-benefits is very limited. Bain et al. (2015) analyze the relevance of ancillary benefits for behaviors aimed at bringing about public and political action involving almost 6200 respondents in 24 countries. The authors focus on health benefits, economic development, as well as benevolence in the community. Especially for the latter two, they find treatment effects on public support for climate protection measures. The effects are as sizable as rating climate change as one of the most important issues that society is facing today.

Bernauer and McGrath (2016) conduct a survey experiment with individuals from the USA and use four different frames. The control frame emphases the risks of climate change. The economic benefit frame focusses on the technological innovations and more economic success. The third frame highlights a potential enhancement of the community spirit that might be attributed to our internal satisfaction block. The final frame emphasizes the health benefits from pollution reduction and more active lifestyles. To the best of our knowledge, this study is the only one that investigates all three blocks of ancillary benefits in a randomized controlled trial. However, the authors find no treatment effects for either of their outcome variables, i.e., policy support, behavioral intentions, and environmental citizenship.

Schwirplies (2018) adds another international dimension to these analyses by investigating the determinants of the acceptance and support of national as well as international climate policies in three countries, namely China, Germany, and the USA. The author finds a high and significantly positive correlation between the belief that contributing to climate protection has a positive impact on the economy and the support of national and international climate policies in all three countries. Further, using the same indicators for internal satisfaction as in Schwirplies and Ziegler (2016), she finds significantly positive correlations, but some variation across the three countries. In China, the feeling of responsibility for climate protection and the desire to act as an example are positively correlated with the support of national and international climate policy. In Germany, the positive interrelation between internal satisfaction and policy support is driven by a good feeling from climate protection and the feeling of responsibility. In the USA, only the support of national climate policy and a good feeling from climate protection are positively correlated. The same indicator plus the desire to act as an example are positively interrelated with the support of international climate policies in the USA. Still, these are correlations, which might be driven by factors that are not included in the econometric model.

One attribute that is often associated with ancillary benefits is the local implementation of climate protection measures and projects. The idea is that local is a proxy for benefits from positive effects on the local economy and air quality. However, the findings are very ambiguous. Stated preferences studies usually find a significantly higher willingness to pay for climate protection measures implemented in the participants' region (e.g., Kaenzig et al. 2013; Schwirplies et al. 2019) or when their region is negatively affected by climate change (e.g., Longo et al. 2012). Field experimental evidence, in contrast, suggests that emphasizing the local engagement has no effect on donation rates and amounts for developing renewable energies (Lange and Schwirplies 2019).

Overall, the findings whether framing public climate protection measures with private benefits will foster support are quite ambiguous. The limited results are mixed depending on the method, survey instrument, and outcome variables. Again, we need more research that, for example, applies experimental methods.

4 Fairness and Equity as Private Ancillary Benefits

The theory on inequality aversion assumes that individuals derive a utility loss from having more or less than others (Fehr and Schmidt 1999; Bolton and Ockenfels 2000). In addition, numerous experimental studies provide evidence that individuals are willing to sacrifice some of their material payoff to obtain a more equitable outcome (e.g., Carlsson et al. 2005). That is, individuals who are averse to earning more than others are willing to propose positive amounts in the dictator and ultimatum game and individuals who are averse to earning less than others are willing to reject low offers in ultimatum games and punish in public goods games.

Similarly, experimental studies on "conditional cooperation" have found that people are more willing to contribute to charities and public goods if they observe, believe, or are informed that others are willing to do the same (e.g., Fischbacher et al. 2001).

In the context of climate change mitigation, individuals might dislike contributing more or less than others. Just as monetary savings, health benefits, and internal satisfaction, perceived fairness and equality in contributions to climate protection would immediately generate the private characteristic X_i. Alternatively, the belief or perception that contributions follow an unfair or unequal distribution might reduce or destroy parts of the private utility component. Still, aspects of fairness and equity are rarely linked to ancillary benefits of climate-friendly behavior and climate policies and empirical evidence remains scarce.

Inspired by disastrous climate change, Tavoni et al. (2011) conduct a public goods lab experiment where inequality in endowments reduces cooperation to prevent a disastrous loss. Nevertheless, if groups succeed they also tend to eliminate inequality over the course of the game. The experiment by Ajzen et al. (2000) shows that the willingness to pay for a public good increases with the perceived fairness of the requested contribution. Johnson (2006) uses the Contingent Valuation method and finds that the share of respondents who are willing to vote for an environmental regulation increases if the producer also pays a part of the associated costs. Similarly, the results in Cai et al. (2010) emphasize the importance and degree of participation of actors that are believed to be responsible for emissions. These findings are in line with Schleich et al. (2016) who compare burden-sharing principles for international climate policy across individuals from China, Germany, and the USA. Individuals in all three countries seem to have a common (normative) understanding of fairness with the highest preference for polluter pays.[7] Most recently, Andor et al. (2018) show that industry exemptions decrease the willingness to accept an increase in the EEG surcharge, while announcing the elimination of the exemptions significantly raises this willingness.

In addition, social approval based on norm-compliant behavior seems to be positively correlated with the share of the population that acts according to these norms (e.g., Rege 2004). Araghi et al. (2014) demonstrate that travelers are more likely to offset their carbon emissions from air traveling if the collective participation rate is high. Blasch and Farsi (2014) provide evidence that carbon offsetting is strongly driven by the adherence to social norms and the expectations about the cooperation of others. Schwirplies and Ziegler (2016) test the correlation of a strong agreement to the statements "my family, friends or colleagues do not contribute to climate protection" and "society expects me to contribute to climate protection" with the willingness to offset emissions or to pay higher prices for climate-friendly products and services among individuals from Germany and the USA. The belief that social environment does not contribute to climate protection decreases the willingness to pay a price premium in Germany and to offset in the

---

[7]However, there is also evidence that individual choose burden-sharing rules in a more self-serving manner, i.e. favoring the rule that is least costly for their country (e.g., Carlsson et al. 2013).

USA. In contrast, expectations from the society increase both the willingness to offset and to pay a price premium in the USA.

Climate policy might also provide a signal about descriptive and injunctive social norms,[8] which then motivate individuals to comply to these norms. External circumstances such as policy interventions may further crowd out or crowd in an individual's moral motivation to privately provide a public good (e.g., Bó et al. 2010). In this respect, Schleich et al. (2018) provide suggestive evidence that beliefs about climate policy are correlated with own contributions. For example, a higher perceived legitimacy and effectiveness of international climate policy are positively correlated with voluntary individual climate protection activities in Germany and the USA. For German respondents the effect of perceived effectiveness of international climate policy varies with internal satisfaction. That is, respondents with a more dominant intrinsic motivation and who at the same time consider international climate policy to be successful are less likely to make own voluntary contributions. Overall, there seems to be some interrelation between own climate action, ancillary benefits, and beliefs about public policy. The latter might certainly have an effect on policy support. Similarly, Rübbelke (2011) argues that supporting adaptation in developing countries may reduce their perception of a lack of fairness in international climate negotiations, which in turn could have the ancillary benefit of facilitating an agreement.

All the results should be interpreted with some caution because the decisions either did not involve any monetary consequences for participants or are conducted with a restricted population in the lab. Still, the findings suggest that perceptions of equity and fairness considerations matter for decisions about private climate protection activities as well as for policy support. If policymakers do not account for them, these factors might even induce undesired behavioral reactions in other areas of consumption. For example, if individuals are forced to pay for a policy instrument, which they consider unfair, they might reduce their climate-friendly activities in other consumption areas to offset this unfairness.

5 Conclusions and Future Research Needs

In this chapter, we focus on private ancillary benefits from climate change mitigation that might work as motivating factors for individuals and their climate-friendly activities and support of climate policies. In contrast to the primary benefits on the climate, private secondary benefits appear to be more attractive to the individual as they directly increase their utility in the short run and are associated with less uncertainty. We review some existing literature on monetary savings,

---

[8]Descriptive norm refers to the perception or belief of how much others actually contribute to climate protection. Injunctive norm refer to the perception or belief of how much oneself or others ought to contribute.

internal satisfaction, health benefits, and fairness in order to derive a conclusion and recommendation whether actors from the public and private sector should lay more emphasis on the secondary private benefits when promoting climate protection measures.

For economic incentives and internal satisfaction, we see the clearest evidence that they have a positive effect on climate-friendly activities and policy support. However, monetary savings have also been shown to be potentially negatively affected by visibility and have limitations in effectiveness (Bolderdijk et al. 2012; Bothner et al. 2019). In terms of internal satisfaction, the framing of the decision might play a role. Regarding health benefits, we find only limited empirical research. Combined with financial incentives Bothner et al. (2019) find promising results, while Perino and Schwirplies (2019) do not see any effect.

Finally, we regard fairness and equity perceptions as another source of private utility that is rarely linked to ancillary benefits. Generally, a distribution of contributions that follows the polluter pays principle seems to be regarded as fair and has the potential to increase climate-friendly activities and policy support (Andor et al. 2018). On average, individuals also seem to be conditional cooperators and more willing to contribute if others do the same. Further, we see suggestive evidence that climate-friendly activities are positively correlated with the perception that climate policies are justified, successful, and fair.

However, for all categories of ancillary benefits the evidence is mixed and limited. Further research in different countries, for a variety of climate protection measures and framings of the decision is needed to draw clear conclusions and recommendations. In addition, many of the studies discussed in this chapter rely on stated preferences methods and intentions, which may result in hypothetical biases if the responses have no monetary consequences for the participants. Future research should rely on experimental methods with the relevant target group or empirical analyses with panel data that elicit actual behavior instead of stated intentions and are able to at least control for time constant unobserved factors. Especially for policy support, revealed preferences are hard to observe, but future research might at least account for the fact that election programs are usually multidimensional and subject to budget constraints.

Evans et al. (2012) bring forward a final warning in laying too much emphasis on private ancillary benefits as they may prevent beneficial spillover effects on other climate-friendly activities. The authors run two experiments promoting car sharing with information on either monetary savings or environmental benefits. They then investigate not only the effect on car sharing, but also the spillover effect on recycling. The authors find no significant difference between recycling activities in the control and the monetary savings group, but a significantly higher recycling rate in the group that received information on environmental benefits. Spillover effects from climate-friendly activities and differences across motives are other interesting and important directions of research for the future.

References

Ajzen I, Rosenthal LH, Brown TC (2000) Effects of perceived fairness on willingness to pay. J Appl Social Pyschol 30(12):2439–2450. https://doi.org/10.1111/j.1559-1816.2000.tb02444.x

Akter S, Brouwer R, Brander L, van Beukering P (2009) Respondent uncertainty in a contingent market for carbon offsets. Ecol Econ 68(6):1858–1863. https://doi.org/10.1016/j.ecolecon.2008.12.013

Andor MA, Frondel M, Sommer S (2018) Equity and the willingness to pay for green electricity in Germany. Nat Energy 3(10):876–881. https://doi.org/10.1038/s41560-018-0233-x

Andreoni J (1995) Cooperation in public-goods experiments: kindness or confusion? Am Econ Rev 85(4):891–904. https://doi.org/10.2307/2118238

Anik L, Norton MI, Ariely D (2014) Contingent match incentives increase donations. J Market Res 51(6):790–801. https://doi.org/10.1509/jmr.13.0432

Araghi Y, Kroesen M, Molin E, van Wee B (2014) Do social norms regarding carbon offsetting affect individual preferences towards this policy? Results from a stated choice experiment. Transport Res D Transport Environ 26(0):42–46. https://doi.org/10.1016/j.trd.2013.10.008

Ariely D, Bracha A, Meier S (2009) Doing good or doing well? Image motivation and monetary incentives in behaving prosocially. Am Econ Rev 99(1):544–555. https://doi.org/10.2307/29730196

Bain PG, Milfont TL, Kashima Y, Bilewicz M, Doron G, Garðarsdóttir RB, Gouveia VV, Guan Y, Johansson L-O, Pasquali C, Corral-Verdugo V, Aragones JI, Utsugi A, Demarque C, Otto S, Park J, Soland M, Steg L, González R, Lebedeva N, Madsen OJ, Wagner C, Akotia CS, Kurz T, Saiz JL, Schultz PW, Einarsdóttir G, Saviolidis NM (2015) Co-benefits of addressing climate change can motivate action around the world. Nat Clim Change 6:154–157. https://doi.org/10.1038/nclimate2814

Bernauer T, McGrath LF (2016) Simple reframing unlikely to boost public support for climate policy. Nat Clim Change 6:680. https://doi.org/10.1038/nclimate2948

Blasch J (2015) Doing good or undoing harm—doing good or undoing harm—framing voluntary contributions to climate change mitigation. Conference paper presented at the EAERE 2015

Blasch J, Farsi M (2014) Context effects and heterogeneity in voluntary carbon offsetting—a choice experiment in Switzerland. J Environ Econ Policy 3(1):1–24. https://doi.org/10.1080/21606544.2013.842938

BMUB (2016) Bundesregierung wirbt für nachhaltigen Konsum: Nationales Programm verabschiedet. http://www.bmub.bund.de/presse/pressemitteilungen/pm/artikel/bundesregierung-wirbt-fuer-nachhaltigen-konsum/. Accessed 13 Dec 2016

Bó PD, Foster A, Putterman L (2010) Institutions and behavior: experimental evidence on the effects of democracy. Am Econ Rev 100(5):2205–2229. https://doi.org/10.1257/aer.100.5.2205

Bohnet I, Frey BS, Huck S (2001) More order with less law: on contract enforcement, trust, and crowding. Am Polit Sci Rev 95(01):131–144. https://doi.org/10.1017/S0003055401000211

Bolderdijk JW, Steg L, Geller ES, Lehman PK, Postmes T (2012) Comparing the effectiveness of monetary versus moral motives in environmental campaigning. Nat Clim Change 3:413. https://doi.org/10.1038/nclimate1767

Bolton GE, Ockenfels A (2000) ERC: a theory of equity, reciprocity, and competition. Am Econ Rev 90(1):166–193. https://doi.org/10.1257/aer.90.1.166

Bothner F, Dorner F, Herrmann A, Fischer H, Sauerborn R (2019) Explaining climate policies' popularity—an empirical study in four European countries. Environ Sci Policy 92:34–45. https://doi.org/10.1016/j.envsci.2018.10.009

Brekke KA, Kverndokk S, Nyborg K (2003) An economic model of moral motivation. J Public Econ 87(9–10):1967–1983. https://doi.org/10.1016/S0047-2727(01)00222-5

Brouwer R, Brander L, Beukering P (2008) "A convenient truth": air travel passengers' willingness to pay to offset their CO2 emissions. Clim Change 90(3):299–313. https://doi.org/10.1007/s10584-008-9414-0

Cai B, Cameron TA, Gerdes GR (2010) Distributional preferences and the incidence of costs and benefits in climate change policy. Environ Resour Econ 46(4):429–458. https://doi.org/10.1007/s10640-010-9348-7

Carlsson F, Daruvala D, Johansson-Stenman O (2005) Are people inequality-averse, or just risk-averse? Economica 72(287):375–396. https://doi.org/10.1111/j.0013-0427.2005.00421.x

Carlsson F, Kataria M, Krupnick A, Lampi E, Löfgren Å, Qin P, Sterner T (2013) A fair share: burden-sharing preferences in the United States and China. Resour Energy Econ 35(1):1–17. https://doi.org/10.1016/j.reseneeco.2012.11.001

Cary JW, Wilkinson RL (1997) Perceived profitability and farmers' conservation behaviour. J Agric Econ 48(1-3):13–21. https://doi.org/10.1111/j.1477-9552.1997.tb01127.x

Clark CF, Kotchen MJ, Moore MR (2003) Internal and external influences on pro-environmental behavior: participation in a green electricity program. J Environ Psychol 23(3):237–246. https://doi.org/10.1016/S0272-4944(02)00105-6

Crumpler H, Grossman PJ (2008) An experimental test of warm glow giving. J Public Econ 92(5–6):1011–1021. https://doi.org/10.1016/j.jpubeco.2007.12.014

Eckel CC, Grossman PJ (2008) Forecasting risk attitudes: an experimental study using actual and forecast gamble choices. J Econ Behav Organ 68(1):1–17. https://doi.org/10.1016/j.jebo.2008.04.006

Evans L, Maio GR, Corner A, Hodgetts CJ, Ahmed S, Hahn U (2012) Self-interest and pro-environmental behaviour. Nat Clim Change 3:122. https://doi.org/10.1038/nclimate1662

Fehr E, Schmidt KM (1999) A theory of fairness, competition, and cooperation. Q J Econ 114(3):817–868. https://doi.org/10.1162/003355399556151

Fischbacher U, Gächter S, Fehr E (2001) Are people conditionally cooperative? Evidence from a public goods experiment. Econ Lett 71(3):397–404. https://doi.org/10.1016/S0165-1765(01)00394-9

Frey BS, Oberholzer-Gee F (1997) The cost of price incentives: an empirical analysis of motivation crowding-out. Am Econ Rev 87(4):746–755

Glazer A, Konrad KA (1996) A signaling explanation for charity. Am Econ Rev 86(4):1019–1028

Harbaugh WT (1998) What do donations buy?: a model of philanthropy based on prestige and warm glow. J Public Econ 67(2):269–284

Holländer H (1990) A social exchange approach to voluntary cooperation. Am Econ Rev 80(5):1157–1167

IPCC (2001) Climate change: mitigation. Contribution of working group III to the Third Assessment Report (TAR) of the Intergovernmental Panel on Climate Change. Cambridge University Press, Cambridge, UK

Johnson LT (2006) Distributional preferences in contingent valuation surveys. Ecol Econ 56(4):475–487. https://doi.org/10.1016/j.ecolecon.2004.11.019

Kaenzig J, Heinzle SL, Wüstenhagen R (2013) Whatever the customer wants, the customer gets? Exploring the gap between consumer preferences and default electricity products in Germany. Energy Policy 53:311–322. https://doi.org/10.1016/j.enpol.2012.10.061

Karlan D, List JA (2007) Does price matter in charitable giving? Evidence from a large-scale natural field experiment. Am Econ Rev 97(5):1774–1793. https://doi.org/10.1257/aer.97.5.1774

Kesternich M, Löschel A, Römer D (2016) The long-term impact of matching and rebate subsidies when public goods are impure: field experimental evidence from the carbon offsetting market. J Public Econ 137:70–78. https://doi.org/10.1016/j.jpubeco.2016.01.004

Kotchen MJ (2006) Green markets and private provision of public goods. J Polit Econ 114(4):816–834. https://doi.org/10.1086/506337

Lange A, Schwirplies C (2019) Private contributions and the regional scope of charities: how donation experiments can inform public policy. Working paper

Lange A, Ziegler A (2015) Offsetting versus mitigation activities to reduce CO2 emissions: a theoretical and empirical analysis for the U.S. and Germany. Environ Resour Econ 66(1):113–133. https://doi.org/10.1007/s10640-015-9944-7

Lange A, Schwirplies C, Ziegler A (2017) On the interrelation between carbon offsetting and other voluntary pro-environmental activities: theory and empirical evidence. Resour Energy Econ 47:72–88. https://doi.org/10.1016/j.reseneeco.2016.11.002

List JA (2011) The market for charitable giving. J Econ Perspect 25(2):157–180. https://doi.org/10.1257/jep.25.2.157

List JA, Price MK (2012) Charitable giving around the world: thoughts on how to expand the pie. CESifo Econ Stud 58(1):1–30. https://doi.org/10.1093/cesifo/ifr023

Longo A, Hoyos D, Markandya A (2012) Willingness to pay for ancillary benefits of climate change mitigation. Environ Resour Econ 51(1):119–140. https://doi.org/10.1007/s10640-011-9491-9

Mayer H, Flachmann C (2016) Umweltökonomische Gesamtrechnungen: Direkte und indirekte CO2-Emissionen in Deutschland 2005–2012. https://www.destatis.de/DE/Publikationen/Thematisch/UmweltoekonomischeGesamtrechnungen/CO2EmissionenPDF_5851305.pdf. Accessed 8 Nov 2016

Meier S (2007) Do subsidies increase charitable giving in the long run? Matching donations in the field. J Eur Econ Assoc 5(6):1203–1222

Menges R, Schroeder C, Traub S (2005) Altruism, warm glow and the willingness-to-donate for green electricity: an artefactual field experiment. Environ Resour Econ 31(4):431–458. https://doi.org/10.1007/s10640-005-3365-y

Miao L, Wei W (2013) Consumers' pro-environmental behavior and the underlying motivations: a comparison between household and hotel settings. Int J Hospit Manag 32.102–112. https://doi.org/10.1016/j.ijhm.2012.04.008

Nyborg K, Rege M (2003) Does public policy crowd out private contributions to public goods. Public Choice 115(3–4):397–418

Nyborg K, Howarth RB, Brekke KA (2006) Green consumers and public policy: on socially contingent moral motivation. Resour Energy Econ 28(4):351–366. https://doi.org/10.1016/j.reseneeco.2006.03.001

Perino G, Schwirplies C (2019) Meaty arguments and fishy associations: field experimental evidence on the impact of reasons to reduce meat consumption on intentions, behavior and satisfaction. Working paper

Pittel K, Rübbelke DTG (2008) Climate policy and ancillary benefits: a survey and integration into the modelling of international negotiations on climate change. Ecol Econ 68(1–2):210–220. https://doi.org/10.1016/j.ecolecon.2008.02.020

Rege M (2004) Social norms and private provision of public goods. J Public Econ Theory 6(1):65–77. https://doi.org/10.1111/j.1467-9779.2004.00157.x

Rondeau D, List JA (2008) Matching and challenge gifts to charity: evidence from laboratory and natural field experiments. Exp Econ 11(3):253–267. https://doi.org/10.1007/s10683-007-9190-0

Rübbelke DTG (2002) International climate policy to combat global warming: an analysis of the ancillary benefits of reducing carbon emissions. Elgar, Cheltenham

Rübbelke DTG (2011) International support of climate change policies in developing countries: strategic, moral and fairness aspects. Ecol Econ 70(8):1470–1480. https://doi.org/10.1016/j.ecolecon.2011.03.007

Schleich J, Dütschke E, Schwirplies C, Ziegler A (2016) Citizens' perceptions of justice in international climate policy: an empirical analysis. Clim Policy 16(1):50–67. https://doi.org/10.1080/14693062.2014.979129

Schleich J, Schwirplies C, Ziegler A (2018) Do perceptions of international climate policy stimulate or discourage voluntary climate protection activities? A study of German and US households. Clim Policy 18(5):568–580. https://doi.org/10.1080/14693062.2017.1409189

Schwirplies C (2018) Citizens' acceptance of climate change adaptation and mitigation: a survey in China, Germany, and the U.S. Ecol Econ 145:308–322. https://doi.org/10.1016/j.ecolecon.2017.11.003

Schwirplies C, Ziegler A (2016) Offset carbon emissions or pay a price premium for avoiding them? A cross-country analysis of motives for climate protection activities. Appl Econ 48(9):746–758. https://doi.org/10.1080/00036846.2015.1085647

Schwirplies C, Dütschke E, Schleich J, Ziegler A (2019) The willingness to offset CO2 emissions from traveling: findings from discrete choice experiments with different framings. Ecol Econ 165. https://doi.org/10.1016/j.ecolecon.2019.106384

Shang J, Croson R (2009) A field experiment in charitable contribution: the impact of social information on the voluntary provision of public goods. Econ J 119(540):1422–1439. https://doi.org/10.1111/j.1468-0297.2009.02267.x

Tavoni A, Dannenberg A, Kallis G, Löschel A (2011) Inequality, communication, and the avoidance of disastrous climate change in a public goods game. Proc Natl Acad Sci USA 108(29):11825–11829. https://doi.org/10.1073/pnas.1102493108

Welsch H, Kühling J (2009) Determinants of pro-environmental consumption: the role of reference groups and routine behavior. Ecol Econ 69(1):166–176. https://doi.org/10.1016/j.ecolecon.2009.08.009

Impure Public Goods and the Aggregative Game Approach

Anja Brumme, Wolfgang Buchholz, and Dirk Rübbelke

1 Introduction

Climate protection has the features of an impure public good from the perspective of individual countries. Country-specific ("private") co-benefits of greenhouse gas mitigation efforts do not only arise in the technical sense, as, e.g., through improved local air quality or through a growth-enhancing technological push in energy production, but also at the psychological level through a specific warm-glow-of-giving effect, i.e., the easing of conscience when actions are taken that reflect a sense of responsibility for mother nature and future generations. To get a realistic understanding of climate protection as a global public good, it therefore becomes necessary to take these ancillary benefits and thus the impurity of this public good into account. For public good theory this, however, represents a big challenge since the treatment of impure public goods seems to require an intricate analysis in the three-dimensional space, where the private co-benefit of a public good appears as the third argument of utility functions besides private consumption and public good supply (see, e.g., Cornes and Sandler 1994, 1996; Kotchen 2005, 2007; Yildirim 2014). In this chapter, we show how the analysis of voluntary provision of an impure public good can be facilitated by tracing the impure public good model back to the standard pure public good model. To this end we construct auxiliary utility functions in which the private co-benefits do no longer appear as a separate variable and then apply the aggregative game approach as devised especially by Cornes and Hartley (2007). In this approach, the Cournot–Nash equilibrium of the voluntary provision

---

A. Brumme (✉) · D. Rübbelke
Technische Universität Bergakademie Freiberg, Freiberg, Germany
e-mail: anja.brumme@vwl.tu-freiberg.de; dirk.ruebbelke@vwl.tu-freiberg.de

W. Buchholz
University of Regensburg, Regensburg, Germany
e-mail: wolfgang.buchholz@ur.de

© Springer Nature Switzerland AG 2020
W. Buchholz et al. (eds.), *Ancillary Benefits of Climate Policy*, Springer Climate,
https://doi.org/10.1007/978-3-030-30978-7_8

game is not found by means of reaction functions but instead by starting from the countries' possible Nash equilibrium positions, which are described by their (income) expansion paths. By adding the aggregate budget constraint as a further condition, it becomes directly possible to determine the level of public good supply in the Cournot–Nash equilibrium.

To keep our argument simple, we visualize the basic idea of our approach through examples in which specific types of preferences, mainly of the Cobb–Douglas type, are assumed. In particular, we show that existence and uniqueness of a Cournot–Nash equilibrium in the voluntary contribution game for an impure public good can be proven in the same way as in the pure public good case. At the same time, however, application of the aggregative game approach also helps to get a better understanding of the differences between the pure and the impure public good model, especially regarding the effects of income changes and the countries' decision to become a contributor to the public good or a complete free rider. In particular, we will show in this chapter why it cannot be expected from the outset that "Warr neutrality," which implies invariance of the Cournot–Nash equilibrium with respect to income redistribution among public good contributors and which is one of the most famous results in public good theory, also holds in the case of impure public goods.

This chapter is organized as follows: In Sect. 2 we first of all describe how expansion paths in the two-dimensional space (with private consumption and public good supply on the axes) can be constructed also in the impure public good model which thus is directly linked to the standard pure public good case. These expansion paths, which are the cornerstone of the aggregative game approach, then are used in Sect. 3 to determine the Cournot–Nash equilibrium in the voluntary provision game for an impure public good in the same way as known from the pure public good model. After conducting some comparative statics exercises in Sect. 4, we explore the effects of income transfers in Sect. 5 highlighting the concomitant differences between the pure and the impure public good model. In Sect. 6, we briefly consider the outcomes that result for an alternative type of preferences. In Sect. 7, we conclude.

2 Impure Public Goods as Pure Public Goods

A country i for which a public good is impure is characterized by its utility function $v_i(c_i, g_i, G)$ where c_i denotes country i's private good consumption and g_i is its contribution to the public good, from which this country's private co-benefits of public good provision result. The unit prices of both goods are equal to unity. Total public good supply is denoted by G. To become able to apply the aggregative game approach (see Cornes and Hartley 2007, and with special reference to environmental problems Cornes 2016) we first of all define for the originally given utility function $v_i(c_i, g_i, G)$ an auxiliary utility function $u_i(c_i, G)$, which—as in the standard case of

Impure Public Goods and the Aggregative Game Approach

a pure public good—only depends on private consumption c_i and total public good supply G. This new utility function is given by

$$u_i (c_i, G) := v_i (c_i, w_i - c_i, G) \tag{1}$$

where w_i is country i's initial endowment. If $v_i(c_i, g_i, G)$ is defined for all $c_i \geq 0$, $g_i \geq 0$ and $G \geq 0$, the auxiliary utility function $u_i(c_i, G)$ is defined for all $G \geq 0$ and for all $c_i \in [0, w_i[$. For the sake of simplification we thus neglect in the description of individual preferences that $g_i < G$ and, equivalently, $w_i < c_i + G$ has to hold in any feasible allocation.

Even though some general analysis of the voluntary provision of impure public goods by means of the aggregative game approach would be possible, the main ideas that underlie this procedure can be more clearly presented by examples. To this end, we assume that country i has the Cobb–Douglas utility function

$$\tilde{v}_i (c_i, g_i, G) = c_i{}^{\tilde{\alpha}_i} g_i{}^{\tilde{\beta}_i} G^{\tilde{\gamma}_i}, \tag{2}$$

which, as an example, is also briefly treated by Yildirim (2014). The same preferences are equivalently represented by

$$v_i (c_i, g_i, G) = c_i{}^{\alpha_i} g_i{}^{\beta_i} G, \tag{3}$$

where $\alpha_i := \tilde{\alpha}_i / \tilde{\gamma}_i$ and $\beta_i := \tilde{\beta}_i / \tilde{\gamma}_i$. If $\beta_i = 0$, the utility function for the impure public good model turns into that of the standard pure public good model $v_i (c_i, G) = c_i{}^{\alpha_i} G$.

For the auxiliary utility function $u_i (c_i, G) = c_i{}^{\alpha_i} (w_i - c_i)^{\beta_i} G$, we now first of all determine the indifference curves and then the expansion paths in a c_i-G diagram. To simplify the exposition, we drop the index "i" in the remainder of this section.

The indifference curve of $u(c, G)$ for a given utility level $\bar{u}$ is given by the condition $c^\alpha (w - c)^\beta G = \bar{u}$. Hence, as a function of private consumption, this indifference curve can be represented as

$$G^{\bar{u}}(c) = \frac{\bar{u}}{c^\alpha (w - c)^\beta}. \tag{4}$$

This function has the derivative

$$\frac{\partial G^{\bar{u}}}{\partial c}(c) = \frac{\bar{u} ((\alpha + \beta) c - \alpha w)}{c^{\alpha+1} (w - c)^{\beta+1}}, \tag{5}$$

which is negative (positive) for $c < \bar{c} := \frac{\alpha w}{\alpha + \beta}$ ($>$) and—as a calculation shows— increasing in c. Each indifference curve attains its minimum at $\bar{c}$ where its value is $G^{\bar{u}} (\bar{c}) = \frac{\bar{u}(\alpha+\beta)^{\alpha+\beta}}{\alpha^\alpha \beta^\beta w^{\alpha+\beta}}$. Since $\lim_{c \to 0} G^{\bar{u}}(c) = \lim_{c \to w} G^{\bar{u}}(c) = \infty$ for any utility level $\bar{u}$ the indifference curve is U-shaped and approximates the G-axis and the vertical line

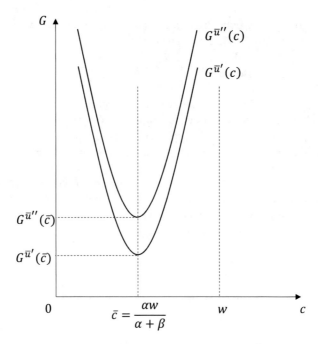

Fig. 1 Indifference curves for the utility function $u(c, G) = c^\alpha (w - c)^\beta G$ with utility levels $\bar{u}'$ and $\bar{u}''$ with $\bar{u}'' > \bar{u}'$

passing through the endowment point $(w, 0)$ as depicted in Fig. 1. If the utility level $\bar{u}$ increases, the indifference curve is shifted in parallel upward.

We now assume that the public characteristic of the impure public good is produced by a summation technology. A country's marginal rate of transformation between the private and the impure public good is equal to some $m > 0$. To be in an equilibrium position for the Nash game of voluntary provision of the impure public good the country's marginal rate of substitution between the private and the impure public good must be equal to this marginal rate of transformation, which is the observation on which the aggregative game approach is essentially based. Formally, this means that

$$mrs = \frac{G\left(\alpha c^{\alpha-1}(w-c)^\beta - \beta c^\alpha (w-c)^{\beta-1}\right)}{c^\alpha (w-c)^\beta} = \frac{G\left(\alpha w - (\alpha+\beta)c\right)}{c(w-c)} = m = mrt \quad (6)$$

has to hold for any Nash equilibrium position (c, G). By solving the quadratic equation, which follows from Eq. (6), the expansion path that connects all these equilibrium positions as a function of G as known from the standard version of the aggregative game approach then is

$$c^e(G) = \frac{(\alpha + \beta)\, G + mw - \sqrt{(\alpha + \beta)^2 G^2 + 2(\beta - \alpha)\, mwG + m^2 w^2}}{2m}, \qquad (7)$$

which looks relatively intricate and is difficult to interpret. Therefore, we consider the inverse of $c^e(G)$. This means that we describe the expansion path as a function of private consumption, which gives

$$G^e(c) = \frac{mc\,(w - c)}{\alpha w - (\alpha + \beta)\, c}. \qquad (8)$$

This function is defined for all $c \in [0, \bar{c}[$ and has $G^e(0) = 0$ and $\lim\limits_{c \to \bar{c}} G^e(c) = \infty$. Its derivative is

$$\frac{\partial G^e}{\partial c}(c) = \frac{m\left(\alpha(w - c)^2 + \beta c^2\right)}{(\alpha w - (\alpha + \beta)\, c)^2}, \qquad (9)$$

which clearly is positive for all c. An additional calculation shows that $G^e(c)$ moreover is convex. This implies that the expansion path $c^e(G)$ is increasing too, but is concave. It is defined for all $G \geq 0$ and has $c^e(0) = 0$ and $\lim\limits_{G \to \infty} c^e(G) = \bar{c} = \frac{\alpha w}{\alpha + \beta}$ (see Fig. 2). As all points of such an expansion path are lying on the left descending parts of the indifference curves, countries are improving their utility when they move outward their expansion paths.

As it directly follows from Eq. (8) the expansion path pivots downward closer to the c-axis if α increases or β and m decrease, i.e., if the preference for private consumption in the original utility function gets stronger, that for the private co-benefits of public good supply gets weaker or if the country becomes relatively less productive in its public good contribution. All these effects clearly are in accordance with intuition. However, taking the derivative of $G^e(c)$ with respect to the country's income w gives

$$\frac{\partial G^e}{\partial w}(c) = \frac{-m\beta c^2}{(\alpha w - (\alpha + \beta)\, c)^2} < 0, \qquad (10)$$

so that an increase of income leads to a downward rotation of the expansion path, closer to the c-axis, which can be regarded as a declining interest in the impure public good.

To explain this effect, which at first sight does not comply with intuition, we represent the country's Cobb–Douglas preferences in an alternative form as $\tilde{v}(c, g, G) = \alpha \ln c + \beta \ln g + \ln G$ so that the auxiliary utility function becomes

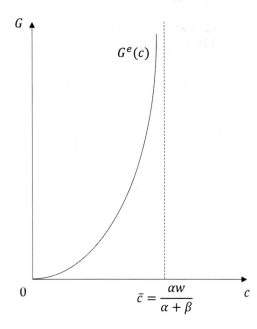

Fig. 2 Expansion path for the utility function $u(c, G) = c^\alpha (w-c)^\beta G$

$\tilde{u}(c, G) = \alpha \ln c + \beta \ln(w-c) + \ln G$. The condition for the country's Nash equilibrium position then is

$$\frac{\partial \tilde{u}}{\partial c} = \frac{\partial \tilde{v}}{\partial c} - \frac{\partial \tilde{v}}{\partial g} = \frac{\alpha}{c} - \frac{\beta}{w-c} = \frac{m}{G} = m\frac{\partial \tilde{v}}{\partial G} = m\frac{\partial \tilde{u}}{\partial G}. \tag{11}$$

Now we keep the level of public good supply G, i.e., the right-hand side of (11), constant and increase income w. For a given level of private consumption c this increases the agent's public good contribution and hence reduces $\frac{\partial \tilde{v}}{\partial g} = \frac{\beta}{w-c}$, i.e., the marginal utility of the private co-benefits at (c, G). Consequently, the left-hand side of (11) increases for the given c. To reestablish identity with the right-hand side, the country's public good contribution must decrease and hence its private consumption c must increase. This, however, means that the country's position must move horizontally to the right to end up on the new expansion path that results for the increased income.

3 Cournot–Nash Equilibrium

Based on the considerations in the previous section the Cournot–Nash equilibrium for voluntary provision of an impure public good can now be determined by the aggregative game approach in the same way as in the case of a pure public good. Thus, if there are n countries $i = 1, \ldots, n$ with contribution productivities m_i and expansion paths $c_i^e(G)$ as given by (7), public good supply $\hat{G}$ in the Cournot–Nash equilibrium is characterized by the familiar condition

$$\hat{G} + \sum_{i=1}^n m_i c_i^e\left(\hat{G}\right) = \sum_{i=1}^n m_i w_i. \qquad (12)$$

Private consumption of country i in the Cournot–Nash equilibrium then is $\hat{c}_i = c_i^e\left(\hat{G}\right)$. Condition (12) reflects the two requirements a Cournot–Nash equilibrium has to satisfy. On the one hand, each single country $i = 1, \ldots, n$ must be in a Nash equilibrium position, i.e., at a point on its expansion path $c_i^e(G)$. On the other hand, the aggregate budget constraint must be fulfilled, i.e., public good supply plus the sum of the weighted private consumption levels must equal the sum of the countries' weighted income levels (where the weighting factors are given by the countries' contribution productivities). Existence and uniqueness of a Cournot–Nash equilibrium is easily confirmed. Since the function $\Psi(G) = G + \sum_{i=1}^n m_i c_i^e(G)$, which corresponds to the left-hand side of (12), is strictly monotone increasing (and continuous) and $\Psi(0) = 0$ and $\lim_{G \to \infty} \Psi(G) = \infty$ hold, it follows from the Intermediate Value Theorem that there exists a unique level of public good supply which satisfies condition (12). For the case of two countries, Fig. 3 visualizes the determination of the Cournot–Nash equilibrium.

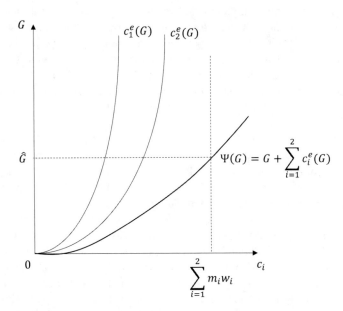

Fig. 3 Determination of the Nash equilibrium for $n = 2$

We exemplify this approach for determining the Cournot–Nash equilibrium for the symmetric case in which both countries have the same preferences $v_i(c_i, g_i, G) = c_i g_i G$, the same income $w_1 = w_2 = w$ and the same contribution productivity $m_1 = m_2 = m$. According to Eq. (7) each country's expansion path is $c^e(G) = \frac{2G + mw - \sqrt{4G^2 + m^2 w^2}}{2m}$ so that the equilibrium condition (12) turns into

$$\hat{G} + 2m \frac{2\hat{G} + mw - \sqrt{4\hat{G}^2 + m^2 w^2}}{2m} = 2mw. \tag{13}$$

Solving the equation that is defined by (13) for $\hat{G}$ gives

$$\hat{G} = \frac{6}{5} mw \tag{14}$$

for public good supply in the Cournot–Nash equilibrium and then (from $\hat{c} = w - \frac{\hat{G}}{2}$)

$$\hat{c} = \hat{c}_1 = \hat{c}_2 = \frac{2}{5} w \tag{15}$$

for each country's private consumption. In the special case with $m = 1$ we have $\hat{G} = \frac{6}{5} w$ and $\hat{c} = \hat{c}_1 = \hat{c}_2 = \frac{2}{5} w$.

In our example with an impure public good and Cobb–Douglas preferences the determination of the Cournot–Nash equilibrium is even less demanding than in the standard case with a pure public good. The reason is that corner solutions in which some countries do not contribute to the public good do not occur in the impure public good case. Rather, as the expansion paths are tangential to the vertical line at $\bar{c}_i$ (and do not cut the vertical line through the endowment point $(w_i, 0)$ as in the case of a standard pure public good) each country i makes a strictly positive contribution to the public good in the Cournot–Nash equilibrium. Independent of the number of countries, country i's contribution is at least $\underline{g}_i = w_i - \bar{c}_i = \frac{\beta w_i}{\alpha + \beta}$, which is decreasing in α and increasing in β and w_i. The country-specific public good contributions will not fall below this threshold if the originally given public good economy is replicated—irrespective of how many times the replication takes place. Thus, interiority of Cournot–Nash equilibrium can be more easily attained for impure than for pure public goods for which interior solutions are rather the exception than the rule (Buchholz et al. 2006).

For other utility functions, however, it is well possible that also in the case of an impure public good some countries do not make a positive contribution to the public good in the Cournot–Nash equilibrium. An example for this will be presented later on in this chapter. Yet, also in this case, which is more like the case of a pure public good, the same methods for determining the contributors and noncontributors as in the case of a pure public good (see, e.g., Andreoni and McGuire 1993; Shrestha and Cheong 2007) can be applied—and a separate procedure (as suggested by Yildirim 2014) is not required. In this respect, hence there is nothing new under the sun for impure public goods.

4 Some Comparative Statics

Based on the equilibrium condition (12) the same method as for a pure public good can also be applied to do some comparative statics exercises in the case of an impure public good. Let us, e.g., assume that for some country j either the preference parameter α_j decreases or the preference parameter β_j increases, which both reflects this country's higher interest in public good provision. Then, as shown before, country j's new expansion path $\tilde{c}_j^e(G)$ lies above the original one $c_j^e(G)$, so that for the public good level $\hat{G}$ in the initial Cournot–Nash equilibrium we now have $\hat{G} + m_j \tilde{c}_j^e \left(\hat{G} \right) + \sum_{i \neq j} m_i c_i^e \left(\hat{G} \right) < \sum_{i=1}^n m_i w_i$. Since the expansion paths are monotone increasing, an increase in public good supply is needed to restore equality in the equilibrium condition (12). All countries except country j then reduce their public good contributions by moving outward their expansion paths (and thus become better off) while country j increases its public good contributions. (As country j's preferences are changing, nothing can be said about its welfare effects.) Such arguments well-known from the standard theory of voluntary provision of a pure public good can also be applied to infer other effects that result from other changes like the increase of a country's contribution productivity or from adding a new country to the economy.

It is more cumbersome to determine the effects of income changes since an income increase for some country j pivots its income expansion path downward, closer to the c-axis, so that in the Nash equilibrium condition (12) both the left-hand and the right-hand side are increased. Therefore, additional considerations are required for the comparative statics analysis of an income change.

To this end, we look at country j's position on the new expansion path after its income has marginally increased but—hypothetically—public good supply remains the same as in the original Cournot–Nash equilibrium, i.e., $\hat{G}$. To simplify the exposition, we normalize the country's contribution productivity to $m_j = 1$. Total differentiation of the equilibrium condition (6) then yields

$$\left. \frac{dc_j}{dw_j} \right|_{G=\hat{G}} = \frac{\alpha_j \hat{G} - \hat{c}_j}{w_i - 2\hat{c}_j + (\alpha_j + \beta_j) \hat{G}} = \frac{\beta_j \hat{c}_j^2}{\alpha_j (w_j - \hat{c}_j)^2 + \beta_j \hat{c}_j^2} < 1. \tag{16}$$

This means that fixing public good supply at $\hat{G}$, private consumption in country j's Nash equilibrium position increases by less than country j's increase in income. Consequently, after the income change, the left-hand side of (12) becomes smaller than the right-hand side. Similar as above, restoring equality in (12) thus requires an increase of public good supply and an increase of all countries' private consumption. Note that for country j this increase consists of two parts—a move to the new expansion path that lies closer to the c-axis and then a move outward this expansion path. For all other countries, clearly only the second effect applies.

5 Non-neutrality of Income Transfers and Matching Schemes

Concerning the effect of income changes on the Cournot–Nash equilibrium a famous result in public good theory is the "neutrality theorem" by Warr (1983). This theorem says that given a linear summation technology for public good provision with identical contribution productivities of all countries, an unconditional redistribution of income between contributing countries does change neither public good supply nor all countries' private consumption in the Cournot–Nash equilibrium. From the perspective of the aggregative game approach this result is based on the fact that in the case of pure public goods the expansion paths are not affected by such an income transfer so that the Nash equilibrium condition (12) remains the same. Yet, in the case of impure public goods this does no longer hold since the income level now has an impact on the utility function and thus on the expansion paths upon which the determination of the Cournot–Nash equilibrium rests. Therefore, it cannot be expected a priori that Warr's neutrality theorem remains valid for impure public goods (see Andreoni 1989, 1990).

One might suspect that this generic non-neutrality could be exploited to improve public good supply by transferring income to countries that have high private co-benefits from public good provision and thus gain twice from their contributions to the public good. This conjecture, however, is not always true as can be shown by the following example.

There are two countries 1 and 2 of which only country 2 has a private co-benefit from public good provision. The preferences of countries 1 and 2 are given by $u_1(c_1, G) = c_1 G$ and $u_2(c_2, g_2, G) = c_2 g_2 G$, respectively, and contribution productivities of both countries are $m = 1$. The relevant expansion path of country 1 then is $c_1{}^e(G) = G$ and that of country 2 is $c_2{}^e(G) = \frac{2G + w_2 - \sqrt{4G^2 + w_2{}^2}}{2}$. The equilibrium condition (12) thus becomes

$$2\hat{G} + \frac{2\hat{G} + w_2 - \sqrt{4\hat{G}^2 + w_2{}^2}}{2} = w_1 + w_2. \tag{17}$$

Equation (17) leads to the quadratic equation $8\hat{G}^2 - 3(2w_1 + w_2)\hat{G} + w_1(w_1 + w_2) = 0$. Solving for $\hat{G}$ yields

$$\hat{G} = \frac{3(2w_1 + w_2) + \sqrt{(2w_1 + w_2)^2 + 8w_2^2}}{16}. \tag{18}$$

We start with the income distribution $w_1 = w_2 = 10$. Inserting these income levels into Eq. (18) gives $\hat{G} = 8.20$. As $\hat{c}_1 = \hat{G} = 8.20 < 10$ an interior Cournot–Nash equilibrium results—since country 2 is at an interior position anyway and has private consumption $\hat{c}_2 = w_1 + w_2 - \hat{c}_1 - \hat{G} = 20 - 2 \cdot 8.20 = 3.60$. Now an amount of income equal to one is transferred from country 1 to country 2 so that the new income distribution is $w_1' = 9$ and $w_2' = 11$. In this case Eq. (15)

gives $\hat{G}' = \hat{c}'_1 = 8.09$ and $\hat{c}'_2 = 3.82$ in the new Cournot–Nash equilibrium after the income redistribution. Country 1 (the one without private co-benefits) which contributes 1.80 to the public good in the initial situation then decreases its public good expenses by $0.89 = 1.80 - 0.91$ after the income transfer. This decrease of country 1's contributions dominates country 2's (the one with co-benefits) increase in public good provision who spends 0.22 of the additional income on private consumption and 0.78 on the public good. In total, public good supply in the Cournot–Nash equilibrium falls by $0.11 = 0.89 - 0.78$.

This example confirms that an income transfer from a country with no co-benefits to a country that has private co-benefits may lead to a decline of public good provision in the Cournot–Nash equilibrium. This result can be traced back to the observation made before at the end of Sect. 2. An increase of income for country 2, for which the public good is impure, decreases its willingness to contribute to the public good, i.e., pushes its expansion path down. The reduced income of country 1 for which the public good is a pure one instead has no impact on its willingness to pay for the public good, i.e., its relevant expansion path is kept unchanged. Therefore, in total the income transfer will reduce public good supply in the Cournot–Nash equilibrium. Conversely, a transfer from country 2 to country 1 without private co-benefits would increase public good supply (see, e.g., Andreoni 1989, p. 1454).

It would be interesting to generalize this analysis and to infer the effects of unconditional income transfers between two countries that both have co-benefits. Yet, in contrast to the case of a pure public goods economy with Cobb–Douglas preferences, in our example of an impure public good it is not possible to specify closed-form expressions for public good supply and private consumption in the Cournot–Nash equilibrium. Rather, one would have to resort to numerical simulations.

Since the seminal works by Guttman (1978) and Boadway et al. (1989), also some kind of conditional transfer schemes under which countries "match" the public good contributions of other countries have attracted much attention in the theory of public goods not only for pure but also for impure public goods (see Rübbelke 2006). When a country i matches the public good of another country j with a matching rate s_i this means that country i commits to spending the additional amount $s_i g_j$ on public good provision if country j has chosen its direct ("flat") public good contribution g_j. This indirect subsidization increases the matched country j's effective marginal rate of transformation between the private and the public good to $\sigma_j = (1 + s_j)m$ (where m again is the common contribution productivity of the countries i and j). Country j's relevant income expansion path then pivots upward further away from the c_j-axis just as if country j's contribution productivity were increased (see Sect. 2).

Matching in general promises to increase public good supply and thus to attenuate the underprovision problem, which is pertinent for voluntary public good provision. In the case of an impure public good, matching works in the same way as in the case of a pure public good, which is also obvious by applying the aggregative game approach. For a visualization in the framework of our example we assume again that there are two countries with the same preferences $v_i(c_i, g_i, G) = c_i g_i G$, the same income $w_1 = w_2 = w$ and the same contribution productivity $m_1 = m_2 = 1$. When there is reciprocal matching and both countries apply the same matching rate s, Eq. (7) then gives $c_i^e(G) = c^e(G) = \frac{2G + \sigma w - \sqrt{4G^2 + \sigma^2 w^2}}{2\sigma}$ as the relevant expansion path of each country where $\sigma = 1 + s$. The equilibrium condition (12) thus becomes

$$\hat{G} + 2\frac{2\hat{G} + \sigma w - \sqrt{4\hat{G}^2 + \sigma^2 w^2}}{2\sigma} = 2w, \tag{19}$$

which is quite similar to condition (13) that referred to the case of varying contribution productivities. Solving (19) for $\hat{G}$ yields

$$\hat{G} = \frac{2(\sigma + 2)}{\sigma + 4} w = \frac{2(s + 3)}{s + 5} w \tag{20}$$

for the level of public good supply in the Cournot–Nash equilibrium with matching. $\hat{G}$ is increasing in σ and hence in the matching rate s. Each country's own flat contribution then is $\hat{g} = \frac{s+3}{(s+5)(s+1)} w$, while it contributes $s\hat{g} = \frac{s(s+3)}{(s+5)(s+1)} w$ indirectly through matching the other country's flat contribution.

Public good supply in the symmetric Pareto optimal solution is $G^* = \frac{4}{3} w$ and private consumption of each country is $c^* = c_1^* = c_2^* = \frac{1}{3} w$. (This follows from combining the Samuelson condition $2mrs = 2\frac{c^*(w - c^*)}{G^*(w - 2c^*)} = 1$ with the feasibility constraint $G^* + 2c^* = 2w$.) This outcome is reached through reciprocal matching if the matching rate that is applied by both countries is $s^* = 1$.

6 An Alternative Type of Preferences

We now briefly contrast the results of the previous sections with those that would be obtained if another type of preferences were supposed. Thus, we assume that country i has the utility function $v_i(c_i, g_i, G) = c_i^{\alpha_i} \cdot (\rho_i g_i + G)^{\beta_i}$ with some $\rho_i > 0$ where we focus on the case $\alpha_i = \beta_i = 1$ and, in a first step, also omit the index "i.". Then we have the auxiliary utility function $u(c, G) = c \cdot (\rho(w - c) + G)$, which is defined for all $c \in [0, w]$ and all $G \geq 0$. If $\bar{u} > \rho w^2$, the indifference curve $G^{\bar{u}}(c) = \frac{\bar{u}}{c} - \rho(w - c)$ is downward sloping everywhere. The expansion path $c^e(G) = \frac{G + \rho w}{1 + 2\rho}$, which is defined by $mrs = \frac{\rho(w - 2c) + G}{c} = 1 = mrt$, is a straight upward sloping line which cuts the vertical line passing through the point $(w, 0)$ at $\widehat{G} = (1 + \rho) w$ (see Fig. 4). This "dropout level" of public good supply is increasing in ρ, i.e., in the country's interest in its private co-benefits of public good provision.

Just as in the case of a pure public good, it therefore becomes possible for such preferences that in the Cournot–Nash equilibrium of the voluntary contribution game of an impure public good a country is a noncontributor. This happens if the

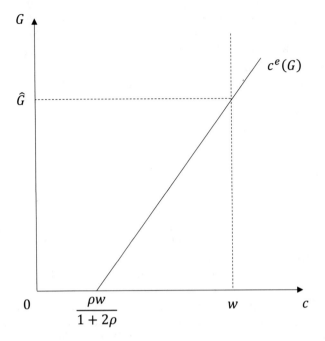

Fig. 4 Expansion path for the utility function $u(c, G) = c \cdot (\rho(w - c) + G)$

other countries provide an amount of the public good that exceeds this complete free rider's dropout level $\widehat{G}$. Likewise, as in the case of a pure public good, an income distribution between two contributing countries now, at least partially, may be neutral. To show this, we assume that there are two countries $i = 1, 2$ that have the same preferences $v(c_i, g_i, G) = c_i \cdot (g_i + G)$ and the same contribution productivity $m = 1$ but possibly different income levels w_1 and w_2. According to (12) an interior Cournot–Nash equilibrium is then characterized by

$$\hat{G} + \frac{\hat{G} + w_1}{3} + \frac{\hat{G} + w_2}{3} = w_1 + w_2, \tag{21}$$

which gives $\hat{G} = \frac{2}{5}(w_1 + w_2)$. Public good supply in an interior Cournot–Nash equilibrium thus only depends on aggregate income but not on its distribution among the two countries.

Yet the reasons for this neutrality are completely different from those in the case of a pure public good since the income redistribution does not leave the expansion paths unchanged. Rather, a transfer from country 1 to country 2 pivots country 1's expansion path downward closer to the c-axis and that of country 2 upward, but these two movements just offset each other. That the expansion paths are changing also implies that neutrality is not complete: Private consumption of country 1, which is $\hat{c}_1 = \frac{7w_1 + 2w_2}{15}$, goes down while private consumption of country 2, which is

$\hat{c}_2 = \frac{2w_1 + 7w_2}{15}$, goes up by the same amount. (By the way, from $\hat{c}_i = \frac{7w_i + 2w_j}{15} < w_i$ $(i, j = 1, 2, i \neq j)$ it follows that an interior Cournot–Nash equilibrium is obtained if $\frac{w_i}{w_1 + w_2} < \frac{4}{5}$ holds for $i = 1, 2$, i.e., if total income is not distributed too unevenly between the two countries as in the pure public good case.)

An easy calculation immediately shows that neutrality regarding public good supply vanishes as soon as the two countries have different preferences, i.e., $v_1(c_1, g_1, G) = c_1 \cdot (\rho_1 g_1 + G)$ and $v_2(c_2, g_2, G) = c_2 \cdot (\rho_2 g_2 + G)$ with $\rho_1 \neq \rho_2$.

7 Conclusions

In this chapter, we have presented an approach by which—based on the construction of an auxiliary utility function—the impure public good model can be considered like a subcase of the standard pure public good model. By pursuing this approach, on the one hand, it becomes easily possible to apply the aggregative game approach to demonstrate existence and uniqueness of the Cournot–Nash equilibrium of the voluntary provision game and to conduct some comparative statics analysis for the impure public good in just the same way as in the case of a pure public good. On the other hand, differences between the impure and the pure public good model come clearly to the fore. According to this, it is more likely in the impure public good case that the Cournot–Nash equilibrium becomes an interior solution without free-riding noncontributors. From the perspective of our approach, it is also obvious a priori that Warr neutrality cannot be expected in the impure public good case: Since an income redistribution between contributing countries changes the auxiliary utility functions and thus the expansion paths, it generally will also change the Cournot–Nash equilibrium.

To visualize the method in this chapter, the analysis has been limited to specific utility functions and normalized prices for the public good and the private co-benefits of public good provision. The elimination of these restrictions and thus the generalization of our approach will be reserved to subsequent work.

References

Andreoni J (1989) Giving with impure altruism: applications to charity and Ricardian equivalence. J Polit Econ 97:1447–1458

Andreoni J (1990) Impure altruism and donations to public goods: a theory of warm-glow giving. Econ J 100:464–477

Andreoni J, McGuire MC (1993) Identifying the free riders: a simple algorithm for determining who will contribute to a public good. J Public Econ 51:447–454

Boadway R, Pestieau P, Wildasin D (1989) Tax-transfer policies and the voluntary provision of public goods. J Public Econ 39:157–176

Buchholz W, Cornes R, Peters W (2006) On the frequency of interior Cournot-Nash equilibria in a public good economy. J Public Econ 8:401–408

Cornes R (2016) Aggregative environmental games. Environ Resour Econ 63:339–365

Cornes R, Hartley R (2007) Aggregative public good games. J Public Econ 9:201–219

Cornes R, Sandler T (1994) The comparative static properties of the impure public good model. J Public Econ 54:403–421

Cornes R, Sandler T (1996) The theory of externalities, public goods and club goods, 2nd edn. Cambridge University Press, New York

Guttman JM (1978) Understanding collective action: matching behavior. Am Econ Rev 68:251–255

Kotchen MJ (2005) Impure public goods and the comparative statics of environmentally friendly consumption. J Environ Econ Manag 49:281–300

Kotchen MJ (2007) Equilibrium existence and uniqueness in impure public good models. Econ Lett 97:91–96

Rübbelke D (2006) Analysis of an international environmental matching agreement. Environ Econ Policy Stud 8:1–31

Shrestha RK, Cheong KS (2007) An alternative algorithm for identifying free riders based on a no-free-rider Nash equilibrium. FinanzArchiv: Public Financ Anal 63:278–284

Warr PG (1983) The private provision of a public good is independent of the distribution of income. Econ Lett 13:207–211

Yildirim H (2014) Andreoni-McGuire algorithm and the limits of warm-glow giving. J Public Econ 114:101–107

Multi-criteria Approaches to Ancillary Effects: The Example of E-Mobility

Stefan Vögele, Christopher Ball, and Wilhelm Kuckshinrichs

Ambitious targets like the ones formulated in the Paris Agreement at the United Nations Climate Change Conference (COP 21) cannot be achieved without a decarbonisation of the transportation sector. Like other policy interventions, policies focusing on this sector will be linked with primary and ancillary effects. In this study, we assess to what extent stakeholders will benefit or suffer from a switch to e-mobility by applying the ancillary benefit approach. Since the attitudes of stakeholders depend on many different factors and the list of factors differs between the stakeholders, an appropriate assessment of a decarbonisation of the transport sector requires the consideration of a broad range of factors including the weighting of options by actors. Using a multi-criteria approach we show that stakeholders, like car users and vehicle manufacturers, will show resistance if they are urged to go for e-mobility. Since the assessment of the characteristics of e-mobility is linked with high uncertainty, we conducted intensive sensitivity analyses. According to these analyses, it is difficult to cause a shift in the attitude of car users towards electric vehicles, since electric vehicles have a lot of disadvantages for the car users (including loss of comfort). According to our assessment, hybrid cars face less resistance since the technology is linked with more benefits/less negative effects for the stakeholders than e-mobility.

S. Vögele (✉) · C. Ball · W. Kuckshinrichs
Institute of Energy and Climate Research - Systems Analysis and Technology Evaluation
(IEK-STE), Forschungszentrum Jülich GmbH, Jülich, Germany
e-mail: s.voegele@fz-juelich.de

© Springer Nature Switzerland AG 2020
W. Buchholz et al. (eds.), *Ancillary Benefits of Climate Policy*, Springer Climate,
https://doi.org/10.1007/978-3-030-30978-7_9

1 Introduction

In principle, each policy intervention, whether on an individual, national or international level, causes multiple direct and indirect effects: Interventions in the area of climate policy are a prominent example of this. Climate change policy aims to reduce GHG emissions as its primary target; however, at the same time, it impacts other kinds of emissions, the production of goods, energy prices and the labour market (see e.g. Groosman et al. 2011; Pittel and Rübbelke 2008; Van Vuuren et al. 2006; Ekins 1996). Since the side effects are not part of the initial goal, they are usually considered as ancillary effects or ancillary benefits (Davis et al. 2000). Depending on the kind and level of the indirect benefits, the affected entities may change their decisions or behaviour patterns. This could have a feedback effect on the entity which initiated the measure—a measure, for example, may fail to deliver the desired outcome due to these unforeseen ancillary effects. In principle, the entity which benefits from the primary target of the measure, namely reductions in CO_2 emissions, and the entity which is impacted by the ancillary effects are not necessarily the same.

Employing a multi-actor multi-criteria approach, we assess the influence of ancillary effects on decision-making processes. We show how ancillary effects can affect investment decisions, taking a complex system of actors and criteria into consideration and using electric mobility as an example. We also aim to illustrate how an ancillary effects approach could enrich multi-criteria analysis.

The meaning of ancillary effects,[1] defined as effects which are related to the primary benefits from a measure, have been highlighted in a broad range of studies (see, e.g. Davis et al. 2000; OECD 2000). The list of identified ancillary effects includes impacts on health, on ecosystems as well as on social (e.g. equity aspects, energy security) and economic factors (e.g. cost, income). Considering the strategies applied to the Clean Development Mechanism, the choice of rate and subsidy focus to facilitate the technological shift in developing countries should consider the primary benefit—global provision of climate protection, as well as private domestic benefits, alleviating poverty and local environmental effects, promoting technological transfer, improving manufacturing and agricultural processes (Rive and Rübbelke 2010). The ancillary effects can have their own side effects—for instance, an intervention to reduce GHG emissions in the electricity sector could lead to the greater use of nuclear power, which itself is associated with environmental externalities in the form of waste and potential accidents (e.g. see Krupnick et al. 2000). Likewise, vehicles emit co-pollutants in addition to GHG emissions (Chavez-Baeza and Sheinbaum-Pardo 2014) and a reduction in these co-pollutants, which include NO_x and SO_x, would represent an ancillary benefit (in this case for health) (Muller 2012; Nemet et al. 2010). Malmgren (2016) also points to ancillary benefits in the form of reduced import dependency and economic

---

[1] Since ancillary impacts can be positive and negative, we prefer to use the term 'ancillary effects' instead of 'ancillary benefits', even if 'ancillary benefits' is the term which is more popular.

possibilities that would emerge from pursuing e-mobility. Thus, the list of effects can be extended by taking second-order effects into consideration (see, e.g. Kirkman et al. 2012; OECD 2000). In particular, economic and social effects are related to (investment) decisions.

The discussion of ancillary benefits can be extended by considering *externalities*, whereby a policy intervention may lead to changes in welfare for a particular group of individuals where these were not considered in the design of the intervention (Krupnick et al. 2000). The effect may be positive if there is a need to appeal to actors not directly interested in the adopted policy measures, such as motivating for an action these who are not convinced in climate change (Bain et al. 2016). In our study, externalities are present where a policy intervenes in the incentives for actors that should adopt a new technology, causing a crossover effect on the actors that are interested in the particulars. For instance, encouraging vehicle manufacturers to favour electric vehicles may have the externality of increasing the trust of car users in the technology. The assessment of the effects of these externalities are implemented in our paper through examining four different scenarios: (1) a scenario without externalities, (2) a scenario in which the government and electric vehicle utilities opt for electric vehicles, (3) a scenario in which all actors other than vehicle manufacturers support electric vehicles and (4) a scenario in which all actors are supportive of electric vehicles.

A well-established and common approach for analysing decisions is the multi-criteria approach (see, e.g. Brans et al. 1986; Pohekar and Ramachandran 2004; Diakoulaki and Karangelis 2007; Kumar et al. 2017). This approach is based on the assumption that decisions result from the evaluation of a set of criteria and the performance of decision alternatives in relation to each of those criteria. A weighting is attached to each criterion, reflecting its importance to the decision maker. The overall performance of an alternative is calculated by summing up the alternative's performance scores on each criterion and multiplying those scores by the weightings attached to each criterion. This overall performance evaluation allows the decision maker to select the most beneficial alternative. Such approaches have been used, e.g. for energy planning problems (see Loken 2007) and appraising transport projects (see, e.g.Macharis and Bernardini 2015).

Since decarbonisation of transport is crucial to meeting the EU's climate change targets, accounting for almost 25% of all EU CO_2 emissions, with road transport making up three quarters of transport emissions (European Commission 2017), we focus our analysis on e-mobility. If transport emissions are not reduced, this sector risks becoming the largest source of emissions, and, assuming the EU follows its decarbonisation targets, transport emissions would threaten to eclipse carbon reductions made in other sectors (European Commission 2017). Therefore, reducing CO_2 emissions is, arguably, the primary motivator behind policies in favour of the deployment of e-mobility.

Policies at the European, national and local level have an impact on the diffusion of electric vehicles. Usmani et al. (2015) argue that policymakers must set a clear vision, accompanied by targets, for the expansion of e-mobility in addition to setting standardised regulations and norms. They also suggest that European policymakers can accelerate the deployment of alternative vehicles by imposing CO_2 standards and by introducing directives in relation to alternative infrastructure to enable stronger diffusion.

In Europe, the diffusion of electric vehicles has been slow so far and this reflects a lack of acceptance, poor understanding of the benefits and costs of such vehicles and the entrenched position of conventional vehicles (Biresselioglu et al. 2018). Moreover, the high price of electric vehicles (Sierzchula et al. 2014), long payback times and the lack of refuelling infrastructure have acted as barriers to diffusion (Hagman et al. 2016). In Germany, electric cars are considered central to decarbonisation and it is predicted that there will be 9 million electric vehicles by 2030, with this dependent on the development of better battery technology (Acatech 2018). Performance has however, so far, lagged behind, and direct financial support for electric vehicles was only introduced in spring 2016 (Truffer et al. 2017).

As regards the German Government's strategy for e-mobility, it places emphasis on making Germany an industrial leader in electric mobility, both co-financing R&D focused on making electric cars more cost and performance competitive and addressing other barriers, namely setting common regulatory standards and the expansion of charging points (Nationale Plattform Elektromobilität 2014). However, e-mobility presents clear and major challenges to the German automobile industry, since electric vehicles are a disruptive innovation. This leads to the risk that knowledge, networks and skills the industry has built up over years are obsolete (Steinhilber et al. 2013). Importance is placed on Germany developing the capacity to produce batteries domestically, even if this production is unprofitable in the short-run (Steinhilber et al. 2013). China's leadership in battery cell production and the concentration of competencies for the large-scale and lower cost production of batteries in Asia is a concern for European manufacturers seeking to establish an industrial strength in electric vehicles (Steinhilber et al. 2013).

Beyond the automobile industry, there must be innovation at the system level and accompanied regulatory changes to enable the wider diffusion of electric vehicles. Augenstein (2015) warns that Germany's National Platform for E-Mobility is dominated by 'regime actors', namely incumbent firms that tend to conceptualise the challenges of e-mobility as largely to do with technology and markets. She suggests that 'regime actors' may neglect issues to do with new mobility patterns, the emergence of new actors and business models, such as car sharing and electric vehicles' involvement in energy storage which may be crucial to the growth of e-mobility. Certain regulatory issues have to be resolved to further the expansion of electric vehicles. For instance, regulators need to determine which actor should be responsible for the charging infrastructure and how a competitive market for recharging services can be ensured (Lo Schiavo et al. 2013).

Theisen (2010) outlines three possible business models for deploying the charging infrastructure which has implications for how this infrastructure is financed. Charging infrastructure could be part of DSOs' assets, it could be run by a separate charging infrastructure operator or, thirdly, it could be operated by a separate e-mobility provider, selling charging services and electricity (Theisen 2010). In the first two models, there is free access to charging infrastructure for retailers, respecting the 'unbundling' concept which leads to more competitive markets. In terms of financing, in the first model, the charging infrastructure is financed jointly by the DSO and the retailer, and so all electricity users finance the infrastructure, whereas, in the second and third models, the cost of financing the charging infrastructure is passed on to users of e-mobility services, via the 'user pays' principle (Theisen 2010). Charging facilities must be accompanied by ICE infrastructure, involving payment systems for electricity charging (Winning 2015) in addition to systems which enable the emergence of an 'intelligent' e-mobility system, based on the local distribution of power for electric vehicles and car sharing (Bundesministerium für Wirtschaft und Energie 2018).

At the consumer level, electric vehicles represent a 'high involvement purchase' and, therefore, the knowledge of car dealerships about electric vehicles is important in overcoming resistance towards adoption (Matthews et al. 2017). Matthews et al. (2017) find that a delay between purchase and delivery can dissuade customers from opting for electric vehicles as can the frequent lack of demonstration vehicles available at dealerships (Matthews et al. 2017). This point about demonstration vehicles is reinforced in findings by Gebauer et al. (2016) who find that the experience of using charging technology has a positive effect on consumer attitudes towards electric vehicles. In contrast, if they are simply given written information about the performance of charging technology, this does not affect their attitudes towards electric vehicles significantly (Gebauer et al. 2016). This highlights the importance of users having the opportunity to experience electric vehicles and their associated technology and car dealerships appear to play an important role in stimulating adoption.

Taking into consideration that certain factors which impact decisions can be as ancillary effects (in the sense that they are determined by other activities or the activities of other actors), a combination of MCDA and ancillary benefits approaches can be beneficial. The chapter is organised as follows: In Sect. 2 we describe the approach we applied for the assessment of the decisions based on ancillary benefits. Results are presented and discussed in Sect. 3. Section 4 concludes.

2 Method

As mentioned in the introduction, measures like climate policies have not only primary effects but also ancillary effects. According to the IPCC, 'ancillary effects' are unintended side effects of a selected measure or policy (IPCC 2001). In principle, 'ancillary effects' consist of a broad range of factors including ecological (e.g. reduction in emissions of local and regional air pollutants), economic (e.g. avoidance cost) and social aspects (e.g. energy security) (see e.g. Davis et al. 2000; Ürge-Vorsatz et al. 2014; Mayrhofer and Gupta 2016; OECD 2000). Given that first-order effects, such as the reduction in GHG emissions, can also trigger ancillary (indirect) effects, such as increases in income or enhancements in health, and that some effects may become apparent following a time lag, the assessment of ancillary effects is challenging.

In this study, we focus on the impacts of a policy measure on the attitudes of different actors. We assume that:

- the measure aims to foster the use of a technology in order to reduce CO_2 emissions
- there is a set of different actors which differ with respect to their preferences
- the actors can choose between different technologies.

The measure to foster a particular technology will be supported by the actors if the actors expect more benefits from using this technology than if they opted for an alternative one. Hence, if the technology fostered by the initial policy measure does not provide more benefits than any alternative, it could be expected that the technology would struggle to gain ground.

In addition to impacts on attitudes towards a technology, we take impacts resulting from externalities, representing feedbacks resulting from activities of other actors, into consideration. Figure 1 gives an overview of our approach. After specifying a policy intervention, we list both the primary effect and the associated ancillary effects. The effects that the intervention is aiming to achieve are defined as the primary benefits. All other effects resulting directly from the selection of a technology are classified as 'first order' ancillary benefits. Effects resulting from interdependencies among actors are captured by implementing additional indicators and defined as 'second order' ancillary benefits. The additional indicators, called externalities, reflect how the attitude of one actor influences the attitude of other actors. For reasons of clarity, we cluster all kinds of benefits to superordinate categories.

In a third step, we describe and specify the benefits, from a technology-specific point of view, by extracting data from studies in combination with expert adjustments. Since we employ an MCDA approach later on, there is no need to assess the benefits using a uniform unit. Hence, it is possible to adjust benefits by using semi-qualitative evaluation scales (e.g. using 0 for not relevant, 1 for very low relevance, 2 for low relevance, ... and 5 for very high relevance).

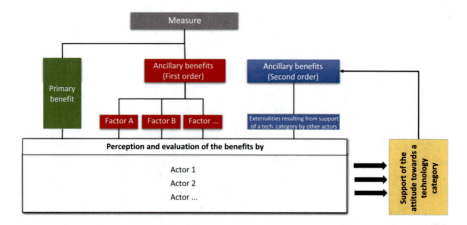

Fig. 1 Overview: primary and ancillary benefits

An assessment of the shapes of benefits from the view point of actors employing MCDA is part of step four. MCDA approaches have been widely used for the assessment of technologies and for scenario comparisons (see e.g. Parkinson et al. 2018; Baležentis and Streimikiene 2017; Wang and Poh 2014; Diakoulaki and Karangelis 2007; Terrados et al. 2009). For the assessment of the effects of an intervention on attitudes, we compare benefits of different technological options. Applying an outranking approach we assume that a decision maker will support the technology with the highest overall performance (see, e.g. Loken 2007; Behzadian et al. 2010).

Regarding the factors which are relevant for the decision process, we employ a hierarchical approach by clustering all factors firstly into superordinate categories. In a second step these categories are disaggregated into subcategories. The chosen assessment approach consists of four subsequent steps:

- Normalisation of the assigned values with the aim of conducting a methodologi- cally reliable comparison of factors with different units.
- Weighting of benefit categories from the perspective of the actors in order to take the relative importance of categories into consideration.
- Weighting of the different kinds of benefits within a category from the perspective of the actors.
- Calculation and aggregation of the weighted values to a composite indicator reflecting the attitude of an actor towards a technology.

For the normalisation, we employ a summation approach: The normalised indicators are calculated by dividing the score of a particular technology, in relation to a particular decision factor by the sum of the score values for all technologies on that particular decision factor.

Equation 1 shows the normalisation of the scored characteristic $x_{i,z}^k$.

$$u_{i,z}^k = \frac{x_{i,z}^k}{\sum_{k=1}^n x_{i,z}^k} \qquad (1)$$

with

$u_{i,z}^k$: normalized value of indicator i of the superordinate categories z assigned for technology k

n: number of technologies under consideration

$x_{i,z}^k$: value of indicator i of the superordinate category z assigned for technology k

In a next step, we introduce weighting factors for the superordinate categories which comprise indicators belonging to e.g. 'economic aspects' or 'ecological aspects'. Within these aggregated categories, we use indicator-specific weighting factors.

Within the superordinate categories as well as in the subcategories, the weighting factors sum up to 1. Taking the weighting factors into consideration, the attitude of actor *a* can be assess as follow:

$$P_k^a = \sum_{i=1}^{m} \sum_{z=1}^{o} w_i^a * v_{i,z}^a * u_{i,z}^k \qquad (2)$$

with

P_k^a: performance index
m: number of superordinate categories
o: number of subcategories
w_i^a: weighting factor of indicator category *cat*, with $\sum_{cat} w_i^a = 1; 0 \leq w_i^a \leq 1$
$v_{i,z}^a$: weighting factor of indicator z within indicator category i, with $\sum_i v_{i,z}^a = 1;$
$\quad 0 \leq v_{i,z}^a \leq 1$

The performance index P_k^a reflects the attitude of actor *a* to technology k.

If an actor can choose between different technologies, he will prefer the one with the highest performance index (OECD/EC/JRC 2008; Nardo et al. 2005).

3 Specification

In the following section, we apply the 'Ancillary Effects/MCDA' approach using the example of e-mobility. The introduction/extension of (an) environmental-friendly mobility system with focus on cars used by private persons is selected as the intervention that triggers the benefits. Hence, we classify reduction of CO_2 emissions as the primary target of the intervention.

We assume that there are different actors which are differently affected by changes in the mobility system. For instance, car users might suffer from a reduction in the vehicles' range whereas utilities might benefit from increases in the demand for electricity. Apart from the reduction of CO_2 emissions, benefits/effects for the different actors are considered as being of secondary interest and are, therefore, classified as ancillary effects.

Applying the two-order approach we differentiate between (1) effects resulting directly from the decision to opt for a technology by a specific group of actors ('first order ancillary effects') and (2) externalities reflecting impacts of an adoption of a technology by one group of actors on other actors ('second order ancillary effects').

Actors, indicators and alternatives

In this study, we consider four different groups of actors:

- Car users: 'car user' represents a private person who is interested in using a car to get to work or to the shops.
- Vehicle manufacturers: 'vehicle manufacturer' corresponds to a company that produces cars.
- Electric utilities: 'electric utility' stands for a company aiming to sell electricity.

- Government: 'government' is used as proxy for decision makers who focus on objectives on a national (or at least regional) political level.

Of course, it is possible to disaggregate the different groups of actors to a greater extent and to expand the range of actors, e.g. by taking oil companies, service station operators and other kinds of stakeholders into consideration. Since our study focuses on showing how multi-criteria analysis can be enriched through the integration of ancillary benefits, we focus on the four representative groups of actors mentioned above.

In principle, the selected groups of actors assess means of transport (and respectively cars) differently: Car users are primarily focused on the cost and comfort aspects, whereas for vehicle manufacturers, profit is the most important factor. Beside profit maximisation, electric utilities are interested in the continued opportunity to sell electricity. For the government, reduction in emissions and promoting employment are key issues.

Table 1 shows characteristics of cars that are of interest to the stakeholder groups. The list starts with environmental factors, namely CO_2 emissions, and emissions with local impacts. CO_2 emissions, measured, e.g. in g/km, are, to some extent, of importance to car users, vehicle manufacturers (because they have to fulfil legal restrictions on their car fleet) and the government (because of national GHG reductions targets). Regarding local emissions (e.g. particulate matters), we assume that they are only of interest to the government.

The second category comprises economic factors: Cost of ownership is one of the key factors car users are interested in. Vehicle manufacturers aim to maximise profit. Thus, we added profit as an indicator. In addition, we assume that structure and level of costs are relevant for vehicle manufacturers. Since tax revenues are (at least partially) related to profits, the indicator 'profit' is considered as also being relevant for the government. Other factors in which the government is interested are: employment, import dependency (with respect to fossil fuels) and electricity supply security. Import dependency and electricity supply security are categorised under 'social/political factors'. Regarding security of electricity supply, we assume that this factor is essentially of interest to electricity utilities.

The car categories considered in this study differ with respect to, e.g. charging time/time needed for refuelling, range and other kinds of comfort aspects (e.g. feedback of the heating system of the car on the range) and thus we added

Table 1 Factors relevant for decisions with respect to e-mobility

Ecological factors	Economic factors	Social/political factors	Comfort/performance	Other factors
– CO_2 emissions – Local emissions	– Cost of ownership – Profit – Employment	– Impact on import dependency – Impact on security of electricity supply	– Charging time – Range – Others	– Complementarity with existing structures – Need for incentives

'comfort/performance' as a category. All indicators assigned to this category are relevant for car users. Since charging time and range can impact electricity demand, we considered both factors as being of interest for electricity utilities too.

Indicators like 'complementarity with existing structures' and 'need for incentives' are allocated to the category others'. 'Complementarity with existing structures' and 'need for incentives' are considered as being relevant for car users, whereas, for vehicle manufacturers, 'complementarity with existing structures' is relevant, and 'need for incentives' is primarily relevant for government. For the consideration of differences in the relevance of indicators, we rate this by using a scale from 0 (not relevant) to 10 (very relevant) and the rating on the relevance scale translates into the weighting attached to the superordinate category and, following this, we attach weighting factors to the subcategories (Fig. 2).

3.1 Alternatives and their Characteristics

Regarding the alternatives available to the actors, we consider the possibility of buying and using:

- an electric car (EV),
- a car with internal combustion engine (ICE) and
- a hybrid car (HEV).

For an appropriate comparison, we assume that the selected cars are compact/medium cars. The specification assigned to the three car categories is based on the latest data for selected cars of this category. As representative example for an 'electric car', we select Nissan Leaf and, we select Toyota Auris 1.2 to represent the typical 'car with internal combustion engine'. For 'hybrid car', we opted for a Toyota Hybrid Comfort (NISSAN Center Europe 2018; Toyota Deutschland 2018).

Based on the list of characteristics in which the actors are interested, we specify the values for the indicators for each car category (Table 2). For indicators which are difficult to quantify, we use a scale ranging from very low/very bad to very high/very good.

3.2 Externalities

We expect that if the government wants to promote the adoption of EV, incentive measures which further the deployment of EV will be established or expanded. This includes measures like subsidies for purchasing EV, special conditions for parking as well as the support of R&D activities with respect to improvements of comfort aspects (e.g. range, charging time) and to reductions in cost (see, e.g. BMUB 2018). In addition, the government might support adjustments to infrastructure (i.e. the electricity grid). Thus, it can be expected that government activities to

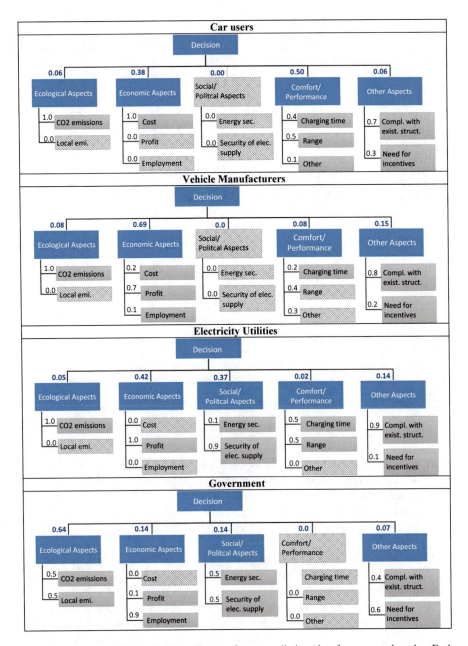

Fig. 2 Indicators and their weighting. (Source: Own compilation (data for car users based on Esch 2016))

Table 2 Characteristics of different car categories

Characteristics	Unit	Electric car	Car with internal combustion	Hybrid car
Ecological factors				
CO_2 emissions[a]	–	63[b]	100	78
Local emissions[a]	–	0	100	79
Economic factors				
Cost of ownership[a]	–	107	100	101
Profit[a]	–	50	100	90
Employment[a]	–	26	100	104
Social/political factors				
Impact on import dependency	–	Very low	Very high	High
Impact on the security of electricity supply	–	Moderate	Very low	Very low
Comfort/performance				
Charging time	–	Very bad	Very good	Very good
Range	km	350	900	1000
Other[c]		Good	Very good	Very good
Other factors				
Complementarity with existing structures		Very low	Very good	Very good
Need for incentives		Very high	Very low	Low

Remarks: [a]Standardised ICE = 100, [b]Calculated based on data on average CO_2-emissions/kWh in Germany, [c]Other factors influencing the comfort of a car (e.g. feedback of the heating system of the car on the range)
Source: Own compilation based on Esch (2016), NISSAN Center Europe (2018), Toyota Deutschland (2018)

promote EV will strengthen the attitude of car users towards EV. In addition, vehicle manufactures and electric utilities will also be encouraged to adopt EV.

Regarding the impacts resulting from electricity utilities favouring EV, we assume that the grid infrastructure will be strengthened and the possibilities for fast charging will be expanded. This could also support the decision of car users to buy an EV.

If the vehicle manufacturers put EV at the top of their agenda, it could be expected that they will invest more R&D in EV leading to lower costs and, hence, increased support, in terms of the attitude of car users towards EV.

An increasing use of EV by car users, on the one hand, results in decreasing costs (because of learning effects in the manufacturing sector) and increases in the profit rates of vehicle manufacturers (with respect to the production of EV). On the other hand, the security of the grid might be impacted more strongly if a rising number of car users use EV. Hence, the decision by car users to invest in EV can affect vehicle manufacturers as well as electricity utilities. Taking into consideration that a greater use of EV also means that the tax revenues linked with oil consumption will drop, the government will also be impacted.

Considering that the attitude of one actor towards a car technology can be positively influenced if other actors promote this technology, the MCDA approach has to be extended. Thus, we include additional factors in the list of indicators reflecting crossover impacts. These impacts are not the primary objective of the actors who promote the technology. Hence, they can be seen as a special kind of additional ancillary benefit that influences the attitude of actors towards a technology and could reinforce diffusion of technologies. Table 3 shows our judgement of the identified crossover impacts.

As dummy indicators, we add 'dum_car users' reflecting the impacts of car users on the other actors, 'dum_vehicle man.' as representative for the effects resulting from the activities of vehicle manufactures, 'dum_elec. utilities' for the integration of crossover effects of the utilities' attitudes and 'dum_goverment' for the assessment of the impacts of the attitude of the government. Each of these indicators has the three states 'EV', 'HEV' and 'ICE'.

We modify Fig. 2 by adding an indicator category called 'External effects' consisting of the dummy variables. With the introduction of the new indicator category all weighting factors on the indicator categories level have to be recalculated.

Regarding the weighting of the external effects, we assume that if the government and the vehicle manufacturers support EV, from the car users' perspective, the cost difference between ICE and EV will be offset. Regarding the impacts of external effects on vehicle manufactures we assume that if the car users opt for EV, differences in the profit rates of ICE and EV will be compensated. The impacts of external effects on the utilities are calibrated by assuming that, through improvements in the charging technology, negative effects on the security of the grid can be minimised.

3.3 Scenarios

In addition to the scenario without externalities, we focus on three additional scenarios. The scenario 'first-order externalities' is computed by taking the external impacts into account which result from the attitudes of the actors towards technologies identified in the scenario 'Without externalities'.

In the third scenario ('second-order externalities') we recalculate the attitudes taking results from the scenario 'first order externalities' into consideration.

In the fourth scenario, we assume that all actor groups favour EV.

4 Results

In the following we present the attitudes of different actor groups assessed by applying MCDA.

Table 3 Crossover impacts

Activity of		Impact on benefits/cost for											
		Car user			Vehicle manufacturers			Electricity utilities			National government		
Actor	Favored car category	EV	HEV	ICE	EV	HEV	ICE	EV	HEV	ICE	EV	HEV	ICE
Car user	EV	0	0	0	1	0	0	1	0	−1	3	1	−1
	HEV	0	0	0	−1	2	1	0	0	0	1	3	−3
	ICE	0	0	0	−2	0	3	−1	0	1	−3	0	3
Vehicle man.	EV	3	1	−3	0	0	0	3	1	−3	3	1	−3
	HEV	1	3	−3	0	0	0	1	1	−2	0	3	−3
	ICE	−3	2	3	0	0	0	−2	0	2	−2	0	3
Elec. utilities	EV	1	0	−1	0	0	0	0	0	0	2	1	−2
	HEV	1	1	−1	0	0	0	0	0	0	0	0	0
	ICE	−2	0	2	0	0	0	0	0	0	0	0	0
Nat. gov.	EV	3	1	−1	1	0	−1	2	0	−1	0	0	0
	HEV	1	2	−1	0	1	−1	0	1	−1	0	0	0
	ICE	−1	0	3	−2	0	2	−2	0	2	0	0	0

Remarks: 0 No impact, 1 weak positive impact, 3 very strong impact, −1 weak negative impact, −3 very strong negative impact
Source: Own compilation

4.1 Attitudes Towards Car Categories

The scenario 'without externalities' indicates that, under the assumed framework, the actor group 'car users' will favour ICE cars. Not surprisingly, vehicle manufacturers also prefer ICE cars, whereas electricity utilities and the government prefer electric vehicles. There are different underlying reasons for the attitude taken by electricity utilities and the government towards EV: The attitude of electricity utilities is mainly influenced by economic aspects whereas the government sets priorities according to environmental factors (Fig. 3).

If we take crossover effects (triggered by the positive attitude of the government and the utilities towards EV cars) into consideration, the attitude of car users will switch from ICE to hybrid cars (because of side effects) (Fig. 4). This change in attitude is linked with changes in crossover effects. Hence, we change the dummy values reflecting the modified attitude of the car users and recalculate the composite indicators for the different actor groups.

The scenario 'second order externalities', based on the modified external effects, i.e. taking into account that car users have switched from ICE to supporting HEV, differs only marginally from the scenario 'first order externalities'. The results confirm the attitude of car users towards HEV as presumed. HEV is also preferred by the vehicle manufactures. Utilities and the government prefer EV. All in all, the standing of ICE cars will become less favourable if externalities are taken into consideration.

In the scenario 'EV', we analyse how the calculated attitude of an actor group towards EV will differ if all actor groups support EV, through their externalities. According to our results, a prioritisation of EV by the government, utilities and car users will especially strongly affect the attitude of the vehicle manufacturers. However, the external effects will not be strong enough to change the order of

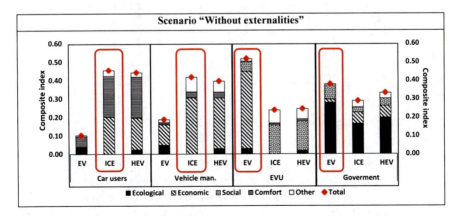

Fig. 3 MCA-Assessment of attitudes (1)

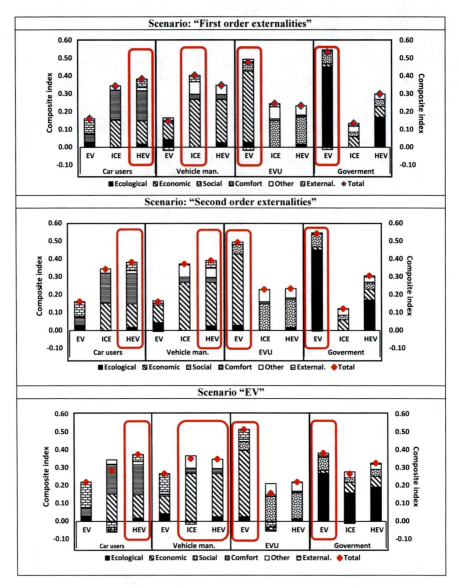

Fig. 4 MCA-Assessment of attitudes (2)

the vehicle manufacturers' preferences significantly, as they do not overcome the original score of the technology on the performance index for this actor.

In particular, restrictions regarding range and charging time limit a switch of the car users to EV. To effect a change in the attitude of these two actor groups, either the characteristics of EV or the weighting of the characteristics have to be changed.

4.2 Sensitivity Analysis

In the following section, we assess the impacts of changes in weighting factors as well as modifications of characteristics of the vehicles on the attitudes of car users and vehicle manufacturers towards EV. In particular, we analyse under which constellation of characteristics and weightings the attitudes of these two actor groups will switch from favouring HEV to EV.

The figure on the left of Table 4 shows the impacts of modifications in the economic and ecological factors on the preferred car technology. According to this figure, the preference of the vehicle manufacturers towards EV will increase if the economic factors become less important or ecological aspects gain in importance. Changes in the meaning of economic factors can result from harmonised profit rates (meaning that the profits from ICE, EV and HEV are more or less the same) or changes in the weighting of these factors by the actor group.

The meaning of ecological factors may change due to increasing pressure on the vehicle manufacturers with respect to a decarbonisation of the vehicle stock or changes in the specific emissions of the electricity mix used for charging EV.

An example for the impacts of modifications in the assumed specific emissions of the electricity mix is presented in the right side of Table 4. The figure is calculated by assuming that no CO_2 emissions are linked to the production of the electricity which is used for charging the EV. As expected, lower specific emissions support the attitude towards EV. In contrast to the calculations presented in Sect. 4.1, vehicle manufacturers will prefer EV if the weighting of ecological and economic factors differ less from the default constellations.

As mentioned, even if all other actor groups prefer EV, car users will prefer HV since HV dominates EV with respect to the comfort/performance aspect. The figure on the left of Table 5 shows the impact of changes in 'comfort/performance' and 'ecological factors' on the attitude of car users towards EV. If 'comfort/performance' is less important, car users could be convinced to use

Table 4 Impact of changes in weighting factors on the attitude of vehicle manufactures

<table>
<tr><th colspan="12">Scenario "EV": CO₂ emissions 73 g/100 km*</th><th colspan="12">Scenario "Variant: EV with zero CO₂ emissions"</th></tr>
<tr><td colspan="2"></td><th colspan="11">ECONOMIC FACTORS (0: not relevant, 10: very important)</th><td colspan="2"></td><th colspan="11">ECONOMIC FACTORS (0: not relevant, 10: very important)</th></tr>
<tr><td colspan="2"></td><th>0</th><th>1</th><th>2</th><th>3</th><th>4</th><th>5</th><th>6</th><th>7</th><th>8</th><th>9</th><th>10</th><td colspan="2"></td><th>0</th><th>1</th><th>2</th><th>3</th><th>4</th><th>5</th><th>6</th><th>7</th><th>8</th><th>9</th><th>10</th></tr>
<tr><td rowspan="11">ECOLOGICAL FACTORS (0: not relevant, 10: very important)</td><td>0</td><td>ICE</td><td>ICE</td><td>ICE</td><td>ICE</td><td>ICE</td><td>ICE</td><td>ICE</td><td>ICE</td><td>ICE</td><td>ICE</td><td>ICE</td><td rowspan="11">ECOLOGICAL FACTORS (0: not relevant, 10: very important)</td><td>0</td><td>ICE</td><td>ICE</td><td>ICE</td><td>ICE</td><td>ICE</td><td>ICE</td><td>ICE</td><td>ICE</td><td>ICE</td><td>ICE</td><td>ICE</td></tr>
<tr><td>1</td><td>HV</td><td>HV</td><td>HV</td><td>HV</td><td>HV</td><td>HV</td><td>HV</td><td>ICE</td><td>ICE</td><td>ICE</td><td>ICE</td><td>1</td><td>ICE</td><td>ICE</td><td>ICE</td><td>ICE</td><td>ICE</td><td>ICE</td><td>ICE</td><td>ICE</td><td>ICE</td><td>ICE</td><td>ICE</td></tr>
<tr><td>2</td><td>HV</td><td>HV</td><td>HV</td><td>HV</td><td>HV</td><td>HV</td><td>HV</td><td>HV</td><td>HV</td><td>HV</td><td>HV</td><td>2</td><td>EV</td><td>EV</td><td>EV</td><td>EV</td><td>EV</td><td>EV</td><td>HV</td><td>HV</td><td>ICE</td><td>ICE</td><td>ICE</td></tr>
<tr><td>3</td><td>HV</td><td>HV</td><td>HV</td><td>HV</td><td>HV</td><td>HV</td><td>HV</td><td>HV</td><td>HV</td><td>HV</td><td>HV</td><td>3</td><td>EV</td><td>EV</td><td>EV</td><td>EV</td><td>EV</td><td>EV</td><td>EV</td><td>EV</td><td>EV</td><td>EV</td><td>EV</td></tr>
<tr><td>4</td><td>EV</td><td>EV</td><td>EV</td><td>HV</td><td>HV</td><td>HV</td><td>HV</td><td>HV</td><td>HV</td><td>HV</td><td>HV</td><td>4</td><td>EV</td><td>EV</td><td>EV</td><td>EV</td><td>EV</td><td>EV</td><td>EV</td><td>EV</td><td>EV</td><td>EV</td><td>EV</td></tr>
<tr><td>5</td><td>EV</td><td>EV</td><td>EV</td><td>EV</td><td>EV</td><td>EV</td><td>EV</td><td>HV</td><td>HV</td><td>HV</td><td>HV</td><td>5</td><td>EV</td><td>EV</td><td>EV</td><td>EV</td><td>EV</td><td>EV</td><td>EV</td><td>EV</td><td>EV</td><td>EV</td><td>EV</td></tr>
<tr><td>6</td><td>EV</td><td>EV</td><td>EV</td><td>EV</td><td>EV</td><td>EV</td><td>EV</td><td>EV</td><td>EV</td><td>EV</td><td>EV</td><td>6</td><td>EV</td><td>EV</td><td>EV</td><td>EV</td><td>EV</td><td>EV</td><td>EV</td><td>EV</td><td>EV</td><td>EV</td><td>EV</td></tr>
<tr><td>7</td><td>EV</td><td>EV</td><td>EV</td><td>EV</td><td>EV</td><td>EV</td><td>EV</td><td>EV</td><td>EV</td><td>EV</td><td>EV</td><td>7</td><td>EV</td><td>EV</td><td>EV</td><td>EV</td><td>EV</td><td>EV</td><td>EV</td><td>EV</td><td>EV</td><td>EV</td><td>EV</td></tr>
<tr><td>8</td><td>EV</td><td>EV</td><td>EV</td><td>EV</td><td>EV</td><td>EV</td><td>EV</td><td>EV</td><td>EV</td><td>EV</td><td>EV</td><td>8</td><td>EV</td><td>EV</td><td>EV</td><td>EV</td><td>EV</td><td>EV</td><td>EV</td><td>EV</td><td>EV</td><td>EV</td><td>EV</td></tr>
<tr><td>9</td><td>EV</td><td>EV</td><td>EV</td><td>EV</td><td>EV</td><td>EV</td><td>EV</td><td>EV</td><td>EV</td><td>EV</td><td>EV</td><td>9</td><td>EV</td><td>EV</td><td>EV</td><td>EV</td><td>EV</td><td>EV</td><td>EV</td><td>EV</td><td>EV</td><td>EV</td><td>EV</td></tr>
<tr><td>10</td><td>EV</td><td>EV</td><td>EV</td><td>EV</td><td>EV</td><td>EV</td><td>EV</td><td>EV</td><td>EV</td><td>EV</td><td>EV</td><td>10</td><td>EV</td><td>EV</td><td>EV</td><td>EV</td><td>EV</td><td>EV</td><td>EV</td><td>EV</td><td>EV</td><td>EV</td><td>EV</td></tr>
</table>

Remarks: * Calculated based on data on average g CO_2-emissions/kWh in Germany, values in red indicates results from default setting

174   S. Vögele et al.

Table 5 Impact of changes in weighting factors on the attitude of car users (A)

Scenario "EV": CO$_2$ emissions 73 g/100 km*												Scenario "Variant I: EV with zero CO$_2$ emissions"											
		ECOLOGICAL FACTORS (0: not relevant, …, 10 very important)													ECOLOGICAL FACTORS (0: not relevant, …, 10 very important)									
		0	1	2	3	4	5	6	7	8	9	10			0	1	2	3	4	5	6	7	8	9	10
Comfort/Performance (0: not relevant, …, 10 very important)	0	HV	HV	HV	HV	EV	EV	EV	EV	EV	EV	EV	Comfort/Performance (0: not relevant, …, 10 very important)	0	HV	HV	EV	EV	EV	EV	EV	EV	EV	EV	EV
	1	HV	HV	HV	HV	HV	EV	EV	EV	EV	EV	EV		1	HV	HV	EV	EV	EV	EV	EV	EV	EV	EV	EV
	2	HV	HV	HV	HV	HV	HV	HV	EV	EV	EV	EV		2	HV	HV	HV	EV	EV	EV	EV	EV	EV	EV	EV
	3	HV	HV	HV	HV	HV	HV	HV	HV	EV	EV	EV		3	HV	HV	HV	EV	EV	EV	EV	EV	EV	EV	EV
	4	HV	HV	HV	HV	HV	HV	HV	HV	HV	EV	EV		4	HV	HV	HV	HV	EV	EV	EV	EV	EV	EV	EV
	5	HV	HV	HV	HV	HV	HV	HV	HV	HV	HV	EV		5	HV	HV	HV	HV	EV	EV	EV	EV	EV	EV	EV
	6	HV	HV	HV	HV	HV	HV	HV	HV	HV	HV	HV		6	HV	HV	HV	HV	HV	EV	EV	EV	EV	EV	EV
	7	HV	HV	HV	HV	HV	HV	HV	HV	HV	HV	HV		7	HV	HV	HV	HV	HV	HV	EV	EV	EV	EV	EV
	8	HV	*HV*	HV	HV	HV	HV	HV	HV	HV	HV	HV		8	HV	*HV*	HV	HV	HV	HV	HV	EV	EV	EV	EV
	9	HV	HV	HV	HV	HV	HV	HV	HV	HV	HV	HV		9	HV	HV	HV	HV	HV	HV	HV	EV	EV	EV	EV
	10	HV	HV	HV	HV	HV	HV	HV	HV	HV	HV	HV		10	HV	HV	HV	HV	HV	HV	HV	EV	EV	EV	EV

Remarks: * Calculated based on data on average g CO$_2$-emissions/kWh in Germany, values in red indicates results from default setting

EV more easily. If ecological aspects become more important (e.g. because of a higher environmental awareness in general), the attitude towards EV will improve.

For the assessment of uncertainties with which EV are associated, we assess the impacts of changes in the specific CO$_2$ emission factors in combination with changes in the range and charging time on the attitude of car users towards EV. The figure on the right of Table 5 is based on the assumption that the electricity used to charge the EV leads to zero CO$_2$ emissions. According to this figure, these modifications are not strong enough to change the attitude towards EV under the default conditions. Only if the attitudes towards comfort/performance or ecological factors change significantly, the conditions could become positive in the sense that car users will favour EV.

Uncertainties concerning the impacts of externalities resulting from attitude patterns of other actors on the attitude of cars users are assessed by modifying the weighting factor for this indicator category. The figure on the left of Table 6 shows how the attitude of the car users will change if the weighting factor for externalities differs. If other actors do not support EV, car users will tend to favour ICE regardless of changes in comfort/performance aspects. With increasing importance

Table 6 Impact of changes in weighting factors on the attitude of car users (B)

Scenario "Variant II": CO$_2$ emissions 73 g/100 km*												Scenario "Variant III: EV with zero CO$_2$ emissions"											
		ECOLOGICAL FACTORS (0: not relevant, …, 10 very important)													ECOLOGICAL FACTORS (0: not relevant, …, 10 very important)									
		0	1	2	3	4	5	6	7	8	9	10			0	1	2	3	4	5	6	7	8	9	10
External. (0: not relevant, …, 10 very important)	0	ICE	ICE	HV	HV	HV	HV	HV	HV	HV	HV	HV	External. (0: not relevant, …, 10 very important)	0	ICE	ICE	ICE	ICE	HV	HV	HV	HV	HV	EV	EV
	1	ICE	HV	HV	HV	HV	HV	HV	HV	HV	HV	HV		1	ICE	HV	HV	HV	HV	HV	HV	HV	HV	EV	EV
	2	HV	HV	HV	HV	HV	HV	HV	HV	HV	HV	HV		2	HV	HV	HV	HV	HV	HV	HV	HV	EV	EV	EV
	3	HV	HV	HV	HV	HV	HV	HV	HV	HV	HV	HV		3	HV	HV	HV	HV	HV	HV	HV	EV	EV	EV	EV
	4	HV	HV	HV	HV	HV	HV	HV	HV	HV	HV	HV		4	HV	HV	HV	HV	HV	HV	HV	EV	EV	EV	EV
	5	HV	*HV*	HV	HV	HV	HV	HV	HV	HV	HV	HV		5	HV	*HV*	HV	HV	HV	HV	EV	EV	EV	EV	EV
	6	HV	HV	HV	HV	HV	HV	HV	HV	HV	HV	HV		6	HV	HV	HV	HV	HV	EV	EV	EV	EV	EV	EV
	7	HV	HV	HV	HV	HV	HV	HV	HV	HV	HV	HV		7	HV	HV	HV	HV	HV	EV	EV	EV	EV	EV	EV
	8	HV	HV	HV	HV	HV	HV	HV	HV	HV	HV	EV		8	HV	HV	HV	HV	EV	EV	EV	EV	EV	EV	EV
	9	HV	HV	HV	HV	HV	HV	HV	HV	HV	EV	EV		9	HV	HV	HV	EV	EV	EV	EV	EV	EV	EV	EV
	10	HV	HV	HV	HV	HV	HV	EV	EV	EV	EV	EV		10	HV	HV	HV	EV	EV	EV	EV	EV	EV	EV	EV

Remarks: * Calculated based on data on average g CO$_2$-emissions/kWh in Germany, values in red indicates results from default setting of externalities (e.g. due to increase in incentives) EV will gain ground. The figure on the right of Table 6 indicates how the situation will change if ecological aspects become more important for car users. The initial situation (marked in red) is based on the assumption that cost differences are compensated by externalities. Since compensation of disadvantages relates to comfort/performance, either the importance of 'comfort/performance' has to be lower or external benefits have to be increased.

Increases in the meaning of ecological aspects support a switch in the attitude of car users towards EV. However, without changing the weighting of 'ecological aspects', a shift in the attitude towards EV does not appear possible.

The results indicate that the introduction/extension of an environmental-friendly mobility system with focus on cars used by private persons will encounter resistance from vehicle manufacturers as well as from car users. According to the results, it should be feasible to cause a shift in the attitude of car users towards HEV. In particular, actions of the government in combination with the car industry can trigger a switch from ICE to HEV. Since EV faces a lot of restrictions, including losses of comfort, it is quite difficult to convince car users to use EV. We assume that the attitude of vehicle manufacturers mainly depends on technology-specific profit rates. Currently, the profit rate for EV is low. According to our calculations (which are based on a lot of simplifications), with increases in the profit rate for EV, the attitude of the vehicle manufacturers towards EV can be influenced significantly. However, we expect that the resulting externalities will not be great enough to compensate the disadvantage car users expect in relation to using EV.

5 Conclusion

In principle, each measure, whether on an individual, national or international level, causes multiple effects. For an assessment of a measure, it is worthwhile to focus not only on the benefits the measure is aiming to achieve, but also on ancillary effects such as impacts on health and changes in income. Usually, different decision makers have different attitudes towards a particular technology. Accordingly, decision makers support measures focusing on technology deployments differently. Using the example of e-mobility, we show that stakeholders like car users and vehicle manufacturers will show resistance, if they are urged to go for e-mobility. In our study, we extend the ancillary benefit approach by taking externalities into consideration. These externalities reflect crossover effects between actor groups. As an example, the car user will benefit if vehicle manufacturers spend more R&D on e-mobility because they favour e-mobility as a technology. Since the assessment of the characteristics of e-mobility is linked with high uncertainty, we conducted intensive sensitivity analyses. These analyses show that it should be feasible to cause a shift in the attitude of car users towards hybrid vehicles but not to electric vehicles, since electric vehicles face a lot of restrictions, including loss of comfort. Regarding the attitude of vehicle manufacturers towards e-mobility, profits rates are a crucial factor.

References

Acatech (2018) Coupling the different energy sectors – options for the next phase of the energy transition. acatech - National Academy of Science and Engineering, Munich

Augenstein K (2015) Analysing the potential for sustainable e-mobility – the case of Germany. Environ Innov Soc Trans 14:101–115

Bain PG et al (2016) Co-benefits of addressing climate change can motivate action around the world. Nat Clim Chang 6(2):154–157

Baležentis T, Streimikiene D (2017) Multi-criteria ranking of energy generation scenarios with Monte Carlo simulation. Appl Energy 185(Part 1):862–871

Behzadian M, Kazemadeh RB, Albadvi A, Aghdasi M (2010) PROMETHEE: a comprehensive literature review on methodologies and applications. Eur J Oper Res 200(1):198–215

Biresselioglu ME, Demirbag Kaplan M, Yilmaz BK (2018) Electric mobility in Europe: a comprehensive review of motivators and barriers in decision making processes. Transp Res Part A Policy Pract 109:1–13

BMUB (2018) General information – Electric mobility. Federal Ministry for the Environment, Nature Conservation, Building and Nuclear Safety (BMUB). https://www.bmu.de. Accessed 24 Oct 2018

Brans JP, Vincke P, Mareschal B (1986) How to select and how to rank projects – the PROMETHEE METHOD. Eur J Oper Res 24(2):228–238

Bundesministerium für Wirtschaft und Energie (2018) IKT für Elektromobilität III: Einbindung von gewerblichen Elektrofahrzeugen in Logistik-, Energie- und Mobilitätsinfrastrukturen. BMWi, Berlin

Chavez-Baeza C, Sheinbaum-Pardo C (2014) Sustainable passenger road transport scenarios to reduce fuel consumption, air pollutants and GHG (greenhouse gas) emissions in the Mexico City metropolitan area. Energy 66:624–634

Davis DL, Krupnick A, Mcglynn G (2000) Ancillary benefits and costs of greenhouse gas mitigation – an overview In: OECD (ed) Ancillary benefits and costs of greenhouse gas mitigation. OECD, Paris, pp 9–49

Diakoulaki D, Karangelis F (2007) Multi-criteria decision analysis and cost–benefit analysis of alternative scenarios for the power generation sector in Greece. Renew Sust Energ Rev 11(4):716–727

Ekins P (1996) The secondary benefits of CO_2 abatement: how much emission reduction do they justify? Ecol Econ 16(1):13–24

Esch F-P (2016) Nutzeranforderungen an Elektrofahrzeuge. Universitäts- und Landesbibliothek Darmstadt, Darmstadt

European Commission (2017) Communication from the Commission to the European Parliament, the Council, the European Economic and Social Committee and the Committee of the Regions: delivering on low emission mobility, a European Union that protects the planet, empowers its consumers and defends its industry and workers. European Commission, COM(2017) 675 Final, Brussels

Gebauer F, Vilimek R, Keinath A, Carbon C-C (2016) Changing attitudes towards e-mobility by actively elaborating fast-charging technology. Technol Forecast Soc Chang 106:31–36

Groosman B, Muller NZ, O'neill-Toy E (2011) The ancillary benefits from climate policy in the United States. Environ Resour Econ 50(4):585–603

Hagman J, Ritzén S, Stier JJ, Susilo Y (2016) Total cost of ownership and its potential implications for battery electric vehicle diffusion. Res Transp Bus Manag 18:11–17

IPCC (2001) Climate change 2001: mitigation. Cambridge University Press, New York

Kirkman GA, Seres S, Haites E, Spalding-Fecher R (2012) Benefits of the clean development mechanism. United Nations Framework Convention on Climate Change, Bonn

Krupnick A, Burtraw D Markandya A (2000) The ancillary benefits and costs of climate change mitigation: a conceptual framework. In: OECD (ed) Ancillary benefits and costs of greenhouse gas mitigation. OECD, Paris, pp 53–93

Kumar A, Sah B, Singh AR, Deng Y, He X, Kumar P, Bansal RC (2017) A review of multi criteria decision making (MCDM) towards sustainable renewable energy development. Renew Sust Energ Rev 69:596–609

Lo Schiavo L, Delfanti M, Fumagalli E, Olivieri V (2013) Changing the regulation for regulating the change: innovation-driven regulatory developments for smart grids, smart metering and e-mobility in Italy. Energy Policy 57:506–517

Loken E (2007) Use of multicriteria decision analysis methods for energy planning problems. Renew Sustain Energy Rev 11(7):1584–1595

Macharis C, Bernardini A (2015) Reviewing the use of multi-criteria decision analysis for the evaluation of transport projects: time for a multi-actor approach. Transp Policy 37:177–186

Malmgren, I. (2016) Quantifying the societal benefits of electric vehicles. In: EV29 Symposium. Montreal, World Electric Vehicle Journal

Matthews L, Lynes J, Riemer M, Del Matto T, Cloet N (2017) Do we have a car for you? Encouraging the uptake of electric vehicles at point of sale. Energy Policy 100:79–88

Mayrhofer JP, Gupta J (2016) The science and politics of co-benefits in climate policy. Environ Sci Policy 57(Supplement C):22–30

Muller NZ (2012) The design of optimal climate policy with air pollution co-benefits. Resour Energy Econ 34(4):696–722

Nardo M, Saisana M, Saltelli A, Tarantola S (2005) Tools for composite indicator building. http://compositeindicators.jrc.ec.europa.eu/Document/EUT%2021682%20EN.pdf. Accessed 24 Feb 2017

Nationale Plattform Elektromobilität (2014) Fortschrittsbericht 2014 – Bilanz der Marktvorbereitung. Nationale Plattform Elektromobilität, Berlin

Nemet GF, Holloway T, Meier P (2010) Implications of incorporating air-quality co-benefits into climate change policymaking. Environ Res Lett 5(1):1–9

Nissan Center Europe (2018) LEAF. https://www.nissan.de. Accessed 24 Oct 2018

OECD (2000) Ancillary benefits and costs of greenhouse gas mitigation. OECD, Paris

OECD/EC/JRC (2008) Handbook on constructing composite indicators: methodology and user guide. OECD Publishing, Paris

Parkinson SC, Makowski M, Krey V, Sedraoui K, Almasoud AH, Djilali N (2018) A multi-criteria model analysis framework for assessing integrated water-energy system transformation pathways. Appl Energy 210:477–486

Pittel K, Rübbelke DTG (2008) Climate policy and ancillary benefits: a survey and integration into the modelling of international negotiations on climate change. Ecol Econ 68(1–2):210–220

Pohekar SD, Ramachandran M (2004) Application of multi-criteria decision making to sustainable energy planning – a review. Renew Sustain Energy Rev 8(4):365–381

Rive N, Rübbelke DTG (2010) International environmental policy and poverty alleviation. Rev World Econ 146(3):515–543

Sierzchula W, Bakker S, Maat K, Van Wee B (2014) The influence of financial incentives and other socio-economic factors on electric vehicle adoption. Energy Policy 68:183–194

Steinhilber S, Wells P, Thankappan S (2013) Socio-technical inertia: understanding the barriers to electric vehicles. Energy Policy 60:531–539

Terrados J, Almonacid G, Perez-Higueras P (2009) Proposal for a combined methodology for renewable energy planning. Application to a Spanish region. Renew Sustain Energy Rev 13(8):2022–2030

Theisen TMRF (2010) Market models for the roll-out of electric vehicle public charging infrastructure. Eurelectric, Union of the Electricity Industry, Brussels

Toyota Deutschland (2018) Auris. https://www.toyota.de. Accessed 24 Oct 2018

Truffer B, Schippl J, Fleischer T (2017) Decentering technology in technology assessment: prospects for socio-technical transitions in electric mobility in Germany. Technol Forecast Soc Chang 122:34–48

Ürge-Vorsatz D, Herrero ST, Dubash NK, Lecocq F (2014) Measuring the co-benefits of climate change mitigation. Annu Rev Environ Resour 39:549–582

Usmani, O., Rösler, H., DE Wilde, H., Straver, K. & Weeda, M. (2015) Policies and good practices to foster electromobility roll-out at the local, national and European level. European Commission, Brussels

Van Vuuren DP, Cofala J, Eerens HE, Oostenrijk R, Heyes C, Klimont Z, Den Elzen MGJ, Amann M (2006) Exploring the ancillary benefits of the Kyoto protocol for air pollution in Europe. Energy Policy 34(4):444–460

Wang Q, Poh KL (2014) A survey of integrated decision analysis in energy and environmental modeling. Energy 77:691–702

Winning I (2015) Deliverable 2.2: Successes encountered in the electromobility policy making process. Green E-motion, European Commission, Brussels

Part III
Ancillary Benefits in Different Sectors and in Adaptation to Climate Change

Ancillary Benefits of Adaptation: An Overview

Elisa Sainz de Murieta

1 Introduction

Climate change is one of the greatest challenges for our societies. Many changes are already undergoing and some of these will continue for centuries. There is evidence that every half a degree of warming matters and in order to avoid intolerable risks, ambitious mitigation and adaptation policies need to be implemented in the following decades (Masson-Delmotte et al. 2018).

The main or primary goal of mitigation policies is emission reduction and climate stabilisation to avoid and reduce climate impacts (Markandya and Rübbelke 2004), while adaptation aims at reducing climate vulnerability (Klein et al. 2014). However, often the benefits of climate policy go beyond its main goal and provide ancillary or secondary benefits. In its Fourth Assessment Report, the IPCC defined ancillary benefits as "the positive effects that a policy or measure aimed at one objective might have on other objectives, irrespective of the net effect on overall social welfare. Co-benefits are often subject to uncertainty and depend on local circumstances and implementation practices, among other factors. Co-benefits are also referred to as *ancillary benefits*". (IPCC 2014: 1762). Following this definition, ancillary benefits and co-benefits of climate policy are used as synonyms in this chapter.

Ancillary benefits are identified in the literature as an important consideration of climate policy. Accounting for these benefits can provide a more favourable cost–benefit ratio of climate options, contributing to maximise the benefits of climate policy implementation (Hallegatte 2009; Krook Riekkola et al. 2011). Co-benefits

---

E. Sainz de Murieta (✉)
Basque Centre for Climate Change (BC3), Leioa, Basque Country, Spain

Grantham Research Institute, London School of Economics (LSE), London, UK
e-mail: elisa.sainzdemurieta@bc3research.org

© Springer Nature Switzerland AG 2020
W. Buchholz et al. (eds.), *Ancillary Benefits of Climate Policy*, Springer Climate,
https://doi.org/10.1007/978-3-030-30978-7_10

are also useful to rank adaptation options, as a criterion for prioritising those with larger positive side effects (de Bruin et al. 2009; UNFCCC 2011).

Ancillary benefits can be significant in magnitude, even exceeding primary benefits (Markandya and Rübbelke 2004). Estimates from Western Europe provide higher benefits than in the US (Pittel and Rübbelke 2008); several studies have found that ancillary benefits of mitigation policy can range from 0.98 (UK) to 6.9 (Germany) times the magnitude of primary benefits. This range in the US varies between 0.1 and 6.7 (e.g. Pearce 2000). A recent study concluded that avoided damages of health impacts due to air pollution in Europe could balance mitigation costs (Schucht et al. 2015). According to the OECD, air pollution-related health costs could reach US$176 billion in 2060 (OECD 2016). Ancillary benefits in developing countries may be even greater than those from industrialised economies (Pittel and Rübbelke 2008).

The importance of ancillary benefits goes beyond quantitative estimations and qualitative, non-market and non-monetary aspects should also be assessed. This is the case, for instance, of human health, biodiversity and ecosystems or water availability. As climate change is expected to affect particularly the most vulnerable, equity and justice considerations, as well as the distributional effects should be considered in the analysis (Markandya and Watkiss 2009). While sometimes these environmental and social issues are difficult to assess in monetary terms, a restrictive quantifiable approach could lead to negative co-effects on the most vulnerable, ecosystems, future generations and, overall, to those whose values cannot always be quantified (Klein et al. 2014).

There is evidence that the willingness to pay for climate policy increases when ancillary benefits are considered (see, for example, Longo et al. 2012; Rodríguez-Entrena et al. 2014). Thus, accounting for ancillary benefits can foster the legitimacy of these policies, in terms of the acceptability by those affected by them (Adger et al. 2005; Krook Riekkola et al. 2011). In fact, co-benefits could be a powerful tool to overcome shifts in climate scepticism or lack of concern and move to action (Bain et al. 2016). Unfortunately, despite the potential of ancillary benefits of climate policy, they have been often ignored or overlooked (Markandya and Rübbelke 2004; Pittel and Rübbelke 2008). In recent years, this topic has gained increasing attention, even though mainly linked to mitigation policies (Dovie 2019).

For many years, most of the climate policy efforts have focused on mitigation, but at present, there is a general agreement that both, mitigation and adaptation policies are needed to face this global challenge (Sainz de Murieta et al. 2014). Adaptation to climate change can provide large direct economic benefits in terms of avoided damages and losses, but beyond direct benefits, adaptation measures offer additional economic, social and environmental co-benefits that, however, are often ignored. The concept of ancillary benefits is traditionally linked to emission reduction policies.

The aim of this chapter is to explore the ancillary benefits of adaptation. An overview of the literature is provided to analyse which are the main co-benefits of adaptation policies. This chapter is organised as follows: Sect. 2 analyses the ancillary benefits of mitigation policies, Sect. 3 focuses on the economic, social and environmental co-benefits of adaptation and Sect. 4 explores the risk of adaptation producing negative ancillary effects. Finally, Sect. 5 outlines the main conclusions.

2 Ancillary Benefits of Mitigation

From an economic perspective, the primary and ancillary benefits of climate mitigation policies present different geographical and temporal characteristics (Altemeyer-Bartscher et al. 2014). First, primary and ancillary benefits occur at different geographical scales. Climate change is a global negative externality, resulting from greenhouse gas (GHG) emissions produced locally and regionally (Rezai et al. 2012). Reducing emissions and stabilising the climate through the implementation of mitigation policies is a global public benefit, as every country and region benefits from emission reductions. Per contra, co-benefits of climate policies that reduce fossil fuel consumption, such as improved air quality and the corresponding health benefits, are enjoyed locally. Thus, while the primary benefits of climate mitigation generate a global public benefit, are enjoyed globally, ancillary benefits are a public domestic effect (Altemeyer-Bartscher et al. 2014).

From a temporal perspective, the occurrence of primary and ancillary benefits also differs. In the case of mitigation policies, the benefits of emission reduction can only be enjoyed in the mid- and long-term future as there is a lag of several decades between the time when emission reduction and climate stabilisation occur. This is due to the historical emission accumulation in the atmosphere and the thermal inertia of the climate system.

On the contrary, ancillary benefits may arise in the short term, even immediately after implementation. This would be the case, for example, of a climate-related transport policy in cities, that can improve air quality almost immediately after such policy is implemented. As explained by Altemeyer-Bartscher et al. (2014), ancillary benefits translate into a fast return of the investment, which in a context of great uncertainty increases the attractiveness of climate policies to decision makers.

The temporal occurrence of primary and ancillary benefits of adaptation is similar to that of emission reduction policies: primary benefits of adaptation will only materialise when future climate change impacts happen. A challenge that is specific to adaptation, which adds to the long-term nature of primary benefits, is the probabilistic nature of some of the impacts that make these benefits difficult to observe. In fact, the primary benefit of adaptation occurs in the future and *if* a certain event happens. In contrast, the ancillary benefits of adaptation take place even in the absence of climate change impacts (Tanner et al. 2016; Wilkinson 2012).

However, the different geographical features of primary and secondary benefits of climate mitigation do not apply in the same way to adaptation. First, adaptation is local by nature (Adger 2001), even if it often has multilevel implications and a wide range of public and private actors might be involved (Lemos and Agrawal 2006). Therefore, both primary and ancillary effects of adaptation will happen at local and regional scales, compared to the global primary effects of mitigation.

From a sectoral perspective, health benefits resulting from reduced air pollution are considered to be the most important type of mitigation co-benefits. This sector concentrates a large share of the academic and public interest (Markandya and Rübbelke 2004; see also Box 1). Other health-related co-benefits of climate mitigation include active transportation, such as cycling or walking (Gould and Rudolph 2015; Haines 2017), reduced morbidity and mortality due to extreme heat (Harlan and Ruddell 2011), or improved health due to shifts to healthy and sustainable diets (Aleksandrowicz et al. 2016).

Another important co-benefit of climate mitigation is biodiversity and ecosystem conservation and restoration, for example, via REDD+ (Reduction of Emissions by Deforestation and forest Degradation) initiatives. REDD+ primary benefit refers to emission reductions avoided deforestation and afforestation (Aaheim and Garcia 2014). Additionally to its mitigation function, REDD+ mechanisms provide a number of ancillary benefits, such as ecosystem and biodiversity protection, poverty reduction, improved rural livelihoods, enhanced governance and employment opportunities (Pasgaard et al. 2016).

Other co-benefits of climate mitigation include improved energy security and efficiency (Forrest et al. 2018; Monforti-Ferrario et al. 2018), higher household comfort in buildings (Ferreira and Almeida 2015), ensuring water availability (Kreye et al. 2014; Pittock et al. 2013; Zhou et al. 2018), reducing traffic noise and congestion (OECD 2000; Shukla and Dhar 2015), and innovation in agriculture (Bustamante et al. 2014).

Co-benefits of adaptation are addressed in detail in the following section.

Box 1 Search for Articles Considering Ancillary Benefits of Climate Adaptation

Objective: Identify articles in the Web of Knowledge focusing on ancillary benefits of climate mitigation and adaptation.

Search method: Systematic review following the approach by Berrang-Ford et al. (2015) in two steps. In Step 1, seven searches were done in the Web of Knowledge following the search criteria as defined in Table 1. An initial database of 1606 studies was obtained in all searches, including articles, books and chapters.

(continued)

Table 1 Inclusion criteria

ID	Search string	Number of documents	Dates
Search 1	Topic = climate change AND co-benefits	748	ALL
Search 2	TS = climate change AND ancillary	316	
Search 3	Topic = ("climate change" AND "adaptation" AND "co-benefits")	155	
Search 4	Topic = climate change AND co-benefits	84	
Search 5	Topic = climate change AND co-benefits	267	
Search 6	TS = ("climate change" AND "adaptation" AND "co-benefits")	10	
Search 7	TS = ("climate change" AND "adaptation" AND "ancillary")	26	

In Step 2, repeated studies were deleted in the different searches and the number of studies decreased to 840, including articles, books and chapters. Next, all document titles were revised, and abstracts when necessary in order to identify those directly focusing on mitigation or adaptation policies and their co-benefits. Some sectors addressed in these studies were also identified (health, air quality, energy, ecosystems).

Results: The final database includes 558 studies, 85% of which addressed climate mitigation (475 studies) and 15% adaptation policies (84 studies). Almost one third (30%) of all studies explicitly mentioned health benefits, and almost half of them referred to air quality. Energy policies are involved in 17% of the studies and there are 97 articles focused on the benefits for biodiversity or ecosystems.

Temporal distribution of the studies: Studies addressing ancillary benefits of climate policy have increased notably since 2015 (see next figure). Based on the evolution of studies, it is possible to identify three phases: the first, before 2008, with less than ten articles focusing on ancillary benefits, were published each year. Only two of these articles mention adaptation. In the second phase, from 2008 to 2014, there is an increase in the attention of the co-benefits of climate policy. In the third phase, from 2015 onwards the

(continued)

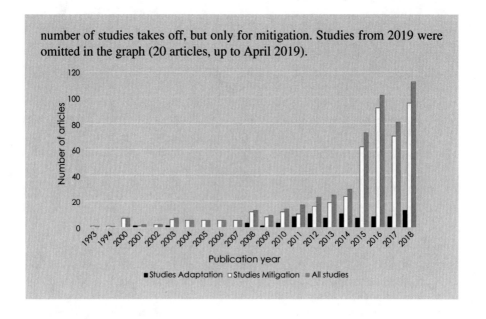

number of studies takes off, but only for mitigation. Studies from 2019 were omitted in the graph (20 articles, up to April 2019).

3 Ancillary Benefits of Adaptation

The primary benefit of adaptation is reducing the vulnerability to climate change but, just as for climate mitigation, adaptation policies can yield important ancillary benefits. In the IPCC's last assessment report, three areas in which ancillary benefits could appear were identified: (1) increasing current adaptive capacity to climate variability, without considering climate change; (2) creating opportunities for the development of new products and services (new insurance products, innovative technological solutions, etc.); and (3) contributing to sustainable development (Klein et al. 2014).

Another classification, sectoral oriented and with a bias towards natural resources, was proposed by Dovie (2019: 737): "(1) development-based natural resource management, (2) integrated water resources management, (3) sustainable agriculture, (4) ecosystem services, (5) biodiversity conservation and (6) bioenergy resource management".

Surminski and Tanner (2016) develop a framework to assess what they name "the triple dividend of resilience", in the context of disaster risk management (DRM). While this framework has been developed as a tool for making the case for DRM investments, the rationale can also be applied in the context of climate change adaptation. The first dividend of resilience is avoiding losses, the second dividend refers to "unlocking the economic potential" and the third dividend focuses on the economic, social and environmental co-benefits of DRM. From a climate change adaptation perspective, the first dividend can be understood as the primary objective of adaptation, i.e. reducing vulnerability and avoiding impacts (damages and losses). The second and third dividends, in contrast, represent co-benefits of adaptation and DRM. If the primary goal of DRM and adaptation is reducing vulnerability, then the economic, environmental and social effects are a secondary benefit of these policies, as analysed in this section.

3.1 The Economic Co-benefits of Adaptation

Adaptation to climate change can provide large direct economic benefits in terms of avoided damages and losses, whose mean net present value has been estimated globally in \$130–\$340 trillion in 2060 (Parry et al. 2009). Beyond direct benefits, adaptation measures offer additional economic co-benefits that, however, are often ignored. Accounting for the ancillary benefits of adaptation, thus, remains an important methodological challenge (Markandya and Watkiss 2009).

One of the economic co-benefits of climate change adaptation comes from *reducing the background risk* generated by climate-related extreme events. Background risk increases risk aversion in individuals, especially after disasters occur, reducing people's risky attitudes (e.g. Ahsan 2014; Cameron and Shah 2015). Risk aversion attitudes may have a negative effect on growth and economic development, due to lower investments in infrastructure, in new economic activities or innovation, as well as avoiding long-term planning (Hallegatte et al. 2016). Reducing background risk by implementing adaptation and disaster risk management policies has the potential to "unlock" entrepreneurship, stimulate long-term investments and innovation, ultimately boosting economic growth (Tanner et al. 2016).

Another important benefit of adaptation is that it can *reduce the present climate vulnerability*, especially of those groups and communities that are currently less prepared to face climate risks, both in developed and developing countries. This contributes to reducing the *adaptation deficit*, which is also closely linked to the *development deficit* (Parry et al. 2009; Preston et al. 2015). Current and future climate risk may also influence *credit ratings*. Rating agencies and investors are increasingly aware of climate change risks, so implementing climate policies can provide important financial co-benefits to firms and institutions at different levels (Rashidi et al. 2019).

Adaptation policies may open a number of opportunities to develop *new products and services*. The New Climate Economy has underlined the need to invest in climate-proof infrastructure, as the current global infrastructure gap requires investments of \$90 trillion by 2030, two-thirds of which concentrates in the global South (NCE 2016). Innovation and new technology may also play an important role. For example, Lobell et al. (2013) analysed the mitigation co-benefits of investing in agricultural adaptation to climate change. According to their results, adaptation in the agriculture sector could even be justified just in terms of the magnitude of the potential emission reduction (annual 0.35 Gt CO_2e) and its cost (\$15 per tonne CO_2e), without considering its primary economic benefits. With regards to risk transfer, there is evidence that climate adaptation may represent an opportunity for new insurance products under climate change (Surminski and Oramas-Dorta 2014).

3.2 Social Co-benefits of Adaptation

Climate adaptation strategies can contribute to improving *social capital*, building adaptive capacity, providing training, strengthening relations and creating networks to respond to climate change (Adger 2010; Pelling and High 2005; Tanner et al. 2016). There are very successful examples of involving grassroot organisations in the design of climate adaptation and disaster risk reduction actions, but quantifying such social co-benefits is challenging (Vorhies and Wilkinson 2016).

As in the case of mitigation, in the literature review, more articles were found addressing the *health* co-benefits of adaptation. Cheng and Berry (2013) identified three types of health-related ancillary benefits of adaptation. The first type refers, precisely, to improved social capital by, for example, building social networks, improving mental health and promoting healthy behaviours (Ebi et al. 2013; Eicher and Kawachi 2011). Increased public health is the second type of co-benefit. Adaptation strategies that provide better health support and infrastructures contribute to current public health. Emergency preparedness, climate-related surveillance and early warning alerts are good examples of this kind of ancillary benefit (Ebi et al. 2013; Harlan and Ruddell 2011).

The third type of health co-benefits of adaptation is related to regional and urban planning and design, such as bike and walkable paths or the increase in green space surface and shaded areas. Additionally to changing transport culture and reduce heat stress and its associated diseases, this type of activity promotes healthy practices physical activity, social interaction and mental health. A systematic review of health benefits of green spaces found that living in a greener environment translates in lower mortality rates and improved mental health (van den Berg et al. 2015). The World Health Organization (2017) reviewed the existing evidence of health benefits of urban green spaces and underlined the role of these for healthy, sustainable and liveable cities. Interestingly, co-benefits are delivered to all population, in particular to socioeconomically disadvantaged groups. From the ecosystem service literature, it is well known that other ancillary benefits of green spaces are related to recreation activities, aesthetic and spiritual values (Dickinson and Hobbs 2017). For many, urban green spaces can be the only places to enjoy nature, and contact with nature has been found to provide multiple benefits, including air quality, physical activity, social cohesion and stress reduction (Hartig et al. 2014).

3.3 Environmental Benefits of Adaptation

Ecosystem-based adaptation and the role of ecosystems in reducing climate risk and vulnerability has received increasing attention in recent years (Arkema et al. 2017;

Tanner et al. 2016). Ecosystem-based adaptation is related to the concept of green infrastructure, which has also attracted a large interest in recent years. In 2013, the European Commission approved its strategy for promoting green infrastructure, which explicitly recognises its links to adaptation and disaster risk reduction, as well as the synergies of these with mitigation and biodiversity conservation (European Commission 2013). The World Bank published a recent report on the opportunities of green infrastructure to face the challenges of this century, including climate change adaptation (Browder et al. 2019).

From an economic perspective, three principles guide the investment in green infrastructure: cost-effectiveness, the "precautionary principle" and the ancillary benefits of ecosystem-based adaptation (Kroeger et al. 2019). Due to its multiple co-benefits, the *cost-effectiveness* of ecosystem-based adaptation can compete with that of traditional infrastructure protection measures and often constitutes a more cost-effective choice (Ojea 2014). For example, a study analysing different hard and green infrastructure options (coastal wetlands, coral and oyster reefs) to prevent coastal risk in the Gulf of Mexico found that nature-based solutions were cost-effective and as competitive as traditional hard infrastructure measures, with benefit to cost ratios above 3.5 (Reguero et al. 2018). A study compared the effectiveness of green, grey and hybrid measures to protect a coastal facility of the company Dow in Texas. The results show that in this case, salt marsh ecosystems did not provide the same level of coastal protection than a levee to the company. However, in the long run, the ancillary benefits of salt marsh ecosystem services generated additional monetary co-benefits compared to the hard infrastructure option. In other words, the private benefits to Dow from ecosystem-based adaptation were limited; however, a mixed marsh-levee option was found to provide the highest private (Dow) and public benefits to the local community (Arkema et al. 2017). A systematic review of the evidence of nature-based solutions' effectiveness for adaptation was carried out by Doswald et al. (2014).

Ecosystem-based adaptation through the conservation and restoration of ecosystems is established on the *precautionary principle* that ecosystems in favourable conservation status are less vulnerable to climate change and other human pressures, and it acknowledges that natural systems have adapted successfully before (Kroeger et al. 2019; Wunder 2013).

Ecosystem-based adaptation can provide multiple environmental and social *co-benefits*, as seen in subsection 3.2, at the same time it reduces climate vulnerability. Nature-based solutions for adaptation yield benefits such as habitat conservation, prevention of erosion, improved water quality, but also mitigation (via carbon sequestration), poverty alleviation and improved livelihoods. Such co-benefits that are not usually produced when opting for hard infrastructures (Munang et al. 2013; Spalding et al. 2014).

Examples of the ancillary environmental co-benefits of adaptation are abundant in the literature. Dovie (2019) explains how increasing forest cover in steep landscapes that present high mudslide and flooding risk can increase resilience and improve adaptive capacity, while at the same time can contribute to biodiversity conservation and even carbon capture. Salt marshes can be considered a flexible, low-regret and low-cost adaptation option, as they have the capacity to accrete (increase their topographic height by capturing more sediment) as a response to certain levels of sea-level rise. Besides this primary effect, salt marshes offer additional co-benefits such as carbon sequestration, water purification, habitat for birds, plants and aquatic fauna or recreation (Sainz de Murieta 2016). Mangroves are considered a highly cost-effective measure to cope with coastal risks, while at the same time support biodiversity and local livelihoods (Marois and Mitsch 2015; McNally et al. 2011).

Many of the ancillary benefits of green spaces in urban areas are related to improved health outcomes as presented in the previous section, but urban nature-based solutions supply other environmental benefits: food security, recreation, water quality or groundwater recharge. Alves et al. (2018) present a comprehensive list of green infrastructure options, the related ecosystem services and potential co-benefits.

4 The Ancillary Costs of Adaptation

The ancillary effects of adaptation policies can be significant, as presented in the previous section. While our focus has been on the positive effect of these policies, they could also create negative externalities or ancillary costs (Chambwera et al. 2014). The concept of co-costs of adaptation is related to that of *maladaptation*, which refers to those actions designed to avoid or reduce vulnerability to climate change "that may lead to increased risk of adverse climate-related outcomes, increased vulnerability to climate change, or diminished welfare, now or in the future" (IPCC 2014).

Barnett and O'Neill (2010: 212) identified five groups of maladaptive actions. The first group encompasses those adaptation options that increase GHG emissions. Examples in this group are the (over)use of air conditioning to reduce exposure to extreme heat or insurance generating riskier behaviours (Surminski 2014).

The second group of maladaptation includes actions that increase the vulnerability of those social groups that are most at risk, by, for example, increasing risks in other (low-income) locations or causing changes in migration patterns (Vorhies and Wilkinson 2016). Adaptation options with high opportunity costs fall into the third group. These are actions that, compared to alternative options, translate into higher economic, environmental or social costs. In order to avoid this risk, Magnan et al. (2016) propose to conduct a cost-benefit analysis before selecting the portfolio of adaptation that would help identify the most cost-effective measures. The fourth group refers to maladaptive policies that reduce the incentive to adapt. The last group includes those adaptation actions that lead to path dependencies that could arise when implementing large infrastructures that are neither robust nor flexible. Failing to account for uncertainty could also lead to these path dependencies and generate missed opportunities for development (McDermott 2016).

In summary, together with estimating the co-benefits of climate change adaptation, potential ancillary costs must be considered in the design and implementation of adaptation policies in order to limit the risk of maladaptation (Magnan et al. 2016).

5 Conclusions

There is a broad consensus that mitigation and adaptation policies generate important ancillary benefits, both quantitatively and qualitatively which can play an important role to foster climate action. However, currently, the academic literature shows an important bias towards the co-benefits of mitigation policies, compared to adaptation. Historically, mitigation has concentrated much of the attention but now it is acknowledged that both, adaptation and mitigation, are needed to face the climate change challenge. Especially after the Paris Agreement was adopted, adaptation and mitigation policies "cannot be uncoupled" (Magnan and Ribera 2016).

Adaptation policies, additionally to reducing vulnerability to climate change, may provide many economic, social and environmental co-benefits. *Economic co-benefits* include: (1) reducing background risk, which can increase entrepreneurship, investments and ultimately economic growth; (2) increasing resilience to current climate variability, which is strongly related to reducing the development deficit; and (3) creating the opportunity space for innovation and the development of new goods and services.

The *social co-benefits* of adaptation include many examples related to human health, such as policies that contribute to improving public health, promote healthy behaviours or the many ancillary benefits of green spaces in urban areas. Recent studies also address the importance of the contribution of adaptation to building the social capital, by creating networks and strengthening community and social interrelations. Studies focused on green infrastructure and ecosystem-based adaptation provide many examples of the *environmental co-benefits* of adaptation, from coastal protection to securing water availability or food-production systems. In all cases, ancillary effects of adaptation must be considered exhaustively to avoid giving rise to negative effects (or co-costs).

In all these cases, co-benefits can be enjoyed in the short term, even immediately after implementation, while the primary benefits of adaptation might only be perceived in the long term or only in case a disaster occurs. Moreover, even if accounting for and quantitatively estimating the magnitude of adaptation co-benefits implies a number of challenges, considering for co-benefits can increase the *legitimacy* of adaptation processes, increase the willingness to adapt of decision makers and boost the implementation of adaptation policies.

Finally, and despite the wide agreement about the importance of ancillary benefits, there is little evidence about the extent to which they are being incorporated into decision-making processes. This could be a policy-relevant question to address in future research.

Acknowledgements This work is supported by the Basque Government through the BERC 2018–2021 programme and by the Spanish Ministry of Science, Innovation and Universities (MICINN) through BC3's María de Maeztu excellence accreditation MDM-2017-0714. E. Sainz de Murieta acknowledges funding from the Basque Government (grant no. POS_2018_2_0027).

References

Aaheim A, Garcia JH (2014) Synergies between adaptation and mitigation and the complexity of REDD+. In: Markandya A, Galarraga I, Sainz de Murieta E (eds) Routledge handbook of the economics of climate change adaptation, Routledge international handbooks. Routledge, Oxon, pp 79–96

Adger WN (2001) Scales of governance and environmental justice for adaptation and mitigation of climate change. J Int Dev 13:921–931

Adger WN (2010) Social capital, collective action, and adaptation to climate change. In: Voss M (ed) Der Klimawandel: Sozialwissenschaftliche Perspektiven. VS Verlag für Sozialwissenschaften, Wiesbaden, pp 327–345. https://doi.org/10.1007/978-3-531-92258-4_19

Adger WN, Arnell NW, Tompkins EL (2005) Successful adaptation to climate change across scales. Glob Environ Chang 15:77–86

Ahsan DA (2014) Does natural disaster influence people's risk preference and trust? An experiment from cyclone prone coast of Bangladesh. Int J Disaster Risk Reduct 9:48–57. https://doi.org/10.1016/j.ijdrr.2014.02.005

Aleksandrowicz L, Green R, Joy EJM, Smith P, Haines A (2016) The impacts of dietary change on greenhouse gas emissions, land use, water use, and health: a systematic review. PLoS One 11:e0165797. https://doi.org/10.1371/journal.pone.0165797

Altemeyer-Bartscher M, Markandya A, Rübbelke DTG (2014) International side-payments to improve global public good provision when transfers are refinanced through a tax on local and global externalities. Int Econ J 28:71–93. https://doi.org/10.1080/10168737.2012.759986

Alves A, Patiño Gómez J, Vojinovic Z, Sánchez A, Weesakul S (2018) Combining co-benefits and stakeholders perceptions into green infrastructure selection for flood risk reduction. Environments 5:29. https://doi.org/10.3390/environments5020029

Arkema KK, Griffin R, Maldonado S, Silver J, Suckale J, Guerry AD (2017) Linking social, ecological, and physical science to advance natural and nature-based protection for coastal communities: advancing protection for coastal communities. Ann N Y Acad Sci 1399:5–26. https://doi.org/10.1111/nyas.13322

Bain PG, Milfont TL, Kashima Y, Bilewicz M, Doron G, Garðarsdóttir RB, Gouveia VV, Guan Y, Johansson L-O, Pasquali C, Corral-Verdugo V, Aragones JI, Utsugi A, Demarque C, Otto S, Park J, Soland M, Steg L, González R, Lebedeva N, Madsen OJ, Wagner C, Akotia CS, Kurz T, Saiz JL, Schultz PW, Einarsdóttir G, Saviolidis NM (2016) Co-benefits of addressing climate change can motivate action around the world. Nat Clim Chang 6:154–157. https://doi.org/10.1038/nclimate2814

Barnett J, O'Neill S (2010) Editorial: maladaptation. Glob Environ Chang 20:211–213. https://doi.org/10.1016/j.gloenvcha.2009.11.004

Berrang-Ford L, Pearce T, Ford JD (2015) Systematic review approaches for climate change adaptation research. Reg Environ Chang 15:755–769. https://doi.org/10.1007/s10113-014-0708-7

Browder G, Ozment S, Rehberger Bescos I, Gartner T, Lange G-M (2019) Integrating green and grey: creating next generation infrastructure. World Bank and World Resources Institute, Washington, DC

Bustamante M, Robledo-Abad C, Harper R, Mbow C, Ravindranat NH, Sperling F, Haberl H, de Siqueira Pinto A, Smith P (2014) Co-benefits, trade-offs, barriers and policies for greenhouse gas mitigation in the agriculture, forestry and other land use (AFOLU) sector. Glob Chang Biol 20:3270–3290. https://doi.org/10.1111/gcb.12591

Cameron L, Shah M (2015) Risk-taking behavior in the wake of natural disasters. J Hum Resour 50:484–515. https://doi.org/10.3368/jhr.50.2.484

Chambwera M, Heal G, Dubeux C, Hallegatte S, Leclerc L, Markandya A, McCarl BA, Mechler R, Neumann JE (2014) Economics of adaptation. In: Field CB, Barros VR, Dokken DJ, Mach KJ, Mastrandrea MD, Bilir TE, Chatterjee M, Ebi KL, Estrada YO, Genova RC, Girma B, Kissel ES, Levy AN, MacCracken S, Mastrandrea PR, White LL (eds) Climate change 2014: impacts, adaptation, and vulnerability. Part a: global and sectoral aspects. Contribution of working group II to the fifth assessment report of the intergovernmental panel of climate change. Cambridge University Press, Cambridge, pp 945–977

Cheng JJ, Berry P (2013) Health co-benefits and risks of public health adaptation strategies to climate change: a review of current literature. Int J Public Health 58:305–311. https://doi.org/10.1007/s00038-012-0422-5

de Bruin K, Dellink RB, Ruijs A, Bolwidt L, van Buuren A, Graveland J, de Groot RS, Kuikman PJ, Reinhard S, Roetter RP, Tassone VC, Verhagen A, van Ierland EC (2009) Adapting to climate change in The Netherlands: an inventory of climate adaptation options and ranking of alternatives. Clim Chang 95:23–45. https://doi.org/10.1007/s10584-009-9576-4

Dickinson DC, Hobbs RJ (2017) Cultural ecosystem services: characteristics, challenges and lessons for urban green space research. Ecosyst Serv 25:179–194. https://doi.org/10.1016/j.ecoser.2017.04.014

Doswald N, Munroe R, Roe D, Giuliani A, Castelli I, Stephens J, Möller I, Spencer T, Vira B, Reid H (2014) Effectiveness of ecosystem-based approaches for adaptation: review of the evidence-base. Clim Dev 6:185–201. https://doi.org/10.1080/17565529.2013.867247

Dovie DBK (2019) Case for equity between Paris climate agreement's co-benefits and adaptation. Sci Total Environ 656:732–739. https://doi.org/10.1016/j.scitotenv.2018.11.333

Ebi KL, Berry P, Campbell-Lendrum D, Corvalan C, Guillemot J (2013) Protecting health from climate change: vulnerability and adaptation assessment. World Health Organization, Geneva

Eicher C, Kawachi I (2011) Social capital and community design. In: Dannenberg AL, Frumkin H, Jackson RJ (eds) Making healthy places: designing and building for health, well-being, and sustainability. Island Press/Center for Resource Economics, Washington, DC, pp 117–128. https://doi.org/10.5822/978-1-61091-036-1_8

European Commission (2013) Green infrastructure strategy—enhancing Europe's natural capital (no. COM(2013) 249 final). European Commission, Brussels

Ferreira M, Almeida M (2015) Benefits from energy related building renovation beyond costs, energy and emissions. Energy Procedia 78:2397–2402. https://doi.org/10.1016/j.egypro.2015.11.199

Forrest K, Tarroja B, Chiang F, AghaKouchak A, Samuelsen S (2018) Assessing climate change impacts on California hydropower generation and ancillary services provision. Clim Chang 151:395–412. https://doi.org/10.1007/s10584-018-2329-5

Gould S, Rudolph L (2015) Challenges and opportunities for advancing work on climate change and public health. Int J Environ Res Public Health 12:15649. https://doi.org/10.3390/ijerph121215010

Haines A (2017) Health co-benefits of climate action. Lancet Planet Health 1:e4–e5. https://doi.org/10.1016/S2542-5196(17)30003-7

Hallegatte S (2009) Strategies to adapt to an uncertain climate change. Glob Environ Chang 19:240–247

Hallegatte S, Bangalore M, Jouanjean M-A (2016) Avoided losses and the development dividend of resilience. In: Surminski S, Tanner T (eds) Realising the "triple dividend of resilience". Springer, Berlin, pp 31–54

Harlan SL, Ruddell DM (2011) Climate change and health in cities: impacts of heat and air pollution and potential co-benefits from mitigation and adaptation. Curr Opin Environ Sustain 3:126–134. https://doi.org/10.1016/j.cosust.2011.01.001

Hartig T, Mitchell R, de Vries S, Frumkin H (2014) Nature and health. Annu Rev Public Health 35:207–228. https://doi.org/10.1146/annurev-publhealth-032013-182443

IPCC (2014) Annex II: glossary. In: Barros VR, Field CB, Dokken DJ, Mastrandrea MD, Mach KJ, Bilir TE, Chatterjee M, Ebi KL, Estrada YO, Genova RC, Girma B, Kissel ES, Levy AN, MacCracken S, Mastrandrea PR, White LL (eds) Climate change 2014: impacts, adaptation, and vulnerability. Part B: regional aspects. Contribution of working group II to the fifth assessment report of the Intergovernmental Panel on Climate Change. Cambridge University Press, Cambridge, pp 1757–1776

Klein RJT, Midgley BL, Preston BL, Alam M, Berkhout FGH, Dow K, Shaw MR (2014) Adaptation opportunities, constraints, and limits. In: Field CB, Barros VR, Dokken DJ, Mach KJ, Mastrandrea MD, Bilir TE, Chatterjee M, Ebi KL, Estrada YO, Genova RC, Girma B, Kissel ES, Levy AN, MacCracken S, Mastrandea PR, White LL (eds) Climate change 2014: impacts, adaptation, and vulnerability. Part A: Global and sectoral aspects. Contribution of working group II to the fifth assessment report of the Intergovernmental Panel on Climate Change. Cambridge University Press, Cambridge, pp 899–943

Kreye M, Adams D, Escobedo F (2014) The value of forest conservation for water quality protection. Forests 5:862–884. https://doi.org/10.3390/f5050862

Kroeger T, Klemz C, Boucher T, Fisher JRB, Acosta E, Cavassani AT, Dennedy-Frank PJ, Garbossa L, Blainski E, Santos RC, Giberti S, Petry P, Shemie D, Dacol K (2019) Returns on investment in watershed conservation: application of a best practices analytical framework to the Rio Camboriú water producer program, Santa Catarina, Brazil. Sci Total Environ 657:1368–1381. https://doi.org/10.1016/j.scitotenv.2018.12.116

Krook Riekkola A, Ahlgren EO, Söderholm P (2011) Ancillary benefits of climate policy in a small open economy: the case of Sweden. Energy Policy 39:4985–4998. https://doi.org/10.1016/j.enpol.2011.06.015

Lemos MC, Agrawal A (2006) Environmental governance. Annu Rev Environ Resour 31:297–325. https://doi.org/10.1146/annurev.energy.31.042605.135621

Lobell DB, Baldos ULC, Hertel TW (2013) Climate adaptation as mitigation: the case of agricultural investments. Environ Res Lett 8:015012. https://doi.org/10.1088/1748-9326/8/1/015012

Longo A, Hoyos D, Markandya A (2012) Willingness to pay for ancillary benefits of climate change mitigation. Environ Resour Econ 51:119–140. https://doi.org/10.1007/s10640-011-9491-9

Magnan AK, Ribera T (2016) Global adaptation after Paris. Science 352:1280–1282. https://doi.org/10.1126/science.aaf5002

Magnan AK, Schipper ELF, Burkett M, Bharwani S, Burton I, Eriksen S, Gemenne F, Schaar J, Ziervogel G (2016) Addressing the risk of maladaptation to climate change: addressing the risk of maladaptation to climate change. Wiley Interdiscip Rev Clim Chang 7:646–665. https://doi.org/10.1002/wcc.409

Markandya A, Rübbelke DTG (2004) Ancillary benefits of climate policy/Sekundäre Nutzen der Klimapolitik. Jahrbücher für Nationalökonomie und Statistik 224:488–503. https://doi.org/10.1515/jbnst-2004-0406

Markandya A, Watkiss P (2009) Potential costs and benefits of adaptation options: a review of existing literature. UNFCCC Technical Paper. FCCC/TP/2009/2 80

Marois DE, Mitsch WJ (2015) Coastal protection from tsunamis and cyclones provided by mangrove wetlands – a review. IntJ Biodivers Sci Ecosyst Serv Manage 11:71–83. https://doi.org/10.1080/21513732.2014.997292

Masson-Delmotte V, Zhai P, Pörtner HO, Roberts D, Skea J, Shukla PR, Pirani A, Moufouma-Okia W, Péan C, Pidcock R, Connors S, Matthews JBR, Chen Y, Zhou X, Gomis MI, Lonnoy E, Maycock T, Tignor M, Waterfield T (eds) (2018) Global warming of 1.5°C. An IPCC special report on the impacts of global warming of 1.5°C above pre-industrial levels and related global greenhouse gas emission pathways, in the context of strengthening the global response to the threat of climate change, sustainable development, and efforts to eradicate poverty. World Meteorological Organization, Geneva, Switzerland

McDermott TKJ (2016) Investing in disaster risk management in an uncertain climate. In: Surminski S, Tanner T (eds) Realising the "triple dividend of resilience". Springer, Berlin, pp 129–150

McNally CG, Uchida E, Gold AJ (2011) The effect of a protected area on the tradeoffs between short-run and long-run benefits from mangrove ecosystems. PNAS 108:13945–13950. https://doi.org/10.1073/pnas.1101825108

Monforti-Ferrario F, Kona A, Peduzzi E, Pernigotti D, Pisoni E (2018) The impact on air quality of energy saving measures in the major cities signatories of the covenant of mayors initiative. Environ Int 118:222–234. https://doi.org/10.1016/j.envint.2018.06.001

Munang R, Thiaw I, Alverson K, Mumba M, Liu J, Rivington M (2013) Climate change and ecosystem-based adaptation: a new pragmatic approach to buffering climate change impacts. Curr Opin Environ Sustain 5:67–71. https://doi.org/10.1016/j.cosust.2012.12.001

NCE (2016) The sustainable infrastructure imperative: financing for better growth and development. The New Climate Economy, London

OECD (ed) (2000) Ancillary benefits and costs of greenhouse gas mitigation, environment. OECD, Paris

OECD (2016) The economic consequences of outdoor air pollution. OECD, Paris

Ojea E (2014) Ecosystem based adaptation. In: Markandya A, Galarraga I, Sainz de Murieta E (eds) Routledge handbook of the economics of climate change adaptation (hardback) – Routledge. Routledge, London

Parry M, Arnell N, Berry P, Dodman D, Fankhauser S, Hope C, Kovats S, Nicholls R, Satterthwhite D, Tiffin R, Wheeler T (2009) Assessing the costs of adaptation to climate change: a review of the UNFCCC and other recent estimates. International Institute for Environment and Development and Grantham Institute for Climate Change, London

Pasgaard M, Sun Z, Müller D, Mertz O (2016) Challenges and opportunities for REDD+: a reality check from perspectives of effectiveness, efficiency and equity. Environ Sci Pol 63:161–169. https://doi.org/10.1016/j.envsci.2016.05.021

Pearce D (2000) Policy framework for the ancillary benefits of climate change policies. In: OECD (ed) Ancillary benefits and costs of greenhouse gas mitigation, environment. OECD, Paris, pp 517–560

Pelling M, High C (2005) Understanding adaptation: what can social capital offer assessments of adaptive capacity? Glob Environ Chang 15:308–319. https://doi.org/10.1016/j.gloenvcha.2005.02.001

Pittel K, Rübbelke DTG (2008) Climate policy and ancillary benefits: a survey and integration into the modelling of international negotiations on climate change. Ecol Econ 68:210–220. https://doi.org/10.1016/j.ecolecon.2008.02.020

Pittock J, Hussey K, McGlennon S (2013) Australian climate, energy and water policies: conflicts and synergies. Aust Geogr 44:3–22. https://doi.org/10.1080/00049182.2013.765345

Preston BL, Mustelin J, Maloney MC (2015) Climate adaptation heuristics and the science/policy divide. Mitig Adapt Strateg Glob Change 20:467–497. https://doi.org/10.1007/s11027-013-9503-x

Rashidi K, Stadelmann M, Patt A (2019) Creditworthiness and climate: identifying a hidden financial co-benefit of municipal climate adaptation and mitigation policies. Energy Res Soc Sci 48:131–138. https://doi.org/10.1016/j.erss.2018.09.021

Reguero BG, Beck MW, Bresch DN, Calil J, Meliane I (2018) Comparing the cost effectiveness of nature-based and coastal adaptation: a case study from the Gulf coast of the United States. PLoS One 13:e0192132. https://doi.org/10.1371/journal.pone.0192132

Rezai A, Foley DK, Taylor L (2012) Global warming and economic externalities. Econ Theory 49:329–351. https://doi.org/10.1007/s00199-010-0592-4

Rodríguez-Entrena M, Espinosa-Goded M, Barreiro-Hurlé J (2014) The role of ancillary benefits on the value of agricultural soils carbon sequestration programmes: evidence from a latent class approach to Andalusian olive groves. Ecol Econ 99:63–73. https://doi.org/10.1016/j.ecolecon.2014.01.006

Sainz de Murieta E (2016) In: University of the Basque Country (ed) Environmental and economic impacts of sea-level rise on the Basque coast (PhD dissertation), Leioa

Sainz de Murieta E, Galarraga I, Markandya A (2014) An introduction to the economics of adaptation to climate change. In: Markandya A, Galarraga I, Sainz de Murieta E (eds)

Routledge handbook of the economics of climate change adaptation, Routledge international handbooks. Routledge, New York, pp 3–26

Schucht S, Colette A, Rao S, Holland M, Schöpp W, Kolp P, Klimont Z, Bessagnet B, Szopa S, Vautard R, Brignon J-M, Rouïl L (2015) Moving towards ambitious climate policies: monetised health benefits from improved air quality could offset mitigation costs in Europe. Environ Sci Pol 50:252–269. https://doi.org/10.1016/j.envsci.2015.03.001

Shukla PR, Dhar S (2015) Energy policies for low carbon sustainable transport in Asia. Energy Policy 81:170–175. https://doi.org/10.1016/j.enpol.2015.02.021

Spalding MD, Ruffo S, Lacambra C, Meliane I, Hale LZ, Shepard CC, Beck MW (2014) The role of ecosystems in coastal protection: adapting to climate change and coastal hazards. Ocean Coast Manag 90:50–57. https://doi.org/10.1016/j.ocecoaman.2013.09.007

Surminski S (2014) The role of insurance in reducing direct risk-the case of flood insurance. IRERE 7:241–278. https://doi.org/10.1561/101.00000062

Surminski S, Oramas-Dorta D (2014) Flood insurance schemes and climate adaptation in developing countries. Int J Disaster Risk Reduct 7:154–164. https://doi.org/10.1016/j.ijdrr.2013.10.005

Surminski S, Tanner T (2016) Realising the "triple dividend of resilience". Springer, Berlin

Tanner T, Surminski S, Wilkinson E, Reid R, Rentschler J, Rajput S, Lovell E (2016) In: Surminski S, Tanner T (eds) The triple dividend of resilience: a new narrative for disaster risk management and development. Springer, Berlin, pp 1–30. Realising the "triple dividend of resilience"

UNFCCC (2011) Assessing the costs and benefits of adaptation options: an overview of approaches. United Nations framework convention on climate change. UNFCCC, Bonn

van den Berg M, Wendel-Vos W, van Poppel M, Kemper H, van Mechelen W, Maas J (2015) Health benefits of green spaces in the living environment: a systematic review of epidemiological studies. Urban For Urban Green 14:806–816. https://doi.org/10.1016/j.ufug.2015.07.008

Vorhies F, Wilkinson E (2016) In: Surminski S, Tanner T (eds) Co-benefits of disaster risk management: the third dividend of resilience. Springer, Berlin, pp 55–72. Realising the "triple dividend of resilience"

WHO (2017) Urban green space interventions and health: a review of impacts and effectiveness. World Health Organization, Copenhagen

Wilkinson E (2012) Transforming disaster risk management: a political economy approach, working and discussion papers. ODI, London

Wunder S (2013) When payments for environmental services will work for conservation. Conserv Lett 6:230–237. https://doi.org/10.1111/conl.12034

Zhou Y, Ma M, Kong F, Wang K, Bi J (2018) Capturing the co-benefits of energy efficiency in China – a perspective from the water-energy nexus. Resour Conserv Recycl 132:93–101. https://doi.org/10.1016/j.resconrec.2018.01.019

Economic Assessment of Co-benefits of Adaptation to Climate Change

Christiane Reif and Daniel Osberghaus

1 Introduction

Especially the last reports by the IPCC clearly show that climate change is already happening and cannot be avoided entirely even with the strongest mitigation efforts (IPCC 2013, 2014a). Thus, adaptation is necessary to reduce climate damage at least to a certain extent, as already emphasized by Pielke (1998). In general, adaptation is the adjustment to "actual or expected climate and its effects" to "moderate or avoid harm or exploit beneficial opportunities" (see, e.g., IPCC 2014b, p. 1251). Hence, the benefits usually expected from adaptation are the reduction of climate-induced adverse effects and the exploitation of climate-induced opportunities. Some adaptation strategies provide additional benefits not covered by this definition. From an economic point of view, such side effects are externalities, which can have a positive but also a negative impact on other unrelated areas that are not internalized (Cornes and Sandler 1986). Of particular interest are those adaptation measures which achieve the effect desired in terms of climate adjustment and at the same time create benefits in areas unrelated to climate change (De Bruin et al. 2009). However, as Hallegatte (2009) emphasizes, adaptation measures might bring about negative impacts on other areas. We address these synergies and conflicts in five different fields. We soften the narrow definition of externalities in "unrelated areas," as policy domains might overlap spatially, temporally, or responsibility-wise. In the following sections, we discuss the linkage between adaptation and

---

C. Reif (✉)
Center for Economics of Materials (CEM), Halle, Germany
e-mail: christiane.reif@imws.fraunhofer.de

D. Osberghaus
ZEW – Leibniz Centre for European Economic Research, Mannheim, Germany
e-mail: osberghaus@zew.de

© Springer Nature Switzerland AG 2020
W. Buchholz et al. (eds.), *Ancillary Benefits of Climate Policy*, Springer Climate,
https://doi.org/10.1007/978-3-030-30978-7_11

mitigation, ecological systems, economic development, acquiring knowledge for decisions under uncertainty, and disaster resilience.

2 Adaptation and Mitigation: Synergies and Conflicts

The discussion on an integrated climate policy defining an ideal mix of mitigation and adaptation is an ongoing debate (e.g., Klein et al. 2005). The difficulty of such a policy is that climate action is complex (Berry et al. 2015). Mitigation is the reduction of greenhouse gases. While its main aim is to reduce the production of CO_2, it might also include carbon sinks, and carbon capture and storage (see, e.g., Zhao et al. 2018). Adaptation encompasses measures to reduce current or expected impacts of climate change. These definitions suggest that adaptation and mitigation are two different actions of climate policy. Thus, a policy might pursue one of the three main strategies: only mitigate, only adapt, or a mixture of both. This raises the question in how far the dichotomy of policies discussed in the literature (Klein et al. 2007; Biesbroek et al. 2009) can be resolved through an integrated policy. Adaptation and mitigation might co-benefit or conflict with each other.

This complementary or substitute character of adaptation and mitigation is discussed in literature (e.g., Ingham et al. 2013; Osberghaus and Baccianti 2013; Watkiss et al. 2015; Yohe and Strzepek 2007) and politics. One can argue that both actions reduce the impacts of climate change, albeit in different ways, and are therefore policy substitutes (e.g., Tol 2005; Buob and Stephan 2011). Dowlatabadi (2007) distinguishes between mitigation, i.e., addressing the drivers, and adaptation, i.e., confronting the impacts of climate change. Complementarity generally remains difficult to define, as the review of Samuelson (1974) shows, and it is the same with defining the link between adaptation and mitigation. There are various approaches with different bases of comparison. Investment costs are most commonly used: (1) Mitigation and adaptation are substitutes, if an increase of mitigation costs leads to an increased demand in adaptation or vice versa; (2) They are complements if they have a joint demand. In this case, a decrease in mitigation (adaptation) costs leads to an increased demand for mitigation and adaptation. This higher demand for adaptation (mitigation) has been attributed to increased marginal productivity (Osberghaus and Baccianti 2013). There is support for both of these perspectives: Tol (2005) argues that mitigation and adaptation compete on resources, which is indicative of a substitutive character. If the costs of one measure increase, it is replaced by the other (Buob and Stephan 2011; Kane and Yohe 2000). This argumentation is based on costs and the assumption that both measures reduce the impact of climate change. Furthermore, investing in one measure might positively influence the other, and hence they reinforce each other (Wilbanks et al. 2007). Put differently, without mitigation efforts, climate change might cross thresholds so that climate change impact is either not addressable with the taken adaptation measures or becomes excessively expensive (see, e.g., Parry 2009). Yohe and Strzepek (2007) view mitigation and adaptation as two complementary angles of risk reduction with mitigation decreasing risk exposure and adaptation lowering the impact. Hamin and Gurran (2009) on the one hand acknowledge that mitigation and adaption policy strategies might be complementary, but on the other hand emphasize that the strategies' goals might differ. The examples presented show us that viewpoints on the linkage between mitigation and adaptation diverge. Thus, mitigation and adaptation are not easily interchangeable. However, no matter whether they have a substitutive or complementary character, they need to be analyzed together to provide policy advice in both cases.

Cost–benefit analysis (CBA) might give guidance on climate policy. However, researchers intensively discuss CBA's limits due to assumptions and data restrictions or the uncertainty about mitigation's potential extent of success (see, e.g., Tol (2005) or van den Bergh (2010) for an evaluation of CBAs). Based on an earth system model, Lowe et al. (2009) and Parry et al. (2009) show that adaptation costs increase steeply or even exponentially with climate change. Tol (2005) points out that a joint analysis is only reasonable for mitigation and facilitative adaptation as decisions on these two are taken on the same political level, i.e., the national one. Another possibility to make an informed decision is integrated assessment models (IAM) with endogenous mitigation and adaptation. Bosello et al. (2013) conclude from the studies based on this approach that mitigation and adaptation are strategic complements. They provide further evidence for this conclusion in their study and specify that the stabilization of emissions can be achieved at a lower cost if adaptation supports mitigation efforts. Warren (2011) stresses that these models do not cover all possible interactions, as for example, within and between sectors or different spatial levels. Weitzman (2009) also discusses the limitations of CBAs and IAMs especially in reflecting catastrophic climate change. He provides a conceptual framework to integrate this aspect in the analyses. However, he points out that even though these analyses are informative, precise predictions are not possible due to uncertainties.

The limits of these different analytical approaches make it difficult to develop an optimal response strategy to climate change. Furthermore, there are differences between mitigation and adaptation in terms of *public or private good character, temporal scale, spatial and political level*. These present a particular challenge to comparing and interrelating actions and thus also to finding the right policy strategy (see, e.g., Berry et al. 2015; Biesbroek et al. 2009; Tol 2005; Wilbanks et al. 2007). In the following, we elaborate on the interlinkages of adaptation and mitigation along these dimensions by summarizing the current approaches in research and politics. As such, we detach our review from a specific definition of complementarity.

2.1 Public Versus Private Good

Mitigation and adaptation often differ in their economic character of the good. In general, mitigation effort is characterized by non-excludability and non-rivalry. The mitigation effort of one actor benefits others, leading to globally better climate conditions, which anybody enjoys at the same time. Thus, mitigation is a global public good associated with a social dilemma situation as the social desire to protect the climate is up against the incentives for individual action to benefit from others' mitigation efforts. In contrast, adaptation measures could represent private or (local) public good characteristics. Individuals' adaptation efforts like insurance or retrofitting measures constitute a private good, as it is only beneficial to the individual itself. Those measures benefiting a group of individuals, like coastal protection, generally form a local public good (see, e.g., Aakre et al. 2010). However, the categorization might also depend on the perspective: For the local government coastal protection is a (local) public good, from a superordinate level such measures could be seen as private good of the specific country or region. As such, different motivational aspects come along with these public or private good characteristics. If adaptation is a private good and we assume a substitutive link of adaptation and mitigation, only investment in adaptation is rational as it is for the private benefit and others cannot exploit own efforts.

However, perspective matters, as the example of afforestation shows, which contributes to mitigation and adaptation goals alike: Forests serve as CO_2-sinks and thus represent a global public good. At the same time, forests help to protect against land- or mudslides in the respective region and therefore are local public goods. Additionally, forests can contribute to the private benefit when harvested or used as recreational area with entrance fees. As such, one measure could jointly produce different types of goods and serve mitigation and adaptation at the same time.

2.2 Temporal Scale

Regarding the time horizon, the point in time of action and effect need to be considered. Mitigation measures which are being currently taken will become effective in the mid- and long-term future (Biesbroek et al. 2009). Concerning adaptation, the time scale of effort and effect depends on the specific measure (Biesbroek et al. 2009). The classification by Fankhauser (1998) supports researchers and policymakers in the distinction of adaptation measures. The most prominent differentiation is the one between reactive and proactive adaptation. Reactive measures are a direct response to the consequences of climate change. They have a rather short-term perspective and often encompass ad hoc support like disaster aid. Proactive measures, in contrast, are taken in the present to reduce expected damage in the future. They include insurance as a private good, but also (local) public goods like flood protection or the setup of disaster management capacities. Fankhauser (1998) further differentiates between adjustment measures and capacity building. Especially capacity building is nowadays a popular subject when it comes to the integration of mitigation and adaptation in the context of sustainable development (see also Sect. 4). These proactive middle- or long-term measures are future oriented and therefore it is uncertain if they will ever pay off.

If mitigation and reactive adaptation are interchangeable, then a decision for either one of these actions also implies a decision on the point in time of action. Proactive adaptation and mitigation might have the same time period. In this case, competition on financial resources is likely when the same political level is responsible for both actions. If the proactive adaptation measure is a private good like insurance, then mitigation efforts co-benefit insurance cost by reducing the risk of an impact.

Another temporal aspect relates to ethical considerations. If we assume adaptation and mitigation to be perfect substitutes, is it fair—intergenerationally speaking—to mainly employ adaptation policies? Delaying mitigation efforts implies that severe climate change impacts become more likely. Although one might argue that the money intended for mitigation efforts could be invested in innovation so that new technologies make it easy to adapt in the future, it is subject to many uncertainties whether this will be sufficient to reduce climate change impacts in the future. Van den Berg (2010) points out that especially the combination of uncertainty and expected extreme events should spur us to stabilize emissions and focus on mitigation. Beside these intergenerational aspects, especially poorer countries strongly depending on the agricultural sector will suffer the most from climate impacts. This also leads to the question as to how the burden of adaptation and mitigation should be shared spatially.

2.3 Spatial and Political Level

On a more detailed *spatial and political level*, spatial planning especially in cities seems to be a promising way to harmonize mitigation, adaptation, and sustainable development objectives (Biesbroek et al. 2009). However, Biesbroek et al. (2009), Laukkonen et al. (2009), and Walsh et al. (2011) stress that the possibly different objectives of vertical policy levels and the related legal responsibilities might become an issue. Furthermore, Moser (2012) also cautions against over-harmonizing policies and ignoring that some policies have a value of their own. As such, spatial planning on the local level refers not only to the initial question on the optimal mixture of adaptation and mitigation policy, but also asks if actions should be initiated top-down or bottom-up. The motivation to take the first step is related to legal responsibilities but also to financial and knowledge capacities. Especially smaller cities or cities in developing countries might not be able to bear the costs neither for adaptation, mitigation, nor for integrated spatial planning. This in turn draws attention to fairness considerations and raises the question of whether transfers are needed to enable the players on specific spatial levels to act. Green funds for adaptation capacity building integrated in the Paris Agreement are one step in this direction. They come with the side effect to nudge beneficiary countries into investing in mitigation. If this leads to an efficient climate policy remains to be seen.

All these characteristics of mitigation and adaptation show the difficulty to generally define and determine whether mitigation and adaptation are complements or substitutes, and if they are co-benefiting or conflicting each other. This calls for a closer look on specific activities to examine whether adaptation and mitigation stimulate or hamper each other on the measure level. Bedsworth and Hanek (2008), and Taylor et al. (2007) provide examples of measures enhancing each other. Measures to store carbon, like re/afforestation, at the same time enhance adaptation or constitute a joint production, protecting humans, species, or nature [see Moser (2012) for a summary on these measures]. Although Moser (2012) recognizes that there are areas in which the synergies of adaptation and mitigation are present, she also gives an overview on areas in which they conflict with each other. For that purpose, she expands the purely financial constraints and differentiates between two types of trade-offs: The first one subsumes the availability of (non-)financial resources and conditions like the legal framework or political enforceability. The second type refers to unwanted environmental, social, and political outcomes. Another perspective taken by Berry et al. (2015) is to analyze interactions horizontally at the sectoral level of measures within and between sectors. They find synergies between sectors in the area of biodiversity or water and conflicts in terms of water quantity and quality. This sectoral viewpoint calls for an ecosystem-integrated climate policy like Warren (2011) already emphasized, which we will further discuss in the next section.

3 Adaptation and Environment: Enhancing Ecological Systems

Since the Millennium Ecosystem Assessment (MA) began its work in 2001, knowledge has been gained on the considerable advances of humanity's depletion of natural resources and on promising actions to reserve the degradation of ecosystems (see Millennium Ecosystem Assessment 2005). While the exploitation of ecosystems is typically private and short-term oriented, the conservation of ecosystems often meets long-term ecological and economic needs of the society as a whole (Turner and Daily 2008). Although the maintenance of ecosystems is challenging, institutions and researchers (e.g., Turner and Daily 2008; Pramova et al. 2012) consider ecosystem services as a promising approach to improve climate resilience of inherently climate vulnerable societies.

Due to climate change, urban areas face more frequent and severe flooding, drought, and heat stress (e.g., Voskamp and Van de Ven 2015). Adaptation measures may support a healthier environment and in turn improve individuals' health (Harlan and Ruddell 2011). For example, green spaces in cities help to prepare against heat stress and at the same time create benefits to the environment and humans apart from climate impacts. However, current urban planning often does not sufficiently consider climate and ecosystem aspects. Voskamp and Van de Ven (2015) promote a radical change in urban planning, which shifts away from pure construction measures to adapt to climate change toward joint measures that incorporate green infrastructure and water management with self-adaptive capacities. At the rural level, adaptation strategies like land or water management co-benefit the ecosystem in the conservation of biodiversity. For example, the protection of a forest might be due to adaptation reasons, i.e., to protect soil, and at the same time the forest as an ecological system is enhanced. Forests represent an ecosystem providing economic and recreational capacities and they contribute to regulating climate at the same time. As such, the environmental co-benefits of adaptation often intersect with co-benefits related to climate change mitigation presented in the previous section. Although this might result in a positive reinforcement of the measures, it has also negative implications, as for example, higher complexity due to the involvement of different policy levels, and unclear sharing of competences, responsibilities, and costs.

Furthermore, ecosystems need to naturally adapt to climate changes and their impacts (e.g., Klein et al. 2005; Goklany 2007), which takes time. This calls for an integrated policy framework of mitigation and adaptation, as discussed above, to buy time for a smooth adaptation rather than preserving an ecosystem at the current status at all costs. Spencer et al. (2017) point out that such a multibenefiting policy needs a thorough design with multidimensional incentives and communication. First attempts to incorporate ecosystem conservation into the adaptation strategies can be found in National Adaptation Programmes of Action (NAPAs). In their review of the proposed projects, Pramova et al. (2012) find only a small share of projects explicitly including ecosystem activities. They conclude that NAPAs have a strong potential for an integrated policy framework but technical, political, and financial resources are needed for further promotion.

While there are opportunities to adapt and enhance ecological systems at the same time, adaptation activities can also endanger and erode ecological systems. These negative side effects are one form of maladaptation (e.g., Scheraga and Grambsch 1998; McCarthy et al. 2001; Barnett and O'Neill 2010). The different definitions of this term have in common that they generally describe adaptation measures that unintentionally increase vulnerability to climate change. It might directly or indirectly affect adaptive capacities immediately or in the long run. Barnett and O'Neill (2010) identify five main dimensions of maladaptation including raising emissions, burdening the most vulnerable, high opportunity costs, decreasing further adaptation efforts, and limiting future options. An example for the very first is the increased usage of air conditioners during heat waves. A decrease in private adaptation measures might occur when the government takes adaptation actions like building a dike, and the incentives to privately insure decrease (Grothmann and Reusswig 2006; Hanger et al. 2018). The different dangers of adaptation measures directly draw up a framework to combat maladaptation in evaluating measures according to these possible side effects. Another way proposed by Hallegatte (2009) is different adaptation strategies like no-regret and reversible strategies, discussed in Sect. 5.

Either way needs thorough and detailed assessment of actions in particular to evaluate not only direct and immediate impacts but also future effects to ensure at least nonnegative side effects of adaptation measures and in the best case taking those actions benefiting ecosystems as well.

4 Adaptation and Economic Development: Using Climate Change as a Window of Opportunity

Economic development may conflict with mitigation targets (e.g., Chuku 2010). Moreover, the countries most affected by adverse economic impacts of climate change are developing countries (e.g., Parry et al. 2007; Furlow et al. 2011). The 17 sustainable development goals demonstrate that we face several global challenges beside climate change. Hence, policies which tackle several problems at the same time are key to cope with the sustainable development challenge. While mitigation might hamper economic development, adaptation to climate change has the potential to support economic development.

The Paris Agreement explicitly points out the role of sustainable development to reduce loss and damage associated with climate change (Art. 8) and additionally international cooperation to strengthen adaptation (Art. 7). This increased awareness of adaptation to climate change may be used to foster programs in sustainable development that have important economic co-benefits beside their adaptation benefits. Furthermore, the implemented adaptation funds help to introduce such co-benefiting programs by providing additional financial resources to developing countries. Moreover, it might allow to design projects in a way to jointly meet sustainable development goals and adaptation targets (e.g., Chuku 2010; Furlow et al. 2011). For example, planning new water management systems should ensure access to clean water and at the same time increase climate resilience. As such, climate change awareness and various adaptation funds provided in the climate change context could create a window of opportunity to support existing or start new programs related to sustainable development goals. This is especially true for developing countries and also partly applies to developed countries. In this perspective, Aakre and Rübbelke (2010) provide a framework for the selection of EU adaptation measures considering efficiency and equity considerations as well.

However, ancillary benefits might not be realized due to three aspects. First, it is not certain if adaptation funds are really additional financial resources or money is just shifted from development aid to adaptation (e.g., Furlow et al. 2011). As such, adaptation measures might still co-benefit other areas but at the same time hamper those measures where the money has been withdrawn. Additionally, development might positively or negatively influence vulnerability (Tol 2005). In turn, these conflicts of interest—development, mitigation, adaptation—might hamper investments. Second, the particular attention to adaptation measures might result in an inefficient allocation of resources. If adaptation is more recognized and popular than other projects, resources could rather be spent on adaptation than on other—perhaps more efficient—projects. This is also strongly related to the third aspect of windfall gains. Projects might be initiated to (primarily) gain profit from the newly introduced funding lines so that adaptation targets might recede into the background. As such, the projects might not even be effective. Furthermore, windfall profits might also lead to financing projects which would have been implemented anyway. Thus, the resources would have been used more beneficially otherwise. As such, an integrated policy with climate-resilient development measures seems the most efficient strategy (Chuku 2010; Furlow et al. 2011). Chuku (2010) provides four criteria such a policy needs to fulfill: long-term environmental effectiveness, equity considerations, cost effectiveness, and the institutional compatibility. However, ex ante and even ex post evaluation is difficult due to data availability (Spencer et al. 2017). These challenges might be the reason why obvious potentials are not fully harnessed in developing countries (Chuku 2010). Capacity building might constitute a starting point to facilitate adaptation and sustainable development alike to overcome the social, resource, and physical barriers raised by Spires et al. (2014). Project-specific reviews, pilot studies, and experience in monitoring development practice help to gain expert knowledge and might be of guidance to harmonize climate and development policy (Furlow et al. 2011; Spencer et al. 2017).

5 Adaptation and Robust Decisions: Dealing with Uncertainties

Although research on climate change has extensively increased during the last decades, climate change is still and will always be subject to uncertainty (see, e.g., Allen et al. 2000). Research has helped to gain knowledge and expand competence: We have a sound understanding on the driving forces of climate change and we know climate change has already begun and will lead to severe impacts without action. However, the complexity of the earth system, the lack of knowledge on long-term policy alignments as well as socioeconomic interrelation make definite projections impossible. Even well-founded science cannot fully eliminate uncertainties regarding the kind of climate impacts, their strength, or the exact time and place they occur.

In complex systems under uncertainty, logic or statistic decision-making is replaced by heuristics. Heuristics simplify decision-making by ignoring part of the information (Gigerenzer and Gaissmaier 2011; Kahneman 2003). In case of decisions under uncertainty, the reason that heuristics are applied is not the lack of rationality or saving effort, but because an optimal solution cannot be determined (Neth and Gigerenzer 2015). Climate policy already deals with this uncertainty. Here, proactive adaptation measures represent special characteristics, as it is unknown if they ever materialize. Hence, adaptation strategies have to be developed that are robust to a wide range of possible future scenarios (Prato 2008). Following these robust strategies may be helpful for dealing with background uncertainty in other domains within the climate change context (such as mitigation policy), and in other contexts as well. Among the range of robust adaptation policies facing uncertainty are "no-regret strategies" and "flexible strategies" also called "real options" (Hallegatte 2009; Anda et al. 2009).

No-regret strategies unite adaptation targets with other overarching policy targets like sustainable development or risk management. As such, these measures create either co-benefits to serve adaptation needs and at the same time benefiting other policy targets like sustainability (described in Sects. 3 and 4) or reduce the investment uncertainty as the ancillary benefits are independent of climate change (see Hallegatte 2009). While in developing countries these co-benefits mainly relate to sustainable development goals like poverty reduction or the sustainable management of natural resources, in developed urban areas it may concern spatial planning policy (see, e.g., Kirshen et al. 2008; de Bruin et al. 2009). No-regret strategies help to harmonize policies, to make informed decisions instead of relying on default options, and to decide quicker on alternatives as the no-regret strategy attribute becomes the crucial decision criteria. This might also allow a better usage of windows of opportunities on short notice (see also Sect. 4). Based on this, one can argue that genuine no-regret strategies would have been already implemented if they were useful in any case. Beside financial (mainly liquidity-related), technical, informational, or legal constraints put forward by Hallegatte (2009), not yet realized no-regret strategies could also be reasoned by the long-time perspective or the political will to take action even in non-climate related areas.

Another kind of robust policies are *flexible strategies* or *real options*. Flexible strategies allow to quickly adjust the strategy to changing requirements or newly arising opportunities (Courtney et al. 1997). Hallegatte (2009) presents flexible adaptation strategies, which include in particular insurance or long-term planning in different sectors. He uses land use planning as one example where a restrictive policy is adjustable to new information on climate damage—the policy may easily be softened or withdrawn. Although real options might present flexible strategies, the underlying idea of real options is more likely to delay decisions. According to the review by Heal and Kristrom (2002), such options incorporate the chance to buy time until new information becomes available. In some cases delaying costly investments may save substantial financial resources if future information reduces uncertainties and shows which kind of investment is actually most beneficial. As such, large investments with an uncertain payoff and sunk costs are avoided and decisions are postponed to the future with a reduced uncertainty of climate damage. Anda et al. (2009) explains that the value of a real option is determined by the opportunity costs and the future benefits from compensation. Although, there might be good reasons to shift adaptation actions into the future due to uncertainty (see, e.g., Dixit and Pindyck 1994; Pindyck 2000; Lin et al. 2007; Heal and Kristrom 2002), Hallegatte (2009) points out that this is no solution for overall climate adaptation policy. This is especially true for those policies with a longer time horizon for investment like urban planning. In this case, delayed actions could simply come too late to prevent climate damage. Furthermore, windows of opportunities (discussed in Sect. 4), which make investment easier, are missed. Additionally, the intergenerational fairness is not explicitly taken into account. Anda et al. (2009) proposes real option analyses to meet these challenges by identifying interim policies, which should be implemented in the meanwhile. This supports decisions to avoid uncertain irreversible damage and sunk costs at the same time.

The future will show in how far these robust strategies will work out for adaptation policies: in particular, the possibility to scale up single local decisions on a broader policy level or how experience is transferable to other domains. At least, climate change as one global challenge with decisions under uncertainty has the potential to build up knowledge on dealing with background uncertainty in other domains.

6 Adaptation and Disaster Resilience: Use Price Signals

This section elaborates on the ancillary effects of adaptation on disaster resilience. Disaster risk reduction and climate change adaptation have broad intersections, and are often seen as two sides of the same coin (Mercer 2010). Here, we focus on the role of price signals, which may be introduced by adaptation policies (in particular by insurance markets) and which have ancillary effects on disaster resilience.

In response to an increasing trend of climate-induced natural disasters, many governments try to strengthen private precautionary measures as part of their adaptation strategies. Examples include the national adaptation strategy of Germany and the UK (Bundesregierung 2011; Department for Environment Food and Rural Affairs 2018). Also the European Union considers the strengthening of private insurance markets as an important part of adaptation to climate change (European Commission 2013).

One reason why insurance markets are brought forward in the debate on adaptation is their potential of triggering and incentivizing risk-reducing behavior amongst the policyholders, i.e., households and businesses (Kunreuther 2015; Porrini and Schwarze 2014). If properly designed, insurance policies reveal risks to the insured by introducing risk-based premiums. They may provide direct financial incentives for risk-mitigating measures by introducing premium rebates for risk reductions. Some risk reduction measures are also made obligatory for policyholders. Hence, insurance markets can be more than just a form of redistribution of damages over time and among the insured, but can contribute to a higher awareness of risks, higher penetration of physical or behavioral risk reduction measures, and finally to a reduction of total damage. In that sense, one can speak of ancillary benefits of insurance markets.

Another form of ancillary benefits may arise if these effects extend to natural hazards, which are not related to climate. Insurance companies sometimes bundle different risks in one policy (in Germany for example, flood insurance is usually bundled with earthquake insurance). Hence, adaptation policies related to insurance markets may increase the resilience toward disasters in general, including non-climatic disasters, such as earthquakes, volcano eruptions, and technological disasters.

However, using insurance markets as an instrument for adaptation may have detrimental effects on distributional issues: Strengthening the role of private insurance markets in the climate risk management may raise affordability concerns for low-income households (Lamond and Penning-Rowsell 2014; Shively 2017). Moreover, some risk-mitigating measures which are economically beneficial in the long run and which are eligible for premium rebates may be unaffordable to poor households due to relatively high up-front costs. Governments should therefore support risk mitigation of low-income households, e.g., by providing refundable subsidies for the investment in risk mitigation measures with high up-front costs. Beside affordability issues, insurance markets may also trigger a false sense of safety. There are extreme events which are not insurable (such as very large inundations due to sea dike breaches), but may falsely be perceived as insured by policyholders. Another caveat of the proposed mechanism is that the practical implementation of incentives for risk reduction measures by differentiated insurance premiums is still at an early stage. But, as Herweijer et al. (2009) point out, using the opportunities to reduce aggregate damage may become decisive for the insurance industry itself to remain economically sustainable.

7 Conclusion

The rather general definition of adaptation leaves room for interpretation in two perspectives. First, it is not clear what the adjustment to climate and its effects precisely entails. A full protection is neither economically nor practically possible due to uncertainty in time, space, and degree of climate impacts. Policymakers and individuals need to make decisions on which and to what extent specific adaptation measures should be carried out and which kind of residual damage is acceptable. Second, climate adaptation is often not easy to distinguish from other adjustments to societal and economic processes. Thus, the outcomes solely driven by adaptation are not easy to assess.

Consequently, research finds ancillary benefits as well as conflicts of adaptation in many domains and sectors. The literature gives numerous positive examples in the field of nature-based adaptation, hence adaptation measures which are positively interlinked with the conservation of ecosystems are key to harmonize policy strategies (Abramowitz et al. 2002; Pramova et al. 2012; EC 2013). Disaster resilience and learning how to deal with uncertainties offer further fields of co-benefits from adaptation. However, the literature review shows a mixed picture of adaptation side effects when it comes to mitigation and economic development. Furthermore, there are obvious difficulties in the quantification of direct and indirect effects of adaptation measures. The complexity and uncertainties of cost–benefit analysis have even increased with growing information and knowledge on possible side effects, notwithstanding the uncertainties inherent to adaptation. Moreover, with a raising awareness of moral responsibilities for the most vulnerable as well as for future generations, qualitative assessments gain in importance. The multi-criteria analysis proposed by de Bruin et al. (2009) is a step in the direction to incorporate co-benefits as one (qualitative) criterion for the assessment of adaptation measures. Another example is the impact metrics by Stadelmann et al. (2015) which disentangles monetary and human benefits. The authors propose to measure the effect of adaptation projects by protected wealth (saved wealth) and protected humans (saved health). Even though progress has been made with the Paris Agreement insisting on the importance of adaptation, it is still a long way to a fully integrated climate policy. Unfortunately, we need to face the reality that so far the ambitions of the global community to curb climate change have not been sufficient to prevent climate change. Thus, climate-induced changes are already clearly noticeable. The current forecasts for climate impacts do not reveal an encouraging picture. Therefore, adaptation becomes increasingly important. A better understanding of ancillary effects—positive as well as negative—is imperative for an integrated and efficient climate policy incorporating temporal scales, spatial and political levels as well as different economic sectors.

References

Aakre S, Rübbelke D (2010) Adaptation to climate change in the European Union: efficiency versus equity considerations. Environ Policy Gov 20(3):159–179

Aakre S, Banaszak I, Mechler R, Rübbelke D, Wreford A, Kalirai H (2010) Financial adaptation to disaster risk in the European Union. Mitig Adapt Strateg Glob Chang 15(7):721–736

Abramovitz J, Banuri T, Girot PO, Orlando B, Schneider N, Spanger-Siegfried E, Switzer J, Hammill A (2002) Adapting to climate change: natural resource management and vulnerability reduction. Background paper to the task force on climate change, adaptation and vulnerable communities. International Institute for Sustainable Development, Manitoba

Allen MR, Stott PA, Mitchell JF, Schnur R, Delworth TL (2000) Quantifying the uncertainty in forecasts of anthropogenic climate change. Nature 407(6804):617

Anda J, Golub A, Strukova E (2009) Economics of climate change under uncertainty: benefits of flexibility. Energy Policy 37(4):1345–1355

Barnett J, O'Neill S (2010) Maladaptation. Glob Environ Chang Hum Pol Dimens 20:211–213

Bedsworth L, Hanak E (2008) Preparing California for a changing climate. PPIC, San Francisco

Berry PM, Brown S, Chen M, Kontogianni A, Rowlands O, Simpson G, Skourtos M (2015) Cross-sectoral interactions of adaptation and mitigation measures. Clim Chang 128(3-4):381–393

Biesbroek GR, Swart RJ, Van der Knaap WG (2009) The mitigation – adaptation dichotomy and the role of spatial planning. Habitat Int 33(3):230–237

Bosello F, Carraro C, De Cian E (2013) Adaptation can help mitigation: an integrated approach to post-2012 climate policy. Environ Dev Econ 18(3):270–290

Bundesregierung (2011) Aktionsplan Anpassung der Deutschen Anpassungsstrategie an den Klimawandel. Berlin, Germany. Retrieved from http://www.bmub.bund.de/fileadmin/bmu-import/files/pdfs/allgemein/application/pdf/aktionsplan_anpassung_klimawandel_bf.pdf

Buob S, Stephan G (2011) To mitigate or to adapt: how to confront global climate change. Eur J Polit Econ 27(1):1–16

Chuku CA (2010) Pursuing an integrated development and climate policy framework in Africa: options for mainstreaming. Mitig Adapt Strateg Glob Chang 15(1):41–52

Cornes R, Sandler T (1986) The theory of externalities, public goods, and club goods. Cambridge University Press, Cambridge

Courtney H, Kirkland J, Viguerie P (1997) Strategy under uncertainty. In: Strategy despite uncertainty: cutting through the fog, Harvard business review, p 1746 de Bruin K, Dellink RB, Ruijs A, Bolwidt L, van Buuren A, Graveland J, de Groot RS, Kuikman PJ, Reinhard S, Rötter RP, Tassone VC, Verhagen A, van Ierland EC (2009) Adapting to climate change in the Netherlands: an inventory of climate adaptation options and ranking of alternatives. Clim Chang 95(1-2):23–45

Department for Environment Food and Rural Affairs (DEFRA) (2018) The national adaptation programme and the third strategy for climate adaptation reporting. Retrieved from https://assets.publishing.service.gov.uk/government/uploads/system/uploads/attachment_data/file/727252/national-adaptation-programme-2018.pdf

Dixit AK, Pindyck RS (1994) Investment under uncertainty. Princeton University Press, Princeton, NJ

Dowlatabadi H (2007) On integration of policies for climate and global change. Mitig Adapt Strateg Glob Chang 12(5):651–663

European Commission (2013) An EU strategy on adaptation to climate change. European Commission, Strasbourg

Fankhauser S (1998) The costs of adapting to climate change. Working paper global environment facility 16, Washington

Furlow J, Smith JB, Anderson G, Breed W, Padgham J (2011) Building resilience to climate change through development assistance: USAID's climate adaptation program. Clim Chang 108:411–421

Gigerenzer G, Gaissmaier W (2011) Heuristic decision making. Annu Rev Psychol 62:451–482

Goklany IM (2007) Integrated strategies to reduce vulnerability and advance adaptation, mitigation, and sustainable development. Mitig Adapt Strateg Glob Chang 12(5):755–786

Grothmann T, Reusswig F (2006) People at risk of flooding: why some residents take precautionary action while others do not. Nat Hazards 38(1-2):101–120

Hallegatte S (2009) Strategies to adapt to an uncertain climate change. Glob Environ Chang 19(2):240–247

Hamin EM, Gurran N (2009) Urban form and climate change: balancing adaptation and mitigation in the US and Australia. Habitat Int 33(3):238–245

Hanger S, Linnerooth-Bayer J, Surminski S, Nenciu-Posner C, Lorant A, Ionescu R, Patt A (2018) Insurance, public assistance, and household flood risk reduction: a comparative study of Austria, England, and Romania. Risk Anal 38(4):680–693

Harlan SL, Ruddell DM (2011) Climate change and health in cities: impacts of heat and air pollution and potential co-benefits from mitigation and adaptation. Curr Opin Environ Sustain 3(3):126–134

Heal G, Kristrom B (2002) Uncertainty and climate change. Environ Resour Econ 22(1-2):3–39

Herweijer C, Ranger N, Ward RET (2009) Adaptation to climate change: threats and opportunities for the insurance industry. Geneva Pap Risk Insur 34(3):360–380

Ingham A, Ma J, Ulph AM (2013) Can adaptation and mitigation be complements? Clim Chang 120(1-2):39–53

IPCC (2013) Climate change 2013: the physical science basis. In: Stocker TF, Qin D, Plattner GK, Tignor MMB, Allen SK, Boschung J, Midgley PM (eds). Cambridge University Press, Cambridge, MA. Retrieved from https://www.ipcc.ch/report/ar5/wg1/

IPCC (2014a) Climate change 2014: impacts, adaptation, and vulnerability. In: Field CB, Barros VR, Dokken DJ, Mach KJ, Mastrandrea MD, Bilir TE, White LL (eds). Cambridge University Press, Cambridge, MA. Retrieved from http://www.ipcc.ch/report/ar5/wg2/

IPCC (2014b) Climate change 2014: mitigation of climate change working group III contribution to the fifth assessment report of the Intergovernmental Panel on Climate Change. In: Edenhofer O, Pichs-Madruga R, Sokona Y, Minx JC, Farahani E, Kadner S, Zwickel T (eds). Cambridge University Press, Cambridge, MA

Kahneman D (2003) Maps of bounded rationality: psychology for behavioral economics. Am Econ Rev 93(5):1449–1475

Kane S, Yohe G (2000) Societal adaptation to climate variability and change: an introduction. In: Societal adaptation to climate variability and change. Springer, Dordrecht, pp 1–4

Kirshen P, Ruth M, Anderson W (2008) Interdependencies of urban climate change impacts and adaptation strategies: a case study of metropolitan Boston USA. Clim Chang 86(1-2):105–122

Klein RJT, Schipper EL, Dessai S (2005) Integrating mitigation and adaptation into climate and development policy: three research questions. Environ Sci Pol 8(6):579–588

Klein R, Sathaye J, Wilbanks T (2007) Challenges in integrating mitigation and adaptation as responses to climate change. Mitig Adapt Strateg Glob Chang 12(5):639–962

Kunreuther H (2015) The role of insurance in reducing losses from extreme events: the need for public-private partnerships. Geneva Pap Risk Insur Issues Pract 40(4):741–762

Lamond J, Penning-Rowsell E (2014) The robustness of flood insurance regimes given changing risk resulting from climate change. Clim Risk Manag 2:1–10

Laukkonen J, Blanco PK, Lenhart J, Keiner M, Cavric B, Kinuthia-Njenga C (2009) Combining climate change adaptation and mitigation measures at the local level. Habitat Int 33(3):287–292

Lin T, Ko C-C, Yeh H-N (2007) Applying real options in investment decisions relating to environmental pollution. Energy Policy 35(4):2426–2432

Lowe JA, Huntingford C, Raper SCB, Jones CD, Liddicoat SK, Gohar LK (2009) How difficult is it to recover from dangerous levels of global warming? Environ Res Lett 4(1):014012

McCarthy JJ, Canziani OF, Leary NA, Dokken DJ, White KS (2001) Climate change 2001: impacts, adaptation, and vulnerability. Contribution of working group II to the third assessment report of the Intergovernmental Panel on Climate Change. Cambridge University Press, Cambridge

Mercer J (2010) Disaster risk reduction or climate change adaptation: are we reinventing the wheel? J Int Dev 22(2):247–264

Millennium Ecosystem Assessment (2005) Ecosystems and human well-being: synthesis. Island Press, Washington, DC

Moser SC (2012) Adaptation, mitigation, and their disharmonious discontents: an essay. Clim Chang 111(2):165–175

Neth H, Gigerenzer G (2015) Heuristics: tools for an uncertain world. In: Emerging trends in the social and behavioral sciences: an interdisciplinary, searchable, and linkable resource. Wiley, New York, NY, pp 1–18

Osberghaus D, Baccianti C (2013) Adaptation to climate change in the Southern Mediterranean, a theoretical framework, a foresight analysis and three case studies, European Commission, FP7, Brussels

Parry M (2009) Closing the loop between mitigation, impacts and adaptation. Clim Chang 96(1):23–27

Parry M, Canziani O, Palutikof J, van der Linden P, Hanson C (2007) Climate change 2007: impacts, adaptation and vulnerability, contribution of working group II to the fourth assessment report of the Intergovernmental Panel on Climate Change. Cambridge University Press, Cambridge

Parry M, Lowe J, Hanson C (2009) Overshoot, adapt and recover. Nature 458(7242):1102

Pielke RA Jr (1998) Rethinking the role of adaptation in climate policy. Glob Environ Chang 8(2):159–170

Pindyck RS (2000) Irreversibilities and the timing of environmental policy. Resour Energy Econ 22(3):233–259

Porrini D, Schwarze R (2014) Insurance models and European climate change policies: an assessment. Eur J Law Econ 38(1):7–28

Pramova E, Locatelli B, Brockhaus M, Fohlmeister S (2012) Ecosystem services in the national adaptation programmes of action. Clim Pol 12(4):393–409

Prato T (2008) Accounting for risk and uncertainty in determining preferred strategies for adapting to future climate change. Mitig Adapt Strateg Glob Chang 13(1):47–60

Samuelson PA (1974) Complementarity: an essay on the 40th anniversary of the Hicks-Allen revolution in demand theory. J Econ Lit 12(4):1255–1289

Scheraga JD, Grambsch AE (1998) Risks, opportunities and adaptation to climate change. Clim Res 11(1):85–95

Shively D (2017) Flood risk management in the USA: implications of national flood insurance program changes for social justice. Reg Environ Chang 17(8):2323–2323

Spencer B, Lawler J, Lowe C, Thompson L, Hinckley T, Kim SH, Voss J (2017) Case studies in co-benefits approaches to climate change mitigation and adaptation. J Environ Plan Manag 60(4):647–667

Spires M, Shackleton S, Cundill G (2014) Barriers to implementing planned community-based adaptation in developing countries: a systematic literature review. Clim Dev 6(3):277–287

Stadelmann M, Michaelowa A, Butzengeiger-Geiyer S, Köhler M (2015) Universal metrics to compare the effectiveness of climate change adaptation projects. Handb Clim Chang Adapt:2143–2160

Taylor A, Downing J, Hassan B, Denton F, Downing TE (2007) Supplementary material to chapter 18: Inter-relationships between adaptation and mitigation. In: Parry ML, Canziani OF, Palutikof JP, van der Linden PJ, Hanson CE (eds) Climate change 2007: impacts, adaptation and vulnerability, contribution of working group II to the fourth assessment report of the Intergovernmental Panel on Climate Change. Cambridge University Press, Cambridge

Tol RSJ (2005) Adaptation and mitigation: trade-offs in substance and methods. Environ Sci Policy 8(6):572–578

Turner RK, Daily GC (2008) The ecosystem services framework and natural capital conservation. Environ Resour Econ 39(1):25–35

van den Bergh JCJM (2010) Safe climate policy is affordable—12 reasons. Clim Chang 101(3–4):339–385

Voskamp IM, van de Ven FHM (2015) Planning support system for climate adaptation: composing effective sets of blue-green measures to reduce urban vulnerability to extreme weather events. Build Environ 83:159–167

Walsh CL, Dawson RJ, Hall JW, Barr SL, Batty M, Bristow AL, Carney S, Dagoumas AS, Ford AC, Harpham C, Tight MR, Watters H, Zanni AM (2011) Assessment of climate change mitigation and adaptation in cities. Proc Inst Civ Eng Urban Des Plann 164(2):75–84

Warren R (2011) The role of interactions in a world implementing adaptation and mitigation solutions to climate change. Philos Trans R Soc A Math Phys Eng Sci 369(1934):217–241

Watkiss P, Benzie M, Klein RJT (2015) The complementarity and comparability of climate change adaptation and mitigation. Wiley Interdiscip Rev Clim Chang 6(6):541–557

Weitzman ML (2009) Some basic economics of extreme climate change. In: Touffut J-P (ed) Changing climate, changing economy. Edward Elgar Publishing, Northhampton, MA

Wilbanks TJ, Leiby P, Perlack R, Ensminger JT, Wright SB (2007) Toward an integrated analysis of mitigation and adaptation: some preliminary findings. Mitig Adapt Strateg Glob Chang 12(5):713–725

Yohe G, Strzepek K (2007) Adaptation and mitigation as complementary tools for reducing the risk of climate impacts. Mitig Adapt Strateg Glob Chang 12(5):727–739

Zhao C, Yan Y, Wang C, Tang M, Wu G, Ding D, Song Y (2018) Adaptation and mitigation for combating climate change–from single to joint. Ecosyst Health and Sustain 4(4):85–94

Ancillary Benefits of Carbon Capture and Storage

Asbjørn Torvanger

1 Introduction

In the context of policies to mitigate climate change, IPCC (2001) defines "Ancillary Benefits" as "the ancillary, or side effects, of policies aimed exclusively at climate change mitigation."[1] This definition focuses on impacts of climate policies, whereas in this chapter we relate ancillary benefits to a technology solely aimed at mitigation of greenhouse gas emissions, namely Carbon Capture and Storage (CCS). We define ancillary benefits of CCS as the positive effects of CCS deployment that are additional to climate change mitigation through reduced carbon dioxide emissions.

Climate change mitigation is the only primary rationale for deployment of CCS. CCS refers to a group of technologies that reduce emissions of carbon dioxide from coal-fired or gas-fired power stations, or from process industries (see Fig. 1). The "CCS value chain" consists of three stages, where the first is capture of carbon dioxide from exhaust gas or from process-related industrial emissions with the help of various physical and chemical methods. To avoid excessive costs only about 90%

---

[1]The full definition of "ancillary benefits" in IPCC (2001) is:

'The ancillary, or side effects, of policies aimed exclusively at climate change mitigation. Such policies have an impact not only on greenhouse gas emissions, but also on resource use efficiency, like reduction in emissions of local and regional air pollutants associated with fossil-fuel use, and on issues such as transportation, agriculture, land-use practices, employment, and fuel security. Sometimes these benefits are referred to as "ancillary impacts" to reflect that in some cases the benefits may be negative. From the perspective of policies directed at abating local air pollution, greenhouse gas mitigation may also be considered an ancillary benefit, but these relationships are not considered in this assessment.'

A. Torvanger (✉)
CICERO Center for International Climate Research, Oslo, Norway
e-mail: asbjorn.torvanger@cicero.oslo.no

© Springer Nature Switzerland AG 2020
W. Buchholz et al. (eds.), *Ancillary Benefits of Climate Policy*, Springer Climate,
https://doi.org/10.1007/978-3-030-30978-7_12

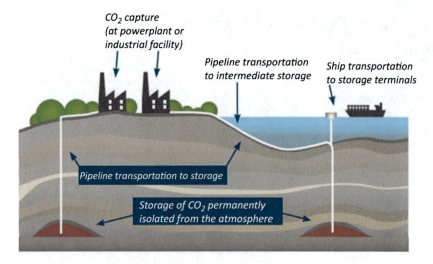

Fig. 1 A schematic overview of carbon capture and storage. Source: CCSP (2019)

of the carbon dioxide in exhaust can be captured in practical terms (IEA GHG 2012). The second stage consists of pressurizing carbon dioxide and transporting the gas in pipelines or with ships to a suitable geological storage site. The third stage consists of injecting the carbon dioxide deep into a geological structure that can sequester the gas for eternity. The injected carbon dioxide is in subcritical form, which is a state between gas and fluid. Sandstone filled with sea water—aquifers—is the geological formation with largest overall potential to store carbon dioxide, but depleted oil and gas reservoirs, and coal-bed seams can also be used. Several studies show that the climate target of limiting man-made warming to 2 °C by 2100 (as well as pursuing 1.5 °C) from the Paris Agreement is certainly more expensive and could even be very difficult without CCS (IPCC 2014, 2018).

About 30 full-scale applications of CCS currently exist or are planned in process industries or the power sector, with the majority in industry (Global CCS Institute 2018; OECD/IEA 2016). We only have long-term experience from a few cases of large-scale geological storage of carbon dioxide. At Sleipner, on the Continental Shelf of Norway, one million ton of carbon dioxide have been injected annually into a geological formation since 1996. Carbon dioxide is separated from the natural gas extracted from this reservoir to make the gas commercial. The integrity of stored carbon dioxide has worked well at this storage site.

According to IPCC (2018), removal of carbon dioxide from the atmosphere is very likely needed to meet the 2 °C maximum warming by 2100 climate target from the Paris Agreement, also referred to as negative emissions, and even more so if the world tries to pursue the 1.5 °C target. The slower we are at reducing emissions of greenhouse gases (GHG), the more dependent we will become on negative carbon dioxide emissions. Two of the major technologies for negative emissions require CCS, namely "Bioenergy combined with CCS" (BECCS) and "Direct Air

Capture" (DAC). Thus, improvements of CCS capture, transportation, and storage technologies contribute to enabling these negative emission technologies.

CCS is expensive in energy and money terms, due to a heavy energy "penalty" capturing carbon dioxide from, e.g., gas-fired power station exhaust, with relatively low concentration of the gas, and since large investments in capture facilities and transportation infrastructure are required. Nevertheless, there is a learning curve and potential to improve the efficiency of carbon dioxide capture technologies and reduce its cost, as well as reducing associated risk—particularly the risk of carbon dioxide leakage from geological sinks. Some learning is likely across different CCS applications, e.g., carbon dioxide capture from the power sector to industry and vice versa, across various process industries, and from process industry applications to BECCS and DAC.

A major challenge for deployment of CCS is the high cost compared to present climate value, e.g., measured as the price of an allowance of one ton of carbon dioxide in the European emissions trading system (which was at 25 Euro in May 2019). This can be compared to an estimated price range of the CCS value chain at 50–100 Euro per ton of carbon dioxide (Rubin et al. 2015). Therefore, industry interest in CCS necessitates a significantly higher value of carbon dioxide mitigation, which again requires stronger climate policies. Much stronger climate policies are required to meet the Paris Agreement target, since the present global emissions trend results in a warming of 3 °C or more by the end of this century (IPCC 2018).

In the following we first briefly describe some possible negative side effects of CCS. Then we review potential ancillary benefits of different CCS applications, starting with various ways to use captured carbon dioxide that can be turned into a commercial value. Next, we assess the importance of CCS for enabling the negative emission technologies BECCS and DAC, followed by a discussion of some new market opportunities that are opened by CCS. The final section examines possible benefits of CCS for petroleum exporting countries, given that this technology can reduce the transfer of petroleum-based royalty from producers to consumers.

2 Negative Side Effects of CCS

Investments in CCS will likely encompass some problematic or even negative side effects. In economic terms CCS has higher risk than for example photovoltaics, since plants are more centralized and larger, and more expensive, and are more exposed to policy risks. Aside from a few ancillary benefits, CCS only has value as part of stringent climate policy. We observe that there is a lack of long-term certainty in climate politics, both at national and international levels. Photovoltaics and wind power that replace fossil fuel reduce carbon dioxide emissions, but in addition generate power. On the other hand, the lifetime of photovoltaics and wind power plants is shorter than for CCS, and power production is intermittent due to variation in sunlight and wind. Therefore, such renewable power sources make backup power necessary, such as gas-fired power plants, storage of energy or imports of power, which adds to the cost. Vögele and Rübbelke (2013) find that the wholesale price of power may drop with investments in CCS compared to renewable energy, foremost because no backup power is needed for CCS. In addition to the current cost–value gap, the policy risk associated with CCS engagement explains industry's reluctance to invest in CCS. Uncertain government support for CCS adds to the risk, since such support would be required to cover the cost–value gap for CCS, and for several years.

Capture of carbon dioxide is energy demanding, where the so-called "energy penalty" can average 29%, which means 8–12% lower energy efficiency for coal-fired power stations (Vögele et al. 2018). This leads to more coal consumption to produce the same amount of power, which generates more air pollution, unless the additional pollutants are scrubbed out (ibid.). The most important air pollutants are sulfur dioxide, nitrous oxides, and particulate matter, which all have negative impacts on health, buildings, and ecosystems.

A non-zero risk for leakage from stored carbon dioxide is unavoidable. A future leakage event can generate expensive future liability claims and make expensive contingency measures necessary (Torvanger et al. 2012). Insufficient regulation of liability for stored carbon dioxide in case of future leakage adds to the leakage risk, both between carbon dioxide transport and storage operators and government, and between countries in case carbon dioxide is exported. Regulation and management of CCS operations should be further advanced, and across countries when carbon dioxide is exported and imported.

There are some negative impacts associated with all measures and technologies to reduce emissions of carbon dioxide, but public skepticism toward CCS is stronger than for most alternative measures, maybe with an exception for nuclear power (Vögele and Rübbelke 2013). A modest public acceptance for CCS is observed among the general public, environmental organizations, and at parliament level in some countries. CCS is perceived as prolonging the fossil era, whereas preferred investments should be in renewable energy and improved efficiency. Furthermore, CCS is seen as an immature and unsafe technology due to the leakage risk from stored carbon dioxide, since escaped gas can harm the health of people living close to storage sites or pipelines, and in addition have harmful effects on animals, plants, and ecosystems.

3 Use of Captured Carbon Dioxide

Captured carbon dioxide can instead of being stored in a geological formation be used as a feedstock for various production processes, so-called Carbon Capture, Use and Storage (CCUS). Carbon dioxide represents a type of "spent" form of carbon (Lim 2016). Most CCUS alternatives require separation of carbon and oxygen atoms and therefore have a high energy cost (Aresta et al. 2013; Scott 2015). CCUS furthermore typically requires hydrogen, which is expensive to produce (Scott

2015). The main alternatives for using carbon dioxide are synthetic fuels, chemicals, plastics, building materials, and Enhanced Oil Recovery (EOR). Broadly, CCUS can be divided into low-energy chemical processes, to produce various chemicals, and high-energy chemical processes, to produce various fuels (Aresta et al. 2013).

3.1 Current and Future Use

One hundred and twenty two megaton of carbon dioxide is currently used for urea and methanol production annually (Armstrong and Styring 2015). Compared to the huge need for CCS in many IPCC scenarios (IPCC 2014), carbon dioxide use likely has a rather small potential, and may be considered as a niche solution in a climate context. Fourteen billion tons of carbon dioxide is annually produced from coal-fired power stations globally, whereas an entirely carbon dioxide-based chemical industry would in the best case only require 300 million tons of carbon dioxide (Scott 2015; Aresta et al. 2013). The market for fuels is likely 12–14 times larger than the market for chemicals (Aresta et al. 2013). Armstrong and Styring (2015) describe a CCUS scenario where 1.3 Gt carbon dioxide could be used every year, amounting to 83% of a CCS target for 2030 cited by IPCC. In 20 years CCUS in combination with renewable energy could easily reduce global carbon dioxide emissions by more than 10%, according to Aresta et al. (2013). The potential of CCUS is reduced due to a trade-off between mitigation potential (highest for building materials) and profitability (high value chemicals) (Extavour and Bunje 2016).

3.2 Mitigation Effect

Captured carbon dioxide used to produce chemicals or synthetic fuels sequesters less carbon dioxide than geological storage or using the carbon dioxide for long lifetime products such as building materials and some plastics, but CCUS still provides a net reduction and a low-carbon pathway for chemical industries (Armstrong and Styring 2015). Short lifetime CCUS can nevertheless reduce carbon dioxide emissions if the products would otherwise be based on fossil oils (ibid.).

3.3 Permanency

Another issue with CCUS is permanency, since many of the products carbon dioxide could be used for have a relatively short expected lifetime before being converted to carbon dioxide again. Carbon dioxide used for "fizz" in drinks, amounting to 20 Mt annually, which quickly returns to the atmosphere, is at the short end of permanency (Lim 2016). A closed carbon loop recycling, however, can be designed for energy storage and liquid fuels, where energy is largely from renewable energy sources, which means that a sizeable volume of carbon dioxide is more or less permanently contained (Extavour and Bunje 2016).

3.4 Synthetic Fuels

The main synthetic fuels that can be produced from a carbon dioxide feedstock are ethanol, methanol, and various hydrocarbons (e.g., DME, which is a clean-burning fuel and a potential diesel substitute) (Centi and Perathoner 2009). Scott (2015) finds that fuels based on a carbon dioxide fuel stock will always cost more than fossil fuels, and therefore require some type of financial support. Efficiency is a challenge when producing such synthetic fuels, since known methods can consume more energy than the resulting fuels can provide (Lim 2016). A zero power price might be needed for synthetic fuels to compete (Scott 2015). One solution is to use surplus power from intermittent renewable energy production, such as power from wind farms and solar photovoltaic plants (Scott 2015; Aresta et al. 2013). Using surplus renewable power for electrolysis will provide a cheap source of hydrogen, which is needed for synthetic fuel production (Lim 2016). Even so, Aresta et al. (2013) find that synthesizing methanol from hydrogen and carbon dioxide from surplus renewable power will only be economically attractive if the carbon price reaches at least 100 Euro per ton of carbon dioxide, so maybe using surplus renewable power directly to produce hydrogen for transportation from electrolysis is a more effective and competitive solution.

A more ambitious concept is to industrialize man-made photosynthesis with the help of artificial trees that capture carbon dioxide, and thereafter convert it to hydrocarbon-based liquid fuels (Centi and Perathoner 2009; Scott 2015). This technology could have a large potential but might take 20–30 years to mature (Scott 2015; Aresta et al. 2013).

3.5 Chemicals and Plastics

Carbon dioxide can be used as feedstock for a number of chemicals, such as cyclic carbonates, solvents, polyurethane, polyols and polymers, and polycarbonates (Scott 2015; Lim 2016). Polyurethane can be used for many plastic products, such as mattresses, adhesives, and coatings. Combining carbon dioxide with ammonia produces urea, which can be used for fertilizers or production of plastics. Surplus power from wind or solar production can also be used to store energy in the form of chemicals, or by electrolysis of water to produce hydrogen, and then combining hydrogen with carbon dioxide to make, e.g., methanol (Scott 2015).

3.6 Building Materials

Carbon dioxide can be locked up in building blocks to be used in construction (Scott 2015). Combining carbon dioxide with calcium produces calcium carbonate, which is used for construction materials and whitening of paper, and fiber cement boards (Lim 2016). Alternatively, carbon dioxide can be absorbed by waste and ash residues from power stations to form calcium carbonate, to be mixed with more carbon dioxide and binders to form building blocks for construction (Scott 2015).

3.7 Enhanced Oil Recovery

Enhanced Oil Recovery (EOR) refers to an oil extraction technology. Carbon dioxide is a well-suited agent when injected to flush out more oil from the tail-production phase of an oil reservoir, giving a value to this application of carbon dioxide. Presently one of the few ways to incentivize CCS is to use carbon dioxide for EOR. As a climate technology the effect of EOR is uncertain when accounting for increased oil production. The net carbon dioxide emissions effect depends on three components: (1) Does any carbon dioxide escape from the reservoir during the EOR phase?; (2) Will the oil reservoir be used for carbon dioxide storage after end of commercial oil production, and thus allow more carbon dioxide storage?; and (3) Will the oil produced replace or be additional to other oil or fossil fuel production, and in case of substitution, which fossil fuels? Therefore, the net effect on carbon dioxide emissions depends on the additional value of storing extra carbon dioxide (the carbon price) beyond what is profitable to inject for oil production (OECD/IEA 2015; Armstrong and Styring 2015; Stewart and Hazeldine 2015). Mac Dowell et al. (2017) find that EOR can facilitate the deployment of carbon dioxide transport infrastructure. However, maximizing carbon dioxide storage implies injecting much more carbon dioxide than what is profitable from an oil recovery perspective, since oil recovery per ton of carbon dioxide is falling with total injected carbon dioxide. They find that optimal oil recovery implies injecting carbon dioxide until 3.3 barrels of oil is recovered per ton of carbon dioxide, whereas maximizing carbon dioxide storage implies continuing to inject carbon dioxide until 1.1 barrels of oil are recovered per ton of carbon dioxide injected.

4 CCS Development Enables Two Negative Emission Technologies

Negative emission technologies are characterized by their potential to remove carbon dioxide from the atmosphere and store the gas in other sinks with large capacity, such as forest biomass (afforestation and reforestation), chemically bound to certain minerals ("mineralization"), in deep ocean sediments ("ocean fertilization"), increased carbon content of soil ("biochar"), and in suitable geological formations [BECCS and "Direct Air Capture" (DAC)] (Fuss et al. 2016; Peters and Geden 2017). These technologies are also referred to as "Carbon Dioxide Removal" (CDR). BECCS is based on biomass from forests and suitable plants that can be combusted for heat and/or power in central plants. DAC depends on machines that are able to pull carbon dioxide from the very small concentration in the atmosphere, and concentrate this to a pure gas. Thus both BECCS and DAC are dependent on CCS to transfer and store carbon dioxide in geological formations. To be viable, both technologies depend on economic and climate efficiency. For BECCS this means efficiency in production of the biomass, transportation of the biomass, processing the biomass to bioenergy, and combustion of the bioenergy for power and heat. In addition, the biomass production depends on sustainable land use, where deforestation is avoided, and biodiversity and food production for the local population are not compromised (Torvanger 2018).

4.1 CCS Learning Across Applications

Since both BECCS and DAC rely on CCS, industry- and power-related CCS development contributes to making BECCS and DAC viable. The transportation and storage components of CCS are basically the same across all these CCS applications. The three main categories of carbon dioxide capture are oxyfuel, precombustion, and postcombustion (IPCC 2005). Efficient capture of carbon dioxide may, however, differ across different industrial applications, such as cement and fertilizer production, when combusting biomass for power and heat, and in the case of direct capture from air. Since oxyfuel and precombustion capture technologies are integrated with the combustion or production process at hand, the learning transferability of these technologies is likely to be less than for postcombustion technologies. Therefore, the transferability of learning efficient carbon dioxide capture in power and industry to BECCS and DAC becomes an important issue. There are few studies on cross-application learning. Reiner (2016) argues that learning through a diversity of CCS demonstration projects is important to lower the cost of CCS, and that greater international coordination would facilitate learning across projects. Torvanger and Meadowcroft (2011) analyze the balance between learning by replication or diversity. Learning by replication, where many resources are allocated to one CCS technology, allows going further down the learning curve, but the risk is that this technology turns out to be "the wrong horse." They instead advocate learning by some level of diversity, where resources are spent on a small portfolio of different CCS technologies, to diversify the risk of betting on only one CCS technology, as well as moving some distance along the learning curves.

The concept of artificial photosynthesis and synthetic fuel production mentioned in section 3.4 has some overlaps with DAC, since in both cases machines are used to capture carbon dioxide from the atmosphere. However, more energy and hydrogen are needed for additional processes to produce synthetic fuels. Nevertheless, some scope for learning across DAC and artificial photosynthesis should be expected.

4.2 BECCS

BECCS is still at an early stage of development, with five operations currently existing at global level. Three BECCS facilities are situated in USA, one in Canada, and one in the Netherlands (OECD/IEA 2016). All are fermentation plants producing ethanol from agricultural products, foremost corn. For two of these, dedicated storage of carbon dioxide in geological formations is ongoing or planned, whereas the other three supply carbon dioxide for EOR. This state of BECCS deployment reflects that carbon dioxide emissions from fermentation are concentrated and less costly to capture, and that EOR gives sufficient value to incentivize CCS in some applications.

Since CCS development is a prerequisite for BECCS viability we may consider some benefits of BECCS as "second-order" ancillary benefits of CCS. The main benefit of BECCS as a negative emission technology is the option to use the technology as a supplement to reduce emissions of carbon dioxide and other GHG, in order to increase the probability of meeting the climate policy target adopted in the Paris Agreement. In this context it is essential that bioenergy for BECCS goes hand in hand with sustainable biomass production. If this is done properly, BECCS could actually facilitate sustainable land management and green development in biomass producing countries, where land use is either allocated to food production, nature conservation (i.e., biodiversity), carbon storage in biomass, production of feedstock for industry (timber), or energy production.

4.3 Direct Air Capture (DAC)

DAC is not a new technology, since variants of this has been applied at small scale in diving equipment, submarines, and space crafts. The idea is to remove carbon dioxide and make air breathable for a prolonged period. This is expensive per ton of carbon dioxide, however, and not viable for the large volumes required for negative emissions in a climate policy context. In addition, the carbon dioxide concentration in air is lower than, e.g., in a submarine, adding to the capture cost. Currently a number of DAC technologies and pilots are explored, but it seems that a lot of development is still needed to reduce the current high costs substantially (Fasihi et al. 2019). Viability of DAC in any case depends on a much higher value of GHG mitigation and negative carbon dioxide emissions.

5 New Industrial Opportunities

Large-scale deployment of CCS will open a new market for producers of carbon dioxide capture technologies, and for countries and companies providing carbon dioxide storage services.

5.1 Production of Carbon Dioxide Capture Equipment

For CCS to have a sizeable impact on global emissions of carbon dioxide, large investments in capture facilities are needed over many years (Torvanger et al. 2013). This opens a huge global market for technology companies that are at the forefront of producing capture technologies at competitive cost and in large volumes. Learning and economies of scale will enable more competition among producers, lower costs over time, and develop equipment with better performance and robustness.

5.2 Geological Storage of Carbon Dioxide

In a scenario with large-scale CCS implementation, some countries lack sufficient storage capacity to handle a sizeable share of their carbon dioxide emissions. Therefore, countries with large storage capacity can develop storage of carbon dioxide for neighboring countries as a business opportunity and charge a fee for this. In this way CCS can turn suitable geological formations into a valuable resource, foremost subsea saline aquifers, and depleted oil and gas reservoirs. In addition to geological capacity to store carbon dioxide, short distance to large emission sources increases the value of this resource. Due to skepticism among local populations close to storage sites, since some risk of leakage is perceived, long distance from storage sites to cities and dense populations adds to the value. As an example in Northern Europe, Norway has sizeable geological capacity under the North Sea bed to store carbon dioxide for countries like Germany and the UK, situated well away from large population concentrations (Allen et al. 2017).

Transportation of carbon dioxide through pipelines or with ships involves a small risk of leakage or accidents. The most problematic risk is due to storage of large volumes of carbon dioxide in geological formations. Even with substantial efforts to identify prospective carbon dioxide storage sites, there will be a non-zero probability of leakage. Consequently, more elaborated rules for efficient and safe transport and storage of carbon dioxide are needed, including allocation of liability or insurance arrangements should problems arise. Countries that store carbon dioxide for other countries must accordingly add a risk component to their fee since a future leakage would imply remediation efforts and costs. Additional costs could be for insurance schemes, physical remediation operations in the future, or purchase of compensating emission allowances. Pawar et al. (2015) note that there has been significant progress in geological carbon dioxide storage risk assessment and management over the last decade, based on 45 field projects and development of regulation frameworks. Risk is broadly divided into site performance, long-term containment, public perception, and market risk. Liability transfer to the state with jurisdiction over the storage site, provided that certain conditions are fulfilled, seems to be the most realistic solution, which is the solution chosen in EU's CCS directive. Garcia and Torvanger (2019) discuss how carbon dioxide mitigation through CCS should be linked to emission trading systems when there is a non-zero risk of leakage from storage sites. They find that CCS-based carbon dioxide mitigation should be discounted, compared to risk-free mitigation alternatives, such as deployment of wind and solar energy.

6 CCS Benefits for Petroleum Exporting Countries

Taxing of carbon dioxide or establishment of an emissions trading system is required to meet a strict climate policy target. Given a significant cost on carbon dioxide emissions, large-scale deployment of CCS would provide benefits for oil and gas exporting countries since more of the royalty from extraction would be kept by producing countries and companies instead of being transferred to consuming countries.

When extracting natural resources, royalty or resource rent is the extra profit gained by the difference between the market value of a limited resource and the extraction cost. For extraction of oil and gas, royalty gives rise to high profits. When the polluter pays principle is applied in the form of a tax on consumers for emissions of carbon dioxide, or when consumers are included in an emissions trading system, some of the royalty will be transferred from producers to consumers. This means that the government of consuming countries receives the revenue from taxing and sale (i.e., auctioning) of emission allowances in an emissions trading system. If CCS is competitive as a climate mitigation technology, and widely deployed to reduce carbon dioxide emissions from power stations and industry, and given a climate policy target such as the 2 °C target adopted in the Paris Agreement, the total cost of meeting this target will be reduced, and a lower tax or emission allowance price on carbon dioxide emissions will suffice. This implies that less of the royalty from oil and gas production is transferred to consumers. Thus, CCS will both secure higher profits for petroleum producers and reduce the conflict, felt by some petroleum producing countries, between income from petroleum extraction and meeting the climate target of the Paris Agreement.

7 Conclusions

Carbon Capture and Storage (CCS) is expensive in energy consumption and costs, thus only making sense as part of meeting an ambitious climate policy target, such as the 2 °C target from the Paris Agreement. Nevertheless, given a broader interpretation of ancillary benefits, CCS generates some positive side effects, foremost when using captured carbon dioxide to produce chemicals and plastics, building materials, and synthetic fuels of commercial value. Carbon dioxide can be used for Enhance Oil Recovery (EOR), which is a valuable technology for efficient oil extraction, but a dubious technology for mitigating carbon dioxide emissions. Learning how to improve efficiency and robustness of CCS contributes to the facilitation of two negative emission technologies: bioenergy with CCS and direct air capture. Deployment of CCS at scale opens new industrial opportunities in production of equipment, as well as for countries with large storage capacity selling geological storage of carbon dioxide as a service to other countries. Finally, CCS deployment at scale will benefit petroleum exporting countries since the technology will reduce the need for taxing of carbon dioxide emissions, and thereby reduce the transfer of some of the royalty from oil extraction from producer to consumer countries.

References

Allen R, Nilsen HM, Andersen O, Lie K-A (2017) Ranking and categorizing large-scale saline aquifer formations based on optimized CO_2 storage potentials and economic factors. Int J Greenhouse Gas Control 65:182–194. https://doi.org/10.1016/j.ijggc.2017.07.023

Aresta M, Dibenedetto A, Angelini A (2013) The changing paradigm in carbon dioxide utilization. J CO2 Util 3–4:65–73

Armstrong K, Styring P (2015) Assessing the potential of utilization and storage strategies for post-combustion carbon dioxide emissions reduction. Front Energy Res 3:1–9. https://doi.org/10.3389/fenrg.2015.00008

CCSP—CLIC Innovation Oy's Carbon Capture and Storage R&D Program (2019) VTT Technical Research Centre of Finland Ltd. http://ccspfinalreport.fi/print

Centi G, Perathoner S (2009) Opportunities and prospects in the chemical recycling of carbon dioxide to fuels. Catal Today 148:191–205

Extavour M, Bunje P (2016) CCUS: utilizing carbon dioxide to reduce emissions, CEP Magazine, June, 52–59

Fasihi M, Efimova O, Breyer C (2019) Techno-economic assessment of CO_2 direct air capture plants. J Clean Prod 224:957–980

Fuss S, Jones CD, Kraxner F, Peters GP, Smith P, Tavoni M, van Vuuren DP, Canadell JG, Jackson RB, Milne J, Moreira JR, Nakicenovic N, Sharifi A, Yamagata Y (2016) Research priorities for negative emissions. Environ Res Lett 11. https://doi.org/10.1088/1748-9326/11/11/115007

Garcia JH, Torvanger A (2019) Carbon leakage from geological storage sites: implications for carbon trading. Energy Policy 127:320–329. https://doi.org/10.1016/j.enpol.2018.11.015

Global CCS Institute (2018) The global status of CCS 2018. https://www.globalccsinstitute.com/resources/global-status-report/

IEA GHG (2012) Carbon dioxide capture at gas-fired power plants, report no. 8. http://hub.globalccsinstitute.com/sites/default/files/publications/103211/co2-capture-gas-fired-power-plants.pdf

IPCC (2001) Climate change 2001, synthesis report: annex B. https://www.ipcc.ch/site/assets/uploads/2018/03/annex.pdf

IPCC (2005) Carbon dioxide capture and storage, special report. Cambridge University Press. https://www.ipcc.ch/report/carbon-dioxide-capture-and-storage/

IPCC (2014) AR5 climate change 2014: mitigation of climate change. The working group III contribution to the IPCC's fifth assessment report. https://www.ipcc.ch/report/ar5/wg3/

IPCC (2018) Global warming of 1.5°C. An IPCC special report on the impacts of global warming of 1.5°C above pre-industrial levels and related global greenhouse gas emission pathways, in the context of strengthening the global response to the threat of climate change, sustainable development, and efforts to eradicate poverty. https://www.ipcc.ch/sr15/

Lim X (2016) How to make the most of carbon dioxide. Nature 526:628–630. https://doi.org/10.1038/526628a

Mac Dowell N, Fennell PS, Shah N, Maitland GC (2017) The role of carbon dioxide capture and utilization in mitigating climate change. Nat Clim Chang 7:243–249. https://doi.org/10.1038/NCLIMATE3231

OECD/IEA (2015) Storing carbon dioxide through enhanced oil recovery—combining EOR with carbon dioxide storage (EOR+) for profit. Insights Series. https://www.iea.org/publications/insights/insightpublications/Storing_carbon dioxide_through_Enhanced_Oil_Recovery.pdf

OECD/IEA (2016) 20 years of carbon capture and storage—accelerating future deployment, OECD/IEA, Paris. https://www.iea.org/publications/freepublications/publication/20YearsofCarbonCaptureandStorage_WEB.pdf

Pawar RJ, Bromhal GS, Carey JW, Foxall W, Korre A, Ringrose PS, Tucker O, Watson MN, White JA (2015) Recent advances in risk assessment and risk management of geologic carbon dioxide storage. Int J Greenhouse Gas Control 40(Special issue):292–311

Peters GP, Geden O (2017) Catalyzing a political shift from low to negative carbon. Nat Clim Chang 7:619–621. https://doi.org/10.1038/nclimate3369

Reiner DM (2016) Learning through a portfolio of carbon capture and storage demonstration projects. Nat Energy 1(1). https://doi.org/10.1038/nenergy.2015.11

Rubin ES, Davison JE, Herzog HJ (2015) The cost of CO_2 capture and storage. Int J Greenhouse Gas Control 40(Special issue):378–400

Scott A (2015) Learning to love carbon dioxide. Chem Eng News 93(45):10–16

Stewart RJ, Haszeldine RS (2015) Can producing oil store carbon? Greenhouse gas footprint of CO_2EOR, offshore North Sea. Environ Sci Technol 49(9):5788–5795. https://doi.org/10.1021/es504600q

Torvanger A (2018) Governance of bioenergy with carbon capture and storage (BECCS): accounting, rewarding, and the Paris agreement. Clim Pol 19(3):329–341

Torvanger A, Meadowcroft J (2011) The political economy of technology support: making decisions about carbon capture and storage and low carbon energy technologies. Glob Environ Chang 21:303–312

Torvanger A, Grimstad A-A, Lindeberg E, Rive N, Rypdal K, Skeie RB, Fuglestvedt J, Tollefsen P (2012) Quality of geological CO_2 storage to avoid jeopardizing climate targets. Clim Chang 114(2):245–260

Torvanger A, Lund MT, Rive N (2013) Carbon capture and storage deployment rates: needs and feasibility. Mitig Adapt Strateg Glob Chang 18:187–205. https://doi.org/10.1007/s11027-012-9357-7

Vögele S, Rübbelke D (2013) Decisions on investments in photovoltaics and carbon capture and storage: a comparison between two different greenhouse gas control strategies. Energy 62:385–392

Vögele S, Rübbelke D, Mayer P, Kuckshinrichs W (2018) Germany's "no" to carbon dioxide capture and storage: just a question of lacking acceptance? Appl Energy 214:205–218

Health Co-benefits of Climate Mitigation Policies: Why Is It So Hard to Convince Policy-Makers of Them and What Can Be Done to Change That?

Anil Markandya and Jon Sampedro

1 Introduction

A reduction in greenhouse gases (GHGs) lowers the risk of climate change but in many cases it also reduces the emissions of local pollutants, which are harmful to health and to the environment. The associated reduction in local pollutants is referred to variously as ancillary benefits or co-benefits.[1] The recognition of such benefits goes back nearly 30 years (perhaps the first reference to them is Glomsrod (1990). Every major review of climate science and policy over that period, including all five assessment cycles of the Intergovernmental Panel on Climate Change (IPCC), have had some discussion of ancillary benefits and commented on the fact that such benefits make the case for climate mitigation action stronger, especially in developing countries, where measures against local pollution are more limited. Yet it remains the case that action on reducing GHGs has not been as aggressive as the targets for climate stabilisation require, and the presence of ancillary benefits has not played a major role in driving the discussion.

This chapter looks at the reasons for this and offers some suggestions on ways to give health co-benefits a great role in climate policy. The next section sets out

---

[1] The two terms are used interchangeably in this chapter.

The earlier part of this chapter draws significantly on joint work with our colleagues at BC3, namely Mikel Gonzalez, Inaki Arto and Cristina Pizarro-Irizar. We gratefully acknowledge their contribution as well all other co-authors of the Markandya et al. (2018) paper referenced here.

A. Markandya (✉) · J. Sampedro
Basque Centre for Climate Change (BC3), Leioa, Spain
e-mail: anil.markandya@bc3research.org; jon.sampedro@bc3research.org

© Springer Nature Switzerland AG 2020
W. Buchholz et al. (eds.), *Ancillary Benefits of Climate Policy*, Springer Climate,
https://doi.org/10.1007/978-3-030-30978-7_13

the evidence on health co-benefits, which are the main type of co-benefits and the focus of the discussion here, relative to the costs of mitigation in different regions of the world. It shows the evidence is more complex than is sometimes imagined. The way in which the two of them stack up depends on a number, especially the way in which the burden of mitigation is shared across countries and regions. The analysis lays out geopolitical differences on climate policy when all such factors are taken into account. It draws extensively on a recent chapter that compares mitigation costs and health co-benefits using a state-of-the-art economic model as well as detailed health benefits based on air quality modelling (Markandya et al. 2018).[2] Section 3 considers specific factors that make the results of the analysis difficult to interpret. One is the role of the value of statistical life in the health co-benefits estimation. A second is the discount rate applied to the mitigation costs and co-benefits when calculating the present value of different policies. A third is the political economy of diffuse benefits versus specific costs. Section 4 discusses how these factors might be addressed in the policy space so as to give health co-benefits a fair chance of influencing climate policy. Section 4 concludes the chapter with ideas for further research to support the analysis presented.

2 Mitigation Costs Versus Health Co-benefits

2.1 Modelling Costs and Health Benefits

In comparing mitigation costs and health co-benefits we need to use a sophisticated economic model that tracks economic output under different constraints on the emissions of GHGs. This is linked to a detailed air quality source–receptor model that takes the emissions arising from different economic activities in different locations and calculates the corresponding concentrations in across the locations. The concentrations in turn are linked to a health impacts module that estimates the health consequences of the concentrations on a given population based on Burnett et al. (2014). Finally, the health consequences (premature mortality and morbidity) are converted into health costs in monetary units. All this takes place over a number of years during which the constraints on emissions apply. Such modelling is complex and has several uncertainties related to the economic modelling, the air dispersion modelling, and populations at risk and the health impacts of the pollutants. Thus the results should be viewed as ranges, ideally with confidence intervals.

The study used in this chapter (Markandya et al. 2018) can be taken as representative of the kind of results one finds in the literature on health co-benefits, certainly as far as mitigation costs are concerned. In that study, the Global Change Assessment Model (GCAM) was used to quantify the GHG pathways and related mitigation costs of the different scenarios. GCAM is an integrated assessment model

---

[2]This chapter uses material prepared for the Markandya et al. (2018) paper but not published in the article.

originally developed by the Joint Global Change Research Institute (JGCRI), which is part of the Pacific Northwest National Laboratory (PNNL). It has been used in most major climate/energy assessments over the last 20 years, including the last IPCC Report. The model is disaggregated into 32 geopolitical regions and operates in 5-year time steps from 2005 to 2100. Details of the model can be found online at https://github.com/JGCRI/gcam-doc. It provides the mitigation cost of different energy and climate policies for each specific region. It also reports the emissions of the main air pollutants (Smith et al. 2005) including organic carbon (OC), black carbon (BC), nitrogen oxides (NOx), non-methane volatile organic compounds (NMVOCs), carbon monoxide (CO) and sulphur dioxide (SO_2) which are main precursor gases of $PM_{2.5}$ and O_3. These emissions are calculated by applying an emission factor (EF) for every pollutant to each technology; consequently, the activity level, such as fuel consumption, drives emissions per period and region. In addition, an "emission control" per activity is also applied. The emission control level generally increases as GDP increases, representing historical trends that, with higher income levels, more stringent pollution control measures will be put into place.

GCAM projections of emissions of air pollutants are passed on to TM5-FASST[3]—an air quality source–receptor model, which translates emission levels into pollutant concentrations, exposure and premature deaths. TM5-FASST is a reduced-form global air quality source–receptor model developed by the European Commission's Joint Research Centre (JRC). It analyses how the emissions of a "source region" affect "receptor points" (grid cells) in terms of concentrations and subsequently of premature deaths and other health impacts.

Given the concentration levels for each region the model calculates the health impacts derived from exposure to O_3 and to $PM_{2.5}$, disaggregating by different causes of death[4] as defined in Forouzanfar et al. (2016). These calculations require baseline mortality rates, which are taken from the World Health Organization (WHO). More details are available on the website of the Institute for Health Metrics and Evaluation and in the aforementioned literature.

These health impacts are then monetized using Value of Statistical Life (VSL) (Lindhjem et al. 2012; OECD 2014) with the valuation extended to incorporate morbidity effects (Hunt et al. 2016).

In order to obtain results on mitigation costs and health benefits some further assumptions are required. The first is to embed the GCAM model in a scenario of future economic and population change. There is a rich literature on such socioeconomic pathways (SSPs) developed by IIASA and others that lays out the evolution of GDP and population by country under different development narratives.[5] The outcomes of different mitigation policies are then studied as

---

[3]Further details of the model can be found in Van Dingenen et al. (2018).

[4]For O_3 coverage is for respiratory disease and for PM it is for ischemic heart disease, chronic obstructive pulmonary disease, stroke, lung cancer, and acute lower respiratory airway infections.

[5]See https://tntcat.iiasa.ac.at/SspDb/dsd?Action=htmlpage&page=about. An important part of the SSP scenario is the emissions factors over time. For details of these see Rao et al. (2017).

deviations from the pathways described under these scenarios. In this case the socioeconomic scenario chosen is SSP2, which is considered a "middle of the road" framework. It has the following features:

- Current trends continue with some progress towards the Sustainable Development Goals, with lower energy and material intensity consumption and lower fossil fuel dependency.
- There is an unequal development rate between low income countries and a persistence of global and in-country inequalities.
- Low level of investment in education prevents low population growth.
- Global governance achieves an intermediate level of environmental protection

Further details can be found in Van Vuuren et al. (2017). This scenario gives the use of different fuels and the associated emissions of GHGs and local pollutants in the base case. We report in how sensitive the results are to the chosen SSP.

When additional controls are introduced they reduce GHG emissions but at a cost in terms of economic output, which is calculated by the model.

The second set of assumptions relates to the distribution of mitigation targets across countries and regions. While the overall cost of mitigation will depend little on this distribution,[6] the cost to individual countries and regions certainly will. While some indication of what burden different countries are willing to bear can be ascertained from their Nationally Determined Contributions (NDCs), these are not firm enough and do not go far enough into the future to calculate national/regional costs. Alternative allocation rules for sharing the burden are discussed in Robiou du Pont et al. (2016), where the authors suggest five distributional approaches, of which three have been selected. They are summarised in Table 1.[7] The implications of each have been examined.

Finally, the analysis looks at three mitigation strategies in terms of emissions and/or temperature increases. These are in addition to a baseline scenario where no climate policy is set and consist of: (a) the Nationally Determined Contributions (hereinafter NDCs), (b) 2 °C stabilisation target and (c) 1.5 °C stabilisation target (both (b) and (c) objectives to be attained by the year 2100).

Following Robiou du Pont et al. (2016), the world is divided into five regions: China, EU-27, India, USA (which covers 60% of global emissions in 2015) and the rest of the world (ROW). Also, following the same literature, the results are presented until 2050.

---

[6]If the reduction is achieved through a global carbon tax the total cost is independent of the distribution of the burden. With less efficient tools for attaining the reduction, the cost can depend to some extent on the distribution.

[7]The two excluded allocations are ones involving very unequal allocations to developed countries. Moreover, to be realized they require huge negative emissions, which is unrealistic.

Health Co-benefits of Climate Mitigation Policies: Why Is It So Hard... 231

Table 1 Mitigation equity criteria

Allocation name	Code	IPCC category	Allocation characteristics
Constant emission ratios	CER	Staged approach	Maintains current ratios of emissions to output, preserves status quo. Also referred to as grandfathering, it is not considered as an equitable option in climate justice and is not supported as such by any party
Capability	CAP	Capability	Countries with high GDP per capita have low emissions allocations
Equal per capita	EPC	Equality	Convergence towards equal annual emissions per person by 2040

Source: http://paris-equity-check.org

2.2 Results from the Analysis

We begin by looking at the global mitigation costs and the global health co-benefits under different scenarios. For the 2 °C target the global costs over the period 2020–2050 range from 0.7 to 1.3% of global GDP, while for the 1.5 °C target the range is 1.3–1.8%.[8] Between the scenarios the lowest costs emerge under the CER or EPC scenario and the highest ones under the CAP scenario.[9] These numbers are similar to those in the 5th IPCC Working Group III Assessment report (Edenhofer et al. 2014), where the values for different years for the 2 °C scenario range from around 0 to 2%. The cost of meeting the NDC reductions is only around 0.2% of global GDP. So going from the NDC scenario to the 2 °C scenario involves an increase in cost of 3.5 to 6.5 times, which, as we discuss later, may be problematic.

Next we compare these mitigation costs with the global health co-benefits associated with the different scenarios. Figure 1 shows the global health co-benefits and mitigation cost for each scenario. Health co-benefits are the difference between the monetized health damage of each policy scenario with respect to the baseline. The figure includes an uncertainty range based on a sensitivity analysis for the value of a statistical life (VSL), which is the metric used to value premature death. It is the variable most influential in determining the health benefits. The lower and the upper VSL values are drawn from the literature (Viscusi and Aldy 2003).

A notable observation in this figure is that at the global level the central value of the health co-benefit is greater than the cost of achieving the mitigation target for all the scenarios for the mean values of the health costs. Some mitigation strategies show co-benefits that are more than double the mitigation cost. The health co-

---

[8] All costs are discounted at 3%. The impact of the choice of the discount rate is discussed later.

[9] Costs are higher under the CAP scenario because under that scenario the total reduction in emissions to 2050 is greatest. Although total emissions reductions to 2100 are the same under all three scenarios, the CAP rule allocates in inverse proportion to GDP. With growing GDP and convergence in GDP per capita over the century the total allocations post 2050 are greater and pre 2050 smaller, making the required reduction pre-2050 larger.

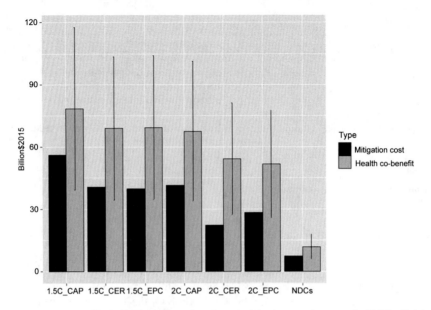

Fig. 1 Cumulative (2020–2050) health co-benefit and mitigation cost by scenario (Trillion$). The discount rate used is 3%. The black bars represent the range of values with lower and upper values of the VSL. Source: Adapted from Markandya et al. (2018)

benefit to mitigation cost ratio ranges from 1.4 (1.5C_CAP) to 2.45 (2C_CER). The sensitivity analysis shows that even when taking the lower bound of VSL, the health co-benefits are very close to the mitigation cost, covering between 70% and 91% of that cost. For the non-equitable allocation of emissions rights (2C_CER), even the lowest estimate of the health co-benefits is higher than the mitigation cost. Note that the higher co-benefits in the CAP scenario do not outweigh the larger policy costs, which results in a lower ratio of co-benefit to cost.

Having looked at the global mitigation costs and global co-benefits the next step is to look at the two components by region/country. In doing so we compare the co-benefits and mitigation costs for the different mitigation pathways by looking at how much of the additional effort of setting a more stringent target is compensated by the additional health co-benefits. Concretely the analysis shows:

- The net benefits of achieving the NDCs or the 2 °C target (following the different defined criteria) against the baseline (no climate policy) scenario
- The net benefits of achieving the 1.5 °C target instead of 2 °C

Table 2 compares, for each of the intermediate steps, the marginal health co-benefits against the marginal mitigation for a range of values of the VSL. A light grey cell indicates that, regardless of the VSL value, the marginal health co-benefits are greater than the marginal mitigation cost. A darker grey colour indicates whether the health co-benefits exceed the extra mitigation cost depending on the VSL value. Finally, if the cell is black, the additional health co-benefits are not sufficient to cover

Health Co-benefits of Climate Mitigation Policies: Why Is It So Hard. . . 233

Table 2 Net marginal benefits by region and scenario (Trillion $)

Scenario	China	EU-27	India	ROW	USA	TOTAL
NDCs	6.36(3.06 ; 9.66)	-2.01(-2.08 ; -1.93)	5.12(2.52 ; 7.72)	-0.72(-0.38 ; -1.06)	-4.42(-4.68 ; -4.16)	4.33(-1.57 ; 10.24)
2C°						
CAP	14.49(0.77 ; 28.21)	-2.70(-3.74 ; -1.67)	26.25(11.18 ; 41.33)	-5.01(-8.29 ; -1.73)	-7.12(-7.76 ; -6.48)	25.91(-7.84 ; 59.67)
CER	14.89(5.39 ; 24.39)	-0.22(-0.60 ; 0.17)	23.40(9.16 ; 37.64)	-4.81(-7.32 ; -2.29)	-1.23(-1.65 ; -0.81)	32.03(4.97 ; 59.10)
EPC	15.22(3.62 ; 26.82)	-1.22(-1.88 ; -0.56)	19.21(8.73 ; 29.70)	-4.42(-7.05 ; -1.79)	-5.33(-5.85 ; -4.81)	23.46(-2.44 ; 49.35)
1.5C°						
CAP	0.27(-1.21 ; 1.75)	-0.27(-0.65 ; 0.12)	3.76(0.98 ; 6.55)	-6.21(-6.83 ; -5.59)	-1.21(-1.37 ; -1.06)	-3.66(-9.08 ; 1.77)
CER	2.08(-1.32 ; 5.47)	-0.60(-1.20 ; -0.01)	3.28(0.93 ; 5.63)	-5.92(-6.76 ; -5.08)	-2.47(-2.70 ; -2.24)	-3.63(-11.05 ; 3.78)
EPC	2.31(-0.05 ; 4.67)	-0.19(-0.68 ; 0.31)	8.40(3.53 ; 13.28)	-3.46(-4.32 ; -2.60)	-0.93(-1.11 ; -0.76)	6.14(-2.63 ; 14.90)

The discount rate used is 3%. The values in brackets show the range of results based on the lower and the upper bounds of the VSL
Note: The first row represents the net marginal benefit of adopting the NDCs and the next three the net marginal benefit of going for the 2 °C stabilisation target. In each case the comparison is made against a no-climate-policy baseline. The last three rows give the net marginal benefits of setting a 1.5 °C policy against the 2 °C. See text for meaning of the colour scale
Source: Markandya et al. (2018)

the additional mitigation cost for any VSL value but they still make a significant contribution to reducing the overall cost.

For the NDC targets the net benefits are positive for China and India but for no other region. For the EU, the USA and the Rest of the World (RoW), mitigation costs exceed the health co-benefits for all values of the VSL. Thus the case for even the NDC targets in most of the world will require appealing to the benefits of reducing climate impacts more widely. Going from the baseline to the 2 °C target, broadly the same applies: China and India have a net gain from the action irrespective of the VSL but other parts of the world do not. Finally going from the 2 °C target to the 1.5 °C target the extra is always fully compensated for India, while for China it depends on the VSL chosen. The results in the other regions suggest that the marginal mitigation cost is generally higher than the marginal co-benefit. Globally, the marginal health co-benefits outweigh the marginal mitigation cost of a 2 °C target depending on the VSL value, except in the case of CER when this holds for all VSL values.

To conclude, the evidence on health co-benefits of mitigation policy is more nuanced than is sometimes realised. The health co-benefits are significant and reduce the overall cost of climate policy but they do not in general exceed those costs. The exceptions are China and India, where they can exceed them, based on the valuations given to the health co-benefits. Of course even if such co-benefits are only a fraction of the mitigation costs, they should make a difference to what policy is adopted. The main reason for seeking a 2 °C or a 1.5 °C stabilisation is to avoid the damage that climate change can do to the economy and society through its impacts on agriculture, on human health and on the infrastructure that underpins all economic activity. Such benefits are notoriously difficult to quantify in monetary terms and have large uncertainties but that does not make them unimportant. On the contrary, risks of major damages if tipping points are exceeded are a strong reason for seeking the stabilisation targets set out in the Paris Agreement.

In the next section we consider why health co-benefits have played a relatively small role in climate policy and go on to look at how that role could be enhanced.

3 Reasons Why Co-benefits Are Problematic for Climate Policy

Issues with the health co-benefits and their role in climate policy have been raised in the following areas:

- Role of valuing premature deaths in overall health co-benefits
- The time profile of the benefits and costs and the role of the discount rate
- Political economy of diffuse benefits versus concentrated costs
- Separating health benefits from mitigation strategies.

We consider the validity of each of these reasons in this section.

3.1 Role of Valuing Premature Deaths in Overall Health Co-benefits

The major share of the health benefits arises from a reduction in premature mortality. These make up 90% of the total. Indeed, the valuation is done such that the premature mortality reductions are valued first and 10% of that value is added on as a proxy for the morbidity or other costs. This 10% figure is an approximation based on various European studies of the two kinds of costs reviewed for the OECD by Hunt et al. (2016). The mortality benefits are derived from a method based on the value of a statistical life (VSL), which has been studied extensively in OECD countries and comes up with range in 2005 US dollars of between $1.8 and $4.5 million (2.3–5.9 million updated to 2018 US Dollars). This range applies to developed countries that are members of the OECD. For countries with a lower level of GDP per capita, the OECD recommends a lower value based on its per capita GDP relative to that of the OECD average.

Problems arise with such an approach in two directions. On the one hand, countries that are less well-off object to loss of life of their citizens being worthless than that of citizens in better off countries. On the other hand, even making this adjustment gives a value for a life saved in a country like India or China that is not reflected in public expenditures to save lives. For example, the average per capita GDP of the OECD countries in 2005 was $30,268 and that of India was $3046. The value of one less premature death in India would then range from $290,000 to \$1 million based on the OECD proposed formula.[10] Given the large number of people affected by air pollution in India, such a range gives a large number to the health benefits. Yet one can argue that public policy in relation to areas such as road safety and public health in India does not value lives saved so highly. There are many actions the government could take that would imply a cost per life saved that is much lower than that but it does not take them.[11]

The other problem with the value of lives saved is that there is no financial flow associated with it. This is not the case with the morbidity savings, which imply less spending by individuals and health agencies. Reducing premature deaths, on the other hand, could even imply an increase in public outlays if the people who are saved are the relatively less healthy and the old (which is partly but not entirely the case with air pollution) and would demand more of public health care if they were alive. Hence government agencies, driven largely by budgetary concerns, will undervalue the billions of dollars of "health benefits" from lives saved compared to the health benefits of less sick people placing demands on public health services.

Both these features of the way that health co-benefits are calculated give rise to values that are too high from a policy perspective. The values are based on the right theoretical foundations but these are not what drives decision-making in the public domain.

3.2 The Time Profile of the Benefits and Costs and the Role of the Discount Rate

The calculations presented earlier are discounted present values, using a 3% discount rate. There is a considerable difference of opinion on what the appropriate discount rate should be (Arrow et al. 2012; Nordhaus 1994; Newell and Pizer 2003). There is no consensus but a broadly held view is that for decisions relating to actions with very long-term consequences, such as climate change, the rate should be lower than the market rate and possibly decline over time, so the rate applied to costs and benefits is lower for later periods than for earlier ones.

In a policy-making context, decisions are often driven disproportionately by present costs and benefits relative to future ones. This reflects the political imperative of immediate consequences of measures taken over the election cycle that extends over a few years. Hence, one could argue that what are relevant to decisions

---

[10]The formula proposed by the OECD for calculating the VSL in country i is: $VSL_i = VSL_{OECD} \left[\frac{GDPPC_i}{GDPPC_{OECD}} \right]^{0.8}$. GDPPC is the per capita GDP. 0.8 is referred to as the elasticity of VSL with respect to GDPPC. The use of a single value such as this has recently been questioned, it being argued that the value should vary regionally. See Viscusi and Masterman (2017).

[11]A lower value to lives saved is arrived at if one uses a formula based on GDP per capita but this has less justification on social grounds, as it is based on the output value of individuals and would by implication place no value on the old and the sick.

Table 3 Ratio of mitigation costs relative to health co-benefits for 2020–2024 and 2020–2050

	2020–2024 (%)	2020–2050 (%)
NDCs	73	59
2 °C CAP	59	60
2C_CER	38	41
2C_EPC	54	53

Source: Own calculations

on climate policy are the mitigation costs and health co-benefits over the next five to ten years, rather than over the three decades taken into account here. One way to focus more on the short term is to raise the discount rate but simulations for higher rates up to 10% do not show a significant change in the relative size of the mitigation cost versus the health co-benefits.[12] Instead, it may be more important to compare the mitigation costs and health co-benefits over the next five years. An analysis of such figures is given in Table 3, which gives the ratio of mitigation costs to health co-benefits over the next five years and over the whole 30-year period 2020–2050. In the NDC scenario the mitigation costs are relatively more important and health benefits less important for the first five years but for the 2 °C scenarios the ratios are quite similar. Thus, this factor does not appear to be a valid reason for underplaying the importance of health co-benefits.

3.3 Political Economy of Diffuse Benefits Versus Concentrated Costs

The health co-benefits of a mitigation policy are diffuse and spread over the whole population while the mitigation costs are concentrated on those sectors that have to switch from a high-carbon production process to a low carbon, and on changes in behaviour of households involving lower consumption of meat, different modes of transport, etc.

To put the health co-benefits in perspective, the 2 °C stabilisation programme for China would reduce mortality rates by something in the order of 2% compared to the baseline mortality rate. For an adult (15–60) baseline annual mortality is around 2 per thousand, so the reduction in that mortality is 0.044 per thousand. The VSL for China in 2018 US dollars is estimated at between $0.47 and $1.19 million (based on the approach described earlier), implying that the value of the reduction in mortality for such persons is between $21 and $53 per year.[13] While this is a useful gain to individuals, it is not going to give them a strong impetus to act. The reason the health

---

[12] A higher discount rate raises the relative importance of health co-benefits in China slightly but reduces it in all other countries/regions. This reflects the relative time profiles of co-benefits relative to mitigation costs, which are influenced by many factors.

[13] The value attached to a reduction in mortality of X% is the VSL times X%.

co-benefits come out as large as they do is that these small benefits apply to the whole population. On the other hand, workers in industries where carbon intensive production is being phased out, such as the coal sector, will face significantly higher costs. Equally, measures to reduce red meat consumption could have high costs in terms of personal preferences.

This asymmetry is reflected in the greater political pressure on decision makers to resist measures that support a low carbon strategy against less pressure to gain from the health co-benefits.

3.4 Separating Health Benefits from Mitigation Strategies

There is, to be sure, a growing realisation of the health consequences of air pollution across the world, especially in developed countries and industrialising economies where pollution levels in some cities has reached a critical phase (World Bank and Institute for Health Metrics and Evaluation 2016). Some data will help show the extent of the problem. According to a recent report from the WHO, one in nine (more than 11%) of today's premature deaths globally is a result of the exposure to air pollution (WHO 2016). Long-term exposure to ambient fine particle air pollution ($PM_{2.5}$), the pollutant that is the greatest contributor to adverse public health impacts, represents 7.6% of total premature global mortality, making it the fifth-ranked global risk for death in 2015 (Cohen et al. 2017). There is also increasing evidence that suggests that morbidity from cardiovascular and respiratory disease and lung cancer is shortening life expectancy. Air pollution exposure reduced average global life expectancy at birth by $\sim$1 year with reductions of $\sim$1.2−1.9 years in polluted countries of Asia and Africa. This statistic is also backed up by the latest estimates from the *Global Burden of Disease Project* (GBD) suggesting that in 2016 over 4.3 million people died prematurely from ambient air pollution.[14] The illnesses associated with air pollution-related premature mortality include lung cancer, heart disease, stroke, acute respiratory infections, and chronic obstructive pulmonary diseases such as bronchitis and emphysema. Recent evidence, however, has increased the range of adverse impacts to include illnesses such as diabetes, mental health and dementia (Bowe et al. 2018; Bishop et al. 2018; Newbury et al. 2019). Moreover, global estimates of mortality have reevaluated (and increased) risks of cardiovascular and cancer illnesses to air pollutants such as $PM_{2.5}$ (Burnett et al. 2018).

These data are now driving a programme to address local air pollution in many countries and while some parts of that programme have a direct link to low carbon actions, other parts are not so low carbon focussed. One measure in the EU, for

---

[14]Household air pollution associated with the use of solid biofuels for cooking and heating in homes also is a major source of premature mortality in the developing world, with an estimated 2.5 million premature deaths in 2016 according to GBD.

example, has been to phase out many diesel vehicles because of their high emissions of NOx. While some of the replacement is to hybrid and electric vehicles, more is to vehicles using gasoline that is less carbon efficient. Another set of measures to reduce local pollution focus on relocating polluting industries so they are further away from concentrations of population.[15] Such relocation does not, of course, reduce GHGs.

4 Possible Measures to Make Health Co-benefits Play a Bigger Role in Climate Policy

The factors described in the previous section make it difficult for the health co-benefits to play their full part in the design of a low carbon strategy at the national and global level. The following are some actions that could be taken to ensure that such co-benefits are given their due importance.

4.1 Treat Health Benefits and Climate Mitigation Gains Equally

There is no reason why climate policy should be the main actor and health impacts of that policy the junior partner. In many emerging economies and even in developed ones, there is at least as much concern about the health impacts of air pollution as there is about meeting GHG reduction targets. One could argue that it makes more sense in some situations to make the GHG reductions of a clean air policy the ancillary benefit, recognising that some clean air policies may have no GHG reductions associated with them (e.g. end-of-pipe measures). This can be done by estimating the GHG reductions associated with a range of clean air interventions and giving a credit to the agency or agencies that implement that action through carbon credits. This will provide a stronger incentive to undertake clean air measures that could be seen as costly.

4.2 Give Less Importance to VSL and More to Morbidity and Economic Costs of Local Air Pollution

The VSL methodology gives large numbers in terms of health co-benefits but, as explained above, those values do not drive decisions in ministries responsible for

---

[15]This can help reduce exposure to pollutant concentrations but account has to be taken of air movement within airsheds.

public health. Such agencies are of course interested in reducing premature deaths but they are also equally interested in the detailed impacts of air pollution on health services and on the economy through absenteeism at work. In the UK, for example, the National Institute for Health Care and Excellence (NICE) uses a rule of thumb for approving medical interventions if the cost per Disability Adjusted Life Year (DALY) is less than £15,000 (US$19,800). While there is no simple link between DALY and VSL, an estimate is made that VSL is equal to 22 DALYs (Cropper and Khanna 2014). This would make the VSL used in the UK equal to about US$436,000, which is much lower than the range recommended by the OECD of US$2.3–5.9 million. Even with this low value, however, many clean air measures will be justified, especially if the full morbidity costs are taken into account. Furthermore, if credit is given for the GHG reduction associated with the measures, the cost per DALY or VSL could fall into the range indicated.

Also missing from the analysis are some other sources of economic costs. Developing countries with high levels of ambient air pollution are not attractive places to live in for local businessmen and expatriates, which also discourages foreign investment and growth. A recent World Bank study has shown that the firm performance is negatively linked to ambient air pollution at lower levels of pollution than expected from previous studies (Soppelsa et al. 2019). Yet the effects of agglomeration economies of large cities on labour productivity growth remain strong. Hence, absent from air pollution these urban areas would contribute significantly to growth. These effects of air pollution have not been taken into account in the calculations reported here but could make a big difference.

4.3 Find Ways to Use Health Co-benefits to Compensate Losers of Mitigation Actions

The problem of diffuse benefits and concentrated costs can be solved if those who bear the costs are compensated in some way from the benefits. We know that in principle this is possible where the health co-benefits are found to be greater than the costs, which holds globally in most cases. The problem is finding the instruments for making the transfers.

One possibility is for low carbon programmes to include compensation for parties that suffer losses, funded through a "clean air dividend". A small charge could be levied on beneficiaries of health co-benefits, reflecting their ability and willingness to pay, and used for this purpose. A second is for the government to use the general budget to support the compensation but adjust the allocation to the health budget downwards to reflect lower health costs. A third, in the case of low income countries where government budgets are inadequate and individuals' ability to pay for clean air very limited, is for donor assistance to include compensation to the losers as part of a "just transition". The last could be of particular relevance in poorer sub-Saharan African countries, where the estimated ambient air pollution health co-benefits of low carbon strategies are small relative to the costs of GHG mitigation. All these alternatives need further discussion and debate to determine which of them are best suited in each country.

5 Conclusions

This chapter has reviewed the evidence on the health co-benefits of different GHG mitigation pathways and compared them to the mitigation costs. The analysis shows that globally the co-benefits exceed the mitigation costs for the central estimates of the co-benefits but that is not true in all countries or regions. Moreover, uncertainty of the health co-benefits, especially the value of reductions in premature death, means that taking the lower bounds of the co-benefits gives a net cost to the pathways in most countries and regions. That of course is not a reason to be despondent. A low carbon strategy has many other benefits by stabilising the climate and the magnitude of the net cost (at 1–2% of GDP) is small when seen as a means of reducing climate impacts overall and when seen as an insurance against the more extreme outcomes if no action is taken.

Nevertheless, it is worth looking at ways in which the health co-benefits can play a full role in determining the right combination of climate stabilisation and clean air policies. It may be better to look at the two sets of policies on an equal footing rather than having health co-benefits as an ancillary to the climate mitigation programmes. It would help if morbidity costs and other economic costs of air pollution were given greater prominence and if measures to compensate the losers of the low carbon strategies were introduced.

References

Arrow KJ et al (2012) How should benefits and costs be discounted in an intergenerational context? RFF DP 12–53. Resources for the Future, Washington, DC

Bishop KC, Ketcham JD, Kuminoff NV (2018) Hazed and confused: the effect of air pollution on dementia. No. w24970. National Bureau of Economic Research, Cambridge, MA

Bowe B, Xie Y, Li T, Yan Y, Xian H, Ziyad Al-Aly Z (2018) The 2016 global and national burden of diabetes mellitus attributable to PM2.5 air pollution. Lancet Planet Health 2:e301–e312

Burnett RT et al (2014) An integrated risk function for estimating the global burden of disease attributable to ambient fine particulate matter exposure. Environ Health Perspect. https://doi.org/10.1289/ehp.1307049

Burnett RT et al (2018) Global estimates of mortality associated with long-term exposure to outdoor fine particulate matter. Proc Natl Acad Sci U S A 115:9592–9597

Cohen AJ et al (2017) Estimates and 25-year trends of the global burden of disease attributable to ambient air pollution: an analysis of data from the global burden of diseases study 2015. Lancet 389(10082):1907–1918

Cropper M, Khanna S (2014) How should the world bank estimate air pollution damages? RFF DP 14–30. Resources for the Future, Washington, DC

Edenhofer O et al (2014) Climate change 2014: mitigation of climate change. IPCC WGII report. Cambridge University Press, Cambridge

Forouzanfar MH, Afshin A, Alexander LT, Anderson HR, Bhutta ZA, Biryukov S, et al (2016) Global, regional, and national comparative risk assessment of 79 behavioural, environmental and occupational, and metabolic risks or clusters of risks, 1990–2015. Lancet [Internet]. [cited 2017 Mar 9]; Available from: https://helda.helsinki.fi/handle/10138/172718

Glomsrod S (1990) Stabilization of emissions of CO_2: a computable general equilibrium assessment, discussion paper no.48. Central Bureau of Statistics, Oslo

Hunt A, Ferguson J, Hurley F, Searl A (2016) Social costs of morbidity impacts of air pollution. Report no 99. OECD Publishing, Paris

Lindhjem H, Navrud S, Biausque V, Braathen N (2012) Mortality risk valuation in environment, health and transport policies. OECD Publishing, Paris

Markandya A, Sampedro J, Smith SJ, van Dingenen R, Pizarro-Irizar C, Arto I, Gonzalez M (2018) Health co-benefits from air pollution and mitigation costs of the Paris agreement: a modelling study. Lancet Planetary Health 2:e126–e133

Newbury JB, Arseneault L, Beevers S, Kitwiroon N, Roberts S, Pariante CM, Kelly FJ, Fisher HL (2019) Association of air pollution exposure with psychotic experiences during adolescence. JAMA Psychiat 76:614–623

Newell RG, Pizer WA (2003) Discounting the distant future: how much do uncertain rates increase valuations? J Environ Econ Manag 46:52–71

Nordhaus WD (1994) Managing the global commons: the economics of climate change. MIT Press, Cambridge

OECD (2014) Cost of air pollution: health impacts of road transport. OECD Publishing, Paris

Rao S et al (2017) Future air pollution in the shared socio-economic pathways. Glob Environ Chang 42:346–358. https://doi.org/10.1016/j.gloenvcha.2016.05.012

Robiou du Pont Y, Jeffery ML, Gütschow J, Rogelj J, Christoff P, Meinshausen M (2016) Equitable mitigation to achieve the Paris agreement goals. Nat Clim Chang 7(1):38–43

Smith SJ, Pitcher H, Wigley TML (2005) Future sulfur dioxide emissions. Clim Chang 73(3):267–318

Soppelsa ME, Gracia NL, Xu LC (2019, April 30) The effects of pollution and business environment on firm productivity in Africa. World Bank Policy Research Working Paper No. 8834. Available at SSRN: https://ssrn.com/abstract=3380727

Van Dingenen R, Dentener F, Crippa M, Leitao J, Marmer E, Rao S et al (2018) TM5-FASST: a global atmospheric source–receptor model for rapid impact analysis of emission changes on air quality and short-lived climate pollutants. Atmos Chem Phys 18(21):16173–16211

Van Vuuren DP, Riahi K, Calvin K, Dellink R, Emmerling J, Fujimori S et al (2017) The shared socio-economic pathways: trajectories for human development and global environmental change. Glob Environ Chang 42:148–152

Viscusi WK, Aldy JE (2003) The value of a statistical life: a critical review of market estimates throughout the world. J Risk Uncertain 27(1):5–76

Viscusi WK, Masterman CJ (2017) Income elasticities and global values of a statistical life. J Benefit-Cost Anal 8:226–250

World Bank and Institute for Health Metrics and Evaluation (2016) The cost of air pollution: strengthening the economic case for action. World Bank, Washington, DC

World Health Organization (WHO) (2016) Ambient air pollution: a global assessment of exposure and burden of disease. WHO, Geneva

Financing Forest Protection with Integrated REDD+ Markets in Brazil

Ronaldo Seroa da Motta, Pedro Moura Costa, Mariano Cenamo, Pedro Soares, Virgílio Viana, Victor Salviati, Paula Bernasconi, Alice Thuault, and Plinio Ribeiro

1 Introduction

It is widely recognized that Brazil has achieved a great degree of greenhouse gas (GHG) abatement. In 2017, GHG emissions in Brazil were 17% lower than in 2000, despite the increase in the economic output of over 70% in the same period (SEEG 2019).

This was achieved mainly due to successful forest management policies that have reduced the deforestation rate by 75% since 2002. The Intended Nationally Determined Contributions (INDC) outlined by Brazil in the UN Framework Convention on Climate Change (UNFCCC) Conference of the Parties (COP21) in Paris in

---

R. Seroa da Motta (✉)
State University of Rio de Janeiro (UERJ), Rio de Janeiro, Brazil
e-mail: ronaldo.motta@uerj.br

P. M. Costa
Bolsa de Valores Ambientais do Rio de Janeiro (BVRio), Rio de Janeiro, Brazil
e-mail: pedro.mouracosta@bvrio.org

M. Cenamo · P. Soares
Instituto de Conservação e Desenvolvimento Sustentável da Amazônia (Idesam), Manaus, Brazil
e-mail: mariano@idesam.org.br; pedro.soares@idesam.org.br

V. Viana · V. Salviati
Fundação Amazonas Sustentável (FAS), Manaus, Brazil
e-mail: virgilio.viana2@fas-amazonas.org; victor.salviati@fas-amazonas.org

P. Bernasconi · A. Thuault
Instituto Centro de Vida (ICV), Cuiabá, Brazil
e-mail: paula.bernasconi@icv.org.br; alice@icv.org.br

P. Ribeiro
Biofílica Investimentos Ambientais (Biofilica), Sao Paulo, Brazil
e-mail: plinio@biofilica.com.br

© Springer Nature Switzerland AG 2020
W. Buchholz et al. (eds.), *Ancillary Benefits of Climate Policy*, Springer Climate,
https://doi.org/10.1007/978-3-030-30978-7_14

December 2015 is even more ambitious and has a commitment to reduce nationwide emissions from forest fire and logging by 90% by 2030.

It is important to highlight that this successful forest policy changes to curb deforestation in Brazil were not initiated to pursue carbon mitigation targets. The main motivation was to protect forest ecosystem services.

Deforestation has negative effects on plant and animal species found in the rainforest. In the case of the Amazon Region, one of the most diverse forests on Earth, such species loss has quite a high social cost.

At least 40,000 plant species, 427 mammals, 1300 birds, 378 reptiles, over 400 amphibians, and around 3000 freshwater fish have been found in the Amazon (Silva 2005). In Brazil alone, between 96,660 and more than 100,000 invertebrate species have been described by scientists (Lewinsohn and Prado 2005).

Many of these species are not found anywhere else in the world. ter Steege et al. (2015) estimated using overlay spatial distribution models with historical and projected deforestation that at least 36% and up to 57% of all Amazonian tree species are likely to qualify as globally threatened under the International Union for Conservation of Nature (IUCN) Red List criteria. If confirmed, these results would increase the number of threatened plant species on Earth by 22%. These authors also indicated how remaining areas of undisturbed and recovering forest provided the last refuge for many species unable to withstand the impact of human activity. Therefore, conserving existing secondary forests may be much cheaper and even more efficient than planting trees to maintain the ecosystem services.

Strassburg et al. (2012) show that adequately funded and broadly implemented carbon-based forest conservation could play a major role in biodiversity conservation, as well as in climate change mitigation.

In addition to non-use values attached to biodiversity, there are important local use values that are also highly dependent on forest management. Malhi et al. (2008) show that evaporation and condensation over the Amazon serve as engines of the global atmospheric circulation, with downstream effects on precipitation across South America and further afield across the Northern Hemisphere. And deforestation exacerbates ecosystem feedbacks such as forest dieback and reduced transpiration.

Paiva et al. (2013), for example, demonstrate that surface waters dominate total water surface variation for the whole Amazon area with a fraction of 56%, followed by soil (27%) and ground water storages (8%).

Floodplains also play a major role in stream flow routing. As stated by Sumila et al. (2017), the Amazon then helps regulate the regional humid climate that affects economic activities that depend on it, such as agricultural productivity and hydropower generation. These effects, according to the authors, are stronger during the onset of the rainy season than when the season is fully developed. Therefore, it leads to less rainfall in the transition months or even a delay in the onset of the rainy season, as deforestation progresses.

The values of these benefits are not priced but are valuable as already pointed out by several studies (Andersen et al. 2002; Seroa da Motta 2005). Strand et al. (2018) have recently estimated economic values for a range of ecosystem services provided by the Brazilian Amazon forest, including extractive production, greenhouse gas mitigation and climate regulation effects on agriculture, cattle raising, and hydroelectricity production. Their highest estimated values range from US\$56.72 $\pm$ 10 ha^{-1} year^{-1} to US\$737 $\pm$ 134 ha^{-1} year^{-1}.

The achieved reduction in the deforestation rate in Brazil has contributed to capturing part of these benefits. It has done so by combining control instruments to improve the monitoring system and strengthening sanction enforcement on forest management, plus restrictions to agricultural expansion in the region, reduced access to agricultural credit and markets, and forging supply chain voluntary actions not to use illegal deforested areas.

However, it has been already pointed out that the incentive power of these control measures has reached its limit and the continuation of the successful pathway may require a more complex set of incentives. Pricing these benefits of ecosystem services in forest conservation must then play a key role in the designing of conservation strategies (Strand et al. 2018; Gibbs et al. 2015; Nepstad et al. 2014; Barreto and Araujo 2012; Hargrave and Kis-Katos 2013).

One instrument is to count on payments for the performance on the reduction emission of forest deforestation and degradation (REDD+). Most tropical forest-rich countries are already designing their REDD+ strategies, while developed countries are in turn requested to scale up respective short- and long-term financing. In most cases, REDD+ strategies will need to draw on customized policy mixes to become effective.

The chapter extends the analysis of a proposal of an Integrated REDD+ approach to fit into Brazil's REDD strategies that was initially presented in Costa et al. (2017). The next section presents the REDD+ framework within the UNFCC and addresses its main additionality concerns. The third section overviews REDD+ markets, and its following section presents details of the proposed Integrated REDD+ approach. The last section concludes the matter.

2 The REDD+ Approach

Since deforestation accounts for more than 15% of global GHG emissions, there is an interest in funding mechanisms to reduce emissions from deforestation and forest degradation. Article 5 of the Paris Agreement (United Nations 2016) reinforces the REDD decisions already agreed under the Climate Convention.

This article contains two paragraphs. The first emphasizes the decision to encourage actions to conserve and improve sinks and reservoirs of greenhouse gases, including forests, as appropriate.

The second paragraph encourages measures to implement and support, together with results-based payments, the guidelines, and decisions that have been already approved for activities related to emission reductions from deforestation and degradation, as well as conservation, sustainable management, sustainable forest management, and increased forest carbon stocks in developing countries. It also includes joint mitigation and adaptation approaches for the integrated and sustainable management of forests, reaffirming the importance of encouraging the other benefits in addition to non-carbon benefits associated with such approaches (Seroa da Motta 2018).

REDD+, however, faces the technical issues that are common to any credit-based offset mechanism. These arise from the potential variability of definitions of what changes in emissions should be credited to the project and how. The validation difficulty is related to baselines, detecting leakage, and the mensuration of carbon sequestration or release from biomass, all of which have been quoted as creating uncertainties about the climatic impacts of REDD+.

Leakage may occur when the funding of reductions in deforestation and degradation in one particular place leads to more trees being cut down in other forest areas. Such possibility may not guarantee additionality, that is, reduced carbon emissions would have occurred even without REDD+ payments. Forest preservation also faces permanence risk when carbon is just temporarily stored in the forests since there is no guarantee that this stored carbon will not be emitted in the future because of economic destructive activities or natural hazards (Moutinho and Guerra 2017; Agrawal et al. 2011; Paoli et al. 2010; Aukland et al. 2003).

Baseline and credit schemes have proven uncertain in relation to baseline definition and leakage when the flow-avoided deforestation is the credit. On the other hand, if the remaining forest stock (i.e., remuneration for any standing biomass) is to be given credit to avoid leakage, it would create an oversupply that could not be priced.

Therefore, the stock and flow method provide a balanced approach when payments are split into stock and flow credits. Payments to forest stocks create the incentives to past protection and flow payments to incentivize avoided deforestation. Carbon balance is made by jurisdiction level where reduced flow of emissions from deforestation and degradation below a historical reference emissions level would be eligible for payments. And carbon balance will have to be deducted from the country's national carbon balance sheet.

With this approach, REDD+ financing has moved away from project-based REDD+ activities toward jurisdictional and nested programs that cover national and sub-national levels (Moutinho and Guerra 2017).

Regarding validation, scientific evidence has been produced to overcome these concerns, and with the improvement of remote sensing and other monitoring technologies, these technical impediments have been mostly overcome (Watson et al. 2010).

Also, the evolution of guidelines and standards such as the Verified Carbon Standard (VCS),[1] the Climate, Communities and Biodiversity (CCB),[2] and the Warsaw

---

[1] www.v-c-s.org/project/vcs-program/

[2] www.climate-standards.org/ccb-standards/governance-of-the-standards/

Framework for REDD are ensuring that REDD+ projects generate comparable climatic benefits while creating other social and environmental co-benefits.[3]

Provided that GHG emission reductions generated by REDD+ initiatives are accounted for in a consistent way, between countries and within projects, programs and national initiatives within countries, REDD+ could become a robust GHG mitigation option.[4]

Developing country Parties implementing REDD+ activities should periodically provide a summary of information on the outcomes of activities related to REDD+, including activities on capacity building, demonstration activities, addressing drivers of deforestation, and mobilization of resources, that is, a transparent REDD + national strategy. This strategy and results should be included in national communications or be provided, on a voluntary basis, via the REDD+ Web Platform.

On December 2, 2015, the Brazilian Ministry of the Environment, through Ordinance No. 370, established the National REDD+ Strategy, designated by the acronym in Portuguese ENREDD+, based on the jurisdictional approach.

ENRRED's main goal is to integrate the governance structures of climate change, forest, and biodiversity-related policies, seeking to promote consistency and synergies among them at the federal, state, and municipal levels.

However, at least in its first phase, funding options do not allow REDD credits to be used as offset in other countries' NDCs. This position is mainly justified on the ground that such an approach generates a very large supply at very low costs, thus reinforcing the identified fears of reduction in the attractiveness of local mitigation actions.

Nevertheless, despite supply and demand management, REDD+ will compete with the sustainable development mechanism (SDM) that has been proposed in Paragraphs 6.4 and 6.7 of the Paris Agreement. This mechanism was based on one Brazilian initiative to replace the Clean Development Mechanism (CDM), and it includes incorporating modalities, procedures, and methodologies to allow for the negotiation of Certified Emissions Reduction units (CERs).[5]

Inspiration from the CDM has shifted to Paragraph 4 (a) of the Paris Agreement (United Nations 2016), which highlights the promotion of GHG emissions mitigation in order to strengthen sustainable development. And Paragraph 4 (b), much like

---

[3] Warsaw Framework for REDD+, http://unfccc.int/land_use_and_climate_change/REDD+/items/8180.php

[4] Indeed, this approach has been considered by the CORSIA (Carbon Offsetting and Reduction Scheme for International Aviation) initiative of the International Civil Aviation Organization (ICAO), i.e., to require that land use projects supplying credits to the scheme must be in conformity with the Warsaw Framework and ensure the harmonization of baselines between projects, jurisdiction, and nations. www.icao.int/environmental-protection/Pages/market-based-measures.aspx

[5] For a more detailed discussion on the differences and similarities between the CDM and the SDM, please refer to Greiner et al. (2017) and Seroa da Motta (2018).

the CDM, states that the mechanism should encourage and facilitate participation in greenhouse gas emissions mitigation by authorized public and private entities.

The Brazilian proposal (Brazil 2014) has also proposed that the mechanism be established to assist mitigation efforts of target countries and assist developing countries in implementing project activities with the aim of reducing GHG emissions or increasing removals by sinks. Consequently, all countries could emit SDM-certified emission reductions, and the scope could cover a wide range of activities, including those associated with sinks (Seroa da Motta 2018; Marcu 2016; IETA 2016).

Paragraph 6.4 (c) seems to confirm this possibility, as it refers to mitigation activities and the reduction of emission levels by the generating country, reaffirming in Paragraph 6 (d) that the SDM should "provide global mitigation in global emissions." Thus, along these lines, it would be possible to integrate a CDS range of activities with the REDD mechanism, as discussed as follows.

3 The REDD+ Market

Increasing urgency required to prevent catastrophic climate change requires the integration of all greenhouse gas (GHG) mitigation options and sectors in parallel. The creation of a carbon price is essential for the decarbonization of energy generation, industrial processes, transportation modes, and consumer patterns.

The process of price formation, however, breaks down if an oversupply of cheap mitigation options is mixed with measures that require higher carbon prices to compete. This has been the dynamic between land use mitigation options, in particular REDD+, reducing emissions from deforestation and forest degradation in relation to industrial and energy options. If REDD+ is included in markets, however, the potential oversupply of this mitigation option could reduce prices to an extent that there could be no financial incentive for promoting action on industrial improvements, energy efficiency, or renewable energy.

The easy solution, adopted since the outset of the Kyoto Protocol in 1997, has been to exclude REDD+ from markets. Land use, however, is still responsible for ca. 24% of global GHG emissions,[6] and as already said for playing a vital role in biodiversity conservation, water flows, and livelihoods. With no carbon pricing to support it, significant levels of deforestation and GHG emissions have occurred from 1997 to date. The exclusion of forests, therefore, resulted in a missed opportunity to create financial incentives for promoting the reduction of deforestation.

Looking forward, it is essential that developing countries secure financial resources to ensure forest protection in order to meet their GHG emission reduction targets under the Paris Agreement (United Nations 2016). The INDCs[7] of many

---

[6]https://www.epa.gov/ghgemissions/global-greenhouse-gas-emissions-data

[7]NDCs (Nationally Determined Contributions) are the GHG emission reduction targets assumed under the Paris Agreement.

tropical forest countries are heavily reliant on the reduction of deforestation and ensuring land use sustainability. In the case of Brazil, for instance, 89% of its NDC emission reductions are expected to derive from reduced deforestation (Piris-Cabezas et al. 2016). However, while government budgets to environmental protection have been severely cut,[8] to date the only source of external funding available to support this effort in Brazil has been ODA transfers, predominantly from Norway and Germany.

At the same time, if REDD+ is adopted, the large volumes of emission reductions with low unit cost would reduce the average costs of abatement and provide an incentive for countries to adopt more ambitious reduction targets.

There is an urgent need to conciliate the tensions and requirements of creating a carbon price for both REDD+ and other sectors of the economy, and secure long-term financial support for the protection of forests at a large scale. This integration of forest protection, production, and wider economic objectives would make it more cost effective and feasible to promote large-scale GHG mitigation and global decarbonization with the participation of all sectors of the economy.

Market concerns, however, remain valid, because of supply and demand unbalances. Land use practices have the potential to generate large volumes of GHG mitigation at low prices (Seymour et al. 2017; Edwards 2016). The implementation of the new Brazilian Forest Code, for instance, has the potential to conserve over 250 million ha of native vegetation in Brazil, storing ca. 100 $GtCO_2e$—the equivalent of 45 years of the European Union's industrial complex operating without caps (Britaldo et al. 2014). Tropical peatlands, in turn, currently store carbon equivalent to 5–9 years of global GHG emissions and their degradation is responsible for ca. 10% of global emissions (Kurnianto et al. 2015).

Global demand for GHG reductions, on the other hand, remains restricted due to the lack of binding commitments of international agreements. While the Paris Agreement points in the right direction, it does not create any binding commitments. California's market has the potential to absorb 225 $MtCO_2e$ of international offsets, and targets in the Carbon Offsetting and Reduction Scheme for International Aviation (CORSIA) could create a demand of 3 ca. Gt by 2035.

Unbounded supply of REDD+ units would result in low prices that could derail efforts to enhance endogenous technical changes in the decarbonization of the industrial, energy, agricultural, and transportation sectors. So, there is a need to control the supply of REDD+ units in markets, to the extent that it does not preclude other mitigation options. If managed properly, the relatively lower cost of REDD+ could enable the adoption of more ambitious deforestation control targets and involve all sectors of the economy.

---

[8]See, for instance, in the case of Brazil, the 2017 reduction of 51% in the budget of the Brazilian Ministry of Environment—www.observatoriodoclima.eco.br/ministerio-do-meio-ambiente-perde-51-da-verba-apos-corte/

4 Integrated REDD+[9]

Given that the inclusion of emission reductions from REDD+ in the same market as other mitigation options could result in undesirable impacts, a possible solution would be to create two distinct but complementary markets. REDD+ units should be negotiated in a pool of other REDD+ units, so as not to affect the pricing of other mitigation options. At the same time, nations or entities should not be allowed to meet their targets solely through the use of REDD+ units, but also by adopting a combined approach.

For instance, a first tranche of a country's target should be met by adopting internal decarbonization measures and/or non-REDD+ offsets, such as public policies and incentives for energy efficiency, renewable energy, improvements in industry and transportation, etc. Only after this "quota" is met could this same country complete its targets using REDD+ offsets.

A similar conditional approach to the conditional use of offsets was adopted by the European Union Emissions Trading System (EU ETS) where the "supplementary" concept required countries to meet only a proportion of their targets using offsets. This proportion celling was justified to keep carbon price incentives to technical change on energy and industrial mitigation options (Trotignon 2011).

The identification of the REDD+ share is also needed and should be the one equalizing the marginal ancillary benefits of REDD+ to the marginal cost of reducing the technological externalities from mitigation options on energy and industrial sources. Measuring these marginal benefit and cost curves are not trivial but proxies and references of them must be the departure points. Political economy factors must enter only to balance their estimation uncertainty (Wang-Helmreich and Kreibich 2019).

Figure 1 shows how integration would work with separate but complementary carbon markets, so that REDD+ units would not affect the price levels of other mitigation mechanisms. They would be supplementary requirements both in terms of domestic abatement measures and external offsets, but also between REDD+ and other types of mitigation mechanisms.

A series of positive impacts could be expected by adopting this approach:

- The separation of markets would not reduce the price of other mitigation options.
- By ensuring that non-REDD+ options receive the necessary financial resources to direct investment in R&D and investment in low-carbon infrastructure, the process of innovation and decarbonization of industrial, transportation, and energy complexes would continue.
- Including REDD+ units as a complement to these measures, however, would reduce the average cost of GHG abatement and enable countries/entities to adopt more ambitious targets.

---

[9]This proposal was initially presented in Costa et al. (2017).

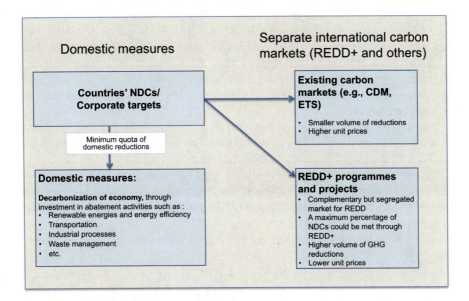

Fig. 1 Integrated REDD+ concept in markets

- The inclusion of REDD+ in markets, at the same time, would ensure access to financial incentives to tackle this important source of emissions.
- Given that the abatement profile of reducing deforestation is frontloaded (as opposed to tree planting or the replacement of energy infrastructure, for instance), it would accelerate the climate benefits associated with mitigation, "buying time" for other measures to enter into force.

Therefore, we propose in this chapter that the design of a future global carbon market regime include REDD+ in its mix, by adopting separate but complementary markets to ensure that forest protection, land use production, and decarbonization of other economic sectors occur in parallel—an integrated protection, production, and decarbonization market—"Integrated REDD+."

At the local scale, the implementation of REDD+ can create tensions with other land use activities. Tree planting schemes envisage carbon finance to finance their plantings. Agriculture, a major deforestation driver, requires finance to intensify and reduce its impacts. Sustainable land use, however, requires the combination and integration of these different measures, which in turn have different carbon benefits, timeframes, and costs.

By allowing all these activities to compete for financial resources in the same playing field, it is unlikely that costlier alternatives would succeed. However, it is the combination of approaches that ensure a sustainable landscape integrated with a productive rural economy.

As proposed for the design of international carbon markets, REDD+ finance flows at the country or jurisdictional level could be conditioned to parallel investments in complementary activities, adopting a "stocks and drivers" approach. That is, for each investment in forest protection, there should be a complementary investment in tackling the drivers of deforestation (predominantly, intensification of agriculture to reduce pressure on land) and/or reforestation of riparian reserves or water catchments.

Figure 2 shows how the integration with other activities within developing countries would ensure combined incentives of forest protection, sustainable agricultural production, and decarbonization of the economy as a whole.

The management of these contributions to land use sustainability at the landscape level could be done by programs themselves or through funds managed by government agencies based on jurisdictional REDD+ programs ensuring this distribution of resources.

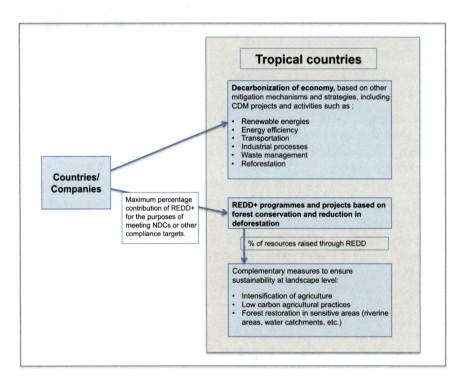

Fig. 2 Integrated REDD+ concept in activities

5 Final Remarks

The urgency in talking about GHG emission mitigation and avoiding climate change requires a concerted and integrated effort involving all sectors of the economy. Given its importance in terms of contribution to GHG emissions, the land use sector must be able to secure financial resources in a large scale to ensure its sustainability and transition to a low-carbon dynamic.

Ancillary benefits are enormous when forest protection contributes in the combat of global warming. The unique biodiversity of tropical forests, as is the case of the Amazon Forest, carries non-use value to the entire world.

In addition to the non-use values attached to biodiversity, water flow use values are also highly dependent on forest management. For example, evaporation and condensation over the Amazon serve as engines of the global atmospheric circulation, with downstream effects on precipitation across South America and further afield across the Northern Hemisphere. In Brazil, it has helped the regional humid climate that affects economic activities that depend on it, such as agricultural production and hydropower generation.

The inclusion of REDD+ into markets is an opportunity to secure such climate and ancillary benefits, directing financial resources to the land use sector.

Since the high volume of reductions and low unit costs of REDD+ units have the potential to create market unbalances and affect the carbon price of other important GHG mitigation activities, we propose herein the creation of a separate market for REDD+ units, associated with a supplementary requirement, so as to ensure the continuation of investments in other mitigation activities.

The creation of a separate REDD+ market has the potential to catalyze the transfer of financial resources to the land use sector, while ensuring that non-REDD+ options continue to receive financial resources for the innovation and decarbonization of industrial, transportation, and energy activities.

Given the low unit cost of REDD+ units, its inclusion in markets would reduce the average cost of GHG abatement and enable recipient countries/entities to adopt more ambitious targets in emission reduction, therefore securing valuable forest ecosystem services.

Ancillary benefits would not only add value to the world as a whole but will also largely benefit national and regional economies in supporting water flows regarding urban and rural supply and hydroelectricity. If so, such gains may attract domestic public and economic support to global warming combat actions.

References

Agrawal A, Nepstad D, Chhatre A (2011) Reducing emissions from deforestation and forest degradation. Annu Rev Environ Resour 36:373–396

Andersen LE et al (2002) The dynamics of deforestation and economic growth in the Brazilian Amazon. Cambridge University Press, Cambridge

Aukland L, Costa PM, Brown S (2003) A conceptual framework and its application for addressing leakage on avoided deforestation projects. Clim Pol 3(2):123–136

Barreto P, Araujo E (2012) O Brasil atingirá sua meta de redução do desmatamento? Imazon, Belém. https://imazon.org.br/publicacoes/1884-2/

Brazil (2014) Views of Brazil on the elements of the new agreement under the convention applicable to all parties. UNFCCC

Britaldo SSF et al (2014) Cracking Brazil's forest code. Science 344:363–364

Costa PM et al (2017) Integrated REDD+ markets: a financial model for forest protection and decarbonization. Brazil REDD Alliance, Rio de Janeiro. https://redd.unfccc.int/uploads/3570_3_redd_brazil_aliance_2C_integrated_redd_proposal.pdf

da Motta S (2018) R. Precificação do carbono: do protocolo de Quioto ao acordo de Paris, Capítulo 14. In: Frangetto FW, Veiga APB, Luedemann G (eds) Legado do MDL: impactos e lições aprendidas a partir da implementação do mecanismo de desenvolvimento limpo no Brasil. IPEA, Brasília

Edwards R (2016) Linking REDD+ to support Brazil's climate goals and implementation of the forest code. Forest Trends Association, Washington

Gibbs HK et al (2015) Brazil's Soy Moratorium. Science 347(6220):377–378

Greiner S et al (2017) CDM transition to Article 6 of the Paris Agreement – options report. Climate Focus, Amsterdam. https://climatefocus.com/publications/cdm-transition-article-6-paris-agreement-options-report

Hargrave J, Kis-Katos K (2013) Economic causes of deforestation in the Brazilian Amazon: a panel data analysis for the 2000s. Environ Resour Econ 54(1):471–494

IETA (2016) A vision for the market provisions of the Paris Agreement. International, Emissions Trading Association, Geneva. https://www.ieta.org/resources/UNFCCC/IETA_Article_6_Implementation_Paper_May2016.pdf

Kurnianto S et al (2015) Carbon accumulation of tropical peatlands over millennia: a modeling approach. Glob Chang Biol 21:431–444

Lewinsohn TM, Prado PI (2005) How many species are there in Brazil? Conserv Biol 19(3):618–624

Malhi Y et al (2008) Climate change, deforestation, and the fate of the Amazon. Science 319:169–172

Marcu A (2016) Carbon market provisions in the Paris Agreement (Article 6). CEPS, Brussels. Special Report, n. 128. https://www.ieta.org/resources/UNFCCC/IETA_Article_6_Implementation_Paper_May2016.pdf

Moutinho P, Guerra R (2017) Programa REDD para early movers – REM: Abordagem de Estoque e Fluxo para a Repartição de Benefícios em Programas de REDD: Conceito e Prática na Implementação de REDD no Estado do Acre. Instituto de Pesquisa Ambiental da Amazonia – IPAM, Brasília

Nepstad DC et al (2014) Slowing Amazon deforestation through public policy and interventions in beef and soy supply chains. Science 344:1118–1123

Paiva et al (2013) Large-scale hydrologic and hydrodynamic modeling of the AmazonRiver basin. Water Resour Res 49:1226–1243

Paoli GD et al (2010) Biodiversity conservation in the REDD. Carbon Balance Manag 5(7):1–9

Piris-Cabezas P et al (2016) Cost-effective emissions reductions beyond Brazil's international target: estimation and valuation of Brazil's potential climate asset, environmental defense fund Washington. http://www.edf.org/sites/default/files/cost-effective-emissions-reductions-brazil.pdf

SEEG (2019) Sistema de Estimativa de Emissões de Gases de Efeito Estufa, Brasília. http://seeg.eco.br/tabela-geral-de-emissoes/

Seroa da Motta R (2005) Custos e benefícios do desmatamento na Amazônia. Ciência & Ambiente 32:73–84

Seymour F et al (2017) Why forests? Why now? The science, economics and politics of tropical forests and climate change. Center for Global Development, Washington. https://www.cgdev.org/sites/default/files/Seymour-Busch-why-forests-why-now-full-book.PDF

Silva et al (2005) The fate of the Amazonian areas of endemism. Conserv Biol 19(3):689–694

Strand et al (2018) Spatially explicit valuation of the Brazilian Amazon Forest's ecosystem services. Nat Sustain 664(1):657–664

Strassburg BBN et al (2012) Impacts of incentives to reduce emissions from deforestation on global species extinctions. Nat Clim Chang 2(5):350–355

Sumila TCA et al (2017) Sources of water vapor to economically relevant regions in Amazonia and the effect of deforestation. J Hydrometeorol 18:1643–1655

ter Steege H et al (2015) Estimating the global conservation status of more than 15,000 Amazonian tree species. Sci Adv 20(10):1–10

Trotignon R (2011) Combining cap-and-trade with offsets: lessons from the EU-ETS. Clim Pol 12(3):273–287

United Nations (2016) Paris Agreement. United Nations Treaty Collection. 8 July 2016. Archived from the original on 21 August 2016. Available at https://treaties.un.org/doc/Publication/MTDSG/Volume%20II/Chapter%20XXVII/XXVII-7-d.en.pdf

Wang-Helmreich H, Kreibich N (2019) The potential impacts of a domestic offset component in a carbon tax on mitigation of national emissions. Renew Sust Energ Rev 101(C):453–460

Watson R et al (2010) IPCC special report on land use, land use change and forestry, a special report of the Intergovernmental Panel on Climate Change. Cambridge University Press, Cambridge

Ancillary Benefits of Climate Policies in the Shipping Sector

Emmanouil Doundoulakis and Spiros Papaefthimiou

1 Introduction

One of the main global challenges that humanity will have to face in the coming years is how to adapt to climate change and mitigate its effects. Four impacts are considered as more prominent and worth emphasising (www.climaterealityproject.org 2019):

- *Rising temperatures*: As temperatures climb around the planet, more heatwaves are expected globally and with higher intensity than those already recorded. Extreme heat can "overpower" all living organisms, which means not only humans but also animals and endangered species.
- *Extreme weather effects*: Climate change has been linked to increased numbers of tornadoes, hurricanes and floods. These adverse effects occur in geographic areas which were not used to face such extreme weather conditions, thus causing unpreceded fatal consequences.
- *Bad air quality*: Apart from the pollution due to burning fossil fuels, a numerous amount of climate change-related processes also impacts air quality, e.g. climate change has been linked to increased numbers of wildfires due to longer lasting dry seasons.
- *Vector-borne diseases*: This term refers to illnesses spread by insects like mosquitoes, fleas, mites, etc. Due to climate change in some areas the warmer and/or more humid conditions allow for some insects to increase their geographic areas thus spreading out diseases (e.g. Lyme, West Nile).

E. Doundoulakis · S. Papaefthimiou (✉)
School of Production Engineering and Management, Technical University of Crete, Chania, Greece
e-mail: spiros@pem.tuc.gr

© Springer Nature Switzerland AG 2020
W. Buchholz et al. (eds.), *Ancillary Benefits of Climate Policy*, Springer Climate,
https://doi.org/10.1007/978-3-030-30978-7_15

Apart from evident costs (especially in economic terms) there can be benefits from the implementation of climate change mitigation policies, either in short and/or in the long run. A climate policy is successful when it provides effects or benefits (primary or ancillary) that make a difference in a positive way. The term "ancillary benefits" may be translated as non-direct, secondary, co-benefits, mainly depending on the relative emphasis given to their relation with the primary effects. To better understand the actual meaning of climate policy ancillary benefits, we can see the reference and the connection between the positive health effects and the air pollution reduction, accompanying a GHG emissions reduction policy. Some other types of ancillary benefits may be considered, like the benefits associated with new technological effects and investments in green renewable technologies, which lead to economic growth of the sector and new employment. Also, afforestation programmes aiming to reduce air pollution concentrations improve the quality of human life and protect the habitat for endangered species.

If a quantitative interpretation of the goals laid out in the Paris Agreement is applied, all energy production sectors based on the combustion of fossil fuel, along with shipping sector, will face substantial challenges in order to successfully decarbonise over the coming years. In this study, we will focus on ancillary benefits of climate policies in the shipping sector and we will provide an overview of the development of the existing regulation of various institutions, analysing the connection of policy measures, towards climate and environmental protection within the shipping sector.

2 Maritime Transportation

The International Maritime Organization (IMO) approved Third IMO GHG Study 2014, providing the latest updated emission estimates for greenhouse gases from ships. According to estimates presented in this study, total shipping emitted 938 million tonnes of CO_2 in 2012, i.e. about 2.6% of the total global carbon dioxide (CO_2) anthropogenic emissions for that year, while in the same year international shipping was estimated to have contributed about 2.2% to the global CO_2 emissions (IMO n.d.). Although international shipping is the most energy-efficient mode of mass transport and only a modest contributor to overall CO_2 emissions, a global approach to further improve its energy efficiency and effective emission control is needed, as sea transport will continue growing apace with world trade.

Emissions from international shipping cannot be attributed to any particular national economy, due to its global nature and complex operation. Therefore, IMO has been energetically pursuing the limitation and reduction of greenhouse gas (GHG) emissions from international shipping, in recognition of the magnitude of the climate change challenge and the intense focus on this topic. Despite this, currently the European Union's (EU) commitment for the reduction of GHG emissions does not include international shipping, making it the only sector of transportation that is not included.

EU has expressed its willingness to widen the enforcement of MARPOL Annex sulphur restrictions to all European seas. Also, concerns were expressed through Strategy for Sustainable Development published on the EU White Paper on Transport Policy (EU Regulation 757 2015) about the impact of maritime transport air quality. This has led to the establishment of the EU Regulation 2016/802 for sulphur content in marine fuels.

The result of the above directive is that all Member States shall take all necessary actions to ensure that sulphur content of marine fuels used in the areas of their territorial seas, exclusive economic zones and pollution control zones does not exceed:

3.50% as from 18/06/2014;
0.50% as from 01/01/2020.

This applies to all vessels of all flags, including those whose journey began outside the European Union. Also during port stays, all vessels calling at EU ports should either use low sulphur fuel (0.1% max) or a shoreside electricity connection staying longer than two hours. Additionally, passenger ships operating on regular services to or from any EU port shall use marine fuel which not exceeds 1.50% by mass, until 1 January 2020. In European designated ECAs, all ships have to burn fuel with a sulphur content of no more than 0.1% since 1 January 2015.

As we can clearly understand from the above, both EU and IMO develop and apply policies to reduce greenhouse gases (GHG) emissions from ships and as a result of this, two similar data collection schemes have been introduced:

EU MRV—Monitoring, Reporting and Verification of **CO_2 emissions** (data collection started 1 January 2018)
IMO DCS—Data Collection System on **fuel consumption** (data collection started 1 January 2019)

EU MRV and IMO DCS requirements are mandatory, and intend to be the first steps in a process to collect and analyse emission data related to the maritime sector. The European Union regulation 2015/757 on the MRV of CO_2 emissions from maritime transport was adopted by the European Council and entered into force July 2015 (EU Regulation 601 2012). This regulation is viewed as the first step of a staged approach for the inclusion of maritime transport emissions into the EU's GHG reduction commitment, alongside the other sectors (energy production, transportation, etc). MRV Regulation covers shipboard CO_2 emissions and requires ships to monitor data on cargo carried and transportation work. The objective of the MRV is to develop a better understanding of fuel consumption and CO_2 emissions from shipping activities within the EU, which could then be used to shape and inform any future GHG monitoring, controlling and mitigation initiatives.

The IMO has addressed ship pollution under the MA
required a gradual decrease of air emissions (NO$_x$, SO$_x$ a
originating from ship engines. In addition, major energy e
for vessels have been proposed through the application of
Design Index (EEDI) and Ship Energy Efficiency Managem

The regulation of air pollution by ships was defined in M.
was firstly adopted in 1997 and revised in 2008, including
of SO$_x$ and NO$_x$ and indirectly Particulate Matter (PM) in I
(ECAs) including EU-territories (such as the Baltic Sea, th
the North Sea). The Marpol Annex VI ECAs are the Baltic
area, the North American area (covering designated coasta
Canada), and the US Caribbean Sea area (around Puerto I
Islands) (Directorate-General for Mobility and Transport 20

When ECAs were firstly introduced, the limit in the conte
fuels was 1%. Nowadays, under the recent IMO regulatio
defined as the maximum sulphur content for marine fuel f
Emission Control Areas since 1 January 2015. This is the on
addresses the control of air emissions from ships in a comp
at present the marine fuels sulphur limit is 3.5%. Marpol
global sulphur cap of 4.5% and in 2012 it was lowered to
2020, IMO will enforce a new 0.5% global sulphur cap o
lowering from the present 3.5% limit (see Fig. 1).

The global fuel sulphur cap is part of the IMO's response
concerns due to harmful air emissions from ships. 2020 as a
at the 70th session of IMO's Marine Environment Protect
held in October 2016.

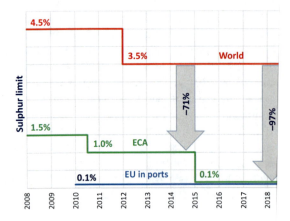

Fig. 1 IMO Marpol Annex VI Sulphur limits timeline

3 EU Regulation on the MRV of GHG Emissions from Maritime Transportation

The EU MRV regulation 2015/757 for maritime transport applies to ships above 5000 gross tonnage (G.T.) and refers to CO_2 emissions released during their voyages, excluding warships, naval auxiliaries, fish-catching or fish-processing ships and government ships used for non-commercial purposes. EU MRV will apply for all ships, regardless of their flag, for voyages:

- Intra-EU
- From the last non-EU port to the first EU port of call (incoming voyages)
- From an EU port to the next non-EU port of call (outgoing voyages).

Ship operators must follow specific monitoring plans to monitor data on per-voyage and annual basis. The monitoring plan, emission reports and the issuance documents of compliance will be accredited by third party verifiers. It is mandatory that verifiers shall be independent of the company or operator of the ship concerned and be accredited by a national accreditation body, according to European Commission (EC) regulation No.765/2008 (EU Regulation 601 2012).

3.1 Monitoring

The actual monitoring of the maritime emissions started in January 2018. Ship owners and operators will not get confirmation of compliance until the first annual report has been satisfactorily verified by their chosen verifier by the end of April 2019. Based on the monitoring plan, for each ship arriving in, or departing from, and for each voyage to or from a port under a Member State's jurisdiction, companies shall monitor the following parameters (EU Regulation 601 2012):

- Port of departure including the date/hour of departure
- Port of arrival including the date/hour of arrival
- For each type of fuel, the amount consumed in total
- Emission factor and quantity of CO_2 emitted
- Distance travelled and time spent at sea
- Cargo carried, transport work.

Reporting on a per-voyage basis is not needed, if both of the below criteria apply during the reporting period:

1. All of the ship's voyages either start from or end at a port of EU region.
2. According to its schedule, the ship performs more than 300 voyages.

In this case, a summarised yearly reporting per ship is needed. Based on the monitoring plan, for each ship and for each calendar year, companies shall monitor the following parameters:

- Amount and emission factor for each type of fuel consumed in total
- Total aggregated CO_2 emitted:

  a. Within the scope of the Regulation
  b. From all voyages between ports under a Member State's jurisdiction
  c. From all voyages which departed from ports or arrived at a port under a Member State's jurisdiction
  d. Which occurred within ports under a Member State's jurisdiction at berth

- Total distance travelled, total time spent at sea
- Total transport work
- Average energy efficiency.

To calculate CO_2 emissions, the following formula is typically applied:

$$CO_2 \text{ emissions} = \text{Fuel consumption} \times CO_2 \text{ emission factor}$$

The fuel consumption includes fuel consumed by main engines, auxiliary engines, gas turbines, boilers and inert gas generators. Ships are using different types of engines which are burning different types of fossil fuels. Fuel consumption at berth shall be calculated for each voyage using one or a combination of the following methods:

1. Bunker Fuel Delivery Note (BDN) and periodic stocktaking of fuel tanks
2. Bunker fuel tank monitoring on board
3. Flow metres for applicable combustion processes, or
4. Direct CO_2 emissions measurements.

The company must define in the monitoring plan which of the above-mentioned methods will be used to calculate fuel consumption for each ship under its responsibility and ensure that once the method has been chosen, it is consistently applied.

For emission factors, default values are used unless the operator decides to use data from the Bunker Fuel Delivery Note (BDN). The BDN is part of the existing legislative requirements for ships to monitor the total amount of fuel bunkered and used for demonstrating compliance with applicable regulations of sulphur emissions. A BDN contains information of the total quantity of fuel bunkered in metric tonnes and density at 15 °C as well as sulphur content. The default values for emission factors are based on the latest available values from Annex VI of the Intergovernmental Panel for Climate Change report (EU Directive 87 2003).

3.2 Reporting

From 2019, by 30 April of each year, companies will have to submit to the EC and to the relevant authorities an emissions report regarding CO_2 emissions and other relevant information for the entire reporting period for each ship under their responsibility which has been accordingly verified. Maritime companies must include in their emissions report the following information:

(a) Data identifying the ship and the company, including:

1. Name of the ship, IMO identification number, port of registry or home port, ice class of the ship, if included in the monitoring plan
2. Technical efficiency of the ship: Energy Efficiency Design Index (EEDI) or the Estimated Index Value (EIV) in accordance with IMO Resolution MEPC.215 (63)
3. Name, address and principal place of business of the shipowner or the managing company, telephone and e-mail details of a contact person

(b) The identity of the verifier that assessed the emissions report
(c) Information on the monitoring method used and the related level of uncertainty
(d) Annual monitoring of the parameters in accordance with the Regulation.

3.3 Verification and Accreditation

In the case that the verifier's assessment identifies non-conformities with the requirements of the regulation, the company revises its monitoring plan accordingly and submits the revised plan for a final assessment by the verifier before the reporting period starts. In particular, the verifier assesses whether the CO_2 emissions and other relevant information included in the emissions report have been determined in accordance with the regulation and the monitoring plan. When the verification assessment concludes with reasonable assurance from the verifier that the emissions report is free from misstatements, the verifier issues a verification report stating that the emissions report has been verified as satisfactory. The verification report specifies all issues relevant to the work carried out by the verifier.

In the case that the verification assessment concludes that the emissions report includes misstatements or non-conformities with the requirements of the regulation, the verifier informs the company thereof in a timely manner. The company then corrects the misstatements or non-conformities so as to enable the verification process to be completed in time and submits to the verifier the revised emissions report and any other necessary information to correct the non-conformities identified. The verifier states whether the initial misstatements or non-conformities have been corrected by the company. If the misstatements or non-conformities are not corrected and, individually or combined, lead to material misstatements, the verifier issues a verification report stating that the emissions report does not comply with the regulation (EU Regulation 601 2012).

3.4 Publication of Information and Report

By 30 June each year, European Commission will make publicly available the information on CO_2 emissions reported. 2019 is the first year that Commission plans to initiate this process and the following information will be publicly available:

(a) Identity of the ship (name, IMO identification number and port of registry or home port)
(b) Technical efficiency of the ship (EEDI or EIV, where applicable)
(c) Annual CO_2 emissions
(d) Annual total fuel consumption for voyages
(e) Annual average fuel consumption and CO_2 emissions per distance travelled of voyages
(f) Annual average fuel consumption and CO_2 emissions per distance travelled and cargo carried on voyages
(g) Annual total time spent at sea in voyages
(h) Method applied for monitoring
(i) Date of issue and the expiry date of the document of compliance
(j) Identity of the verifier that assessed the emissions report
(k) Any other information monitored and reported on a voluntary basis.

The Commission will publish an annual report on CO_2 emissions and other relevant information from maritime transport, including aggregated results, aiming at informing the public and allowing for an assessment of CO_2 emissions and energy efficiency of maritime transport, per size, type of ships, activity, etc. (IMO DCS (Data Collection System) 2019).

3.5 EU MRV Against IMO DCS

Whilst the EU scheme has focused on CO_2 emissions from shipping in the EU area, the IMO scheme covers emissions from shipping globally. It should be noted that it is not yet decided if, how and when the two schemes will converge. Both schemes have overall as objective to mitigate climate change. The outcome of both schemes will be annual reports stating CO_2 emissions per vessel (EU MRV) or aggregated fuel consumption (IMO DCS). The logical consequence would be that considering the experience from EU MRV and IMO DCS schemes, IMO and EU will further decide on setting targets with respect to GHG emission levels from international shipping. Table 1 depicts an overview of the requirements from the two schemes in terms of scope and reporting.

Ancillary Benefits of Climate Policies in the Shipping Sector

Table 1 Comparison of EU MRV and IMO DCS (MEPC 72 2019)

	EU MRV	IMO DCS
Applicability:		
	Ships > 5000 GT calling any EU ports	Ships ≥ 5000 GT trading globally
First reporting period:		
	• 2018 (01/01–31/12) • Reporting to verifier by end of Jan 2019	• 2019 (1 Jan–31 Dec) • Reporting to verifier by end of March 2020
Monitoring plan:		
	• Separate document describing the methodology for data collection and reporting • Predefined format published by the European Commission (EC) • Subject to verification by an independent and accredited verifier • The deadline for submission of monitoring plan was 31 Aug 2017	• Data collection and reporting methodology shall be described as Part II in an integrated part of the Ship Energy Efficiency Management Plan (SEEMP) • Conformation of compliance by Flag/Recognised Organization (RO) • Deadline for submission of SEEMP Part II was 31 Dec 2018
Reporting details:		
	• Amount and emission factor for each type of fuel consumed in total • Total CO_2 emitted and additionally differentiated to aggregated CO_2 emitted (trips to and from EU ports, trips between EU ports, at berth) • Total transport work (time at sea and in port, cargo carried) • Average energy efficiency	• Period of calendar year for which the data is submitted • Distance travelled • Amount of each type of fuel consumed in total • Hours underway under own propulsion • DWT to be used as cargo proxy
Reporting to:		
	European Commission: • Company reports annual emissions to the EMSA data base ("Thetis MRV") • Annual report to be verified by an accredited verifier	Flag state: • Annual emission report to be verified by Flag Admin • Flag State or RO reports to IMO data base
Disclosure:		
	EC will make data publicly available	Individual ship data will be kept confidential

In November 2017, EU decided that international shipping will not be incorporated into the EU Emissions Trading System (ETS) as part of the wider overhaul it is undertaking due to the existing ETS for CO_2 emissions. This decision was a result of intensive negotiations between EU Member States, the European Parliament, the European Commission and shipping stakeholders.

In conjunction with the European Community Shipowners' Associations (ECSA), International Chamber of Shipping (ICS) has consistently argued that the application of a regional EU ETS to all ships calling at EU ports regardless of flag would have been completely inappropriate and would have led to serious market distortion. Many ships would have simply diverted to non-EU ports (including potentially a post-Brexit United Kingdom) in order to minimise cost exposure to the EU system. Additionally, as had happened several years ago when the EU tried unsuccessfully to impose ETS on international aviation, the unilateral application to shipping could generate trade disputes with China and other Asian nations.

This EU decision does not remove the pressure from IMO. Notwithstanding the industry's doubts about the real CO_2 reductions that can be delivered via Market-Based Measures (MBM), the only appropriate forum to have this debate is IMO. The terms of the EU political agreement are that continued exclusion from some form of regional MBM may be dependent on IMO adopting some kind of alternative measure by 2023, which is understood to mean that the EU believes there should indeed be a global MBM.

Moreover, the EC will be required to make an annual report to the European Parliament and EU Member States on progress being made by IMO. In effect, this could mean that if at any time the EC deems progress insufficient, it may seek to justify the need to continue working on unilateral measures. Nevertheless, the EU decision in 2017 seems to represent a recognition that IMO is the best forum in which to have the debate about the appropriateness or otherwise of applying an MBM to shipping.

In November 2017, ICS and ECSA submitted detailed comments to an EC consultation on the possible alignment of its MRV Regulation with the global CO_2 Data Collection System (IMO DCS) which is up and running through 2019. The EU had previously underlined its willingness to consider this alignment in order to help persuade non-EU governments to agree to the establishment of the IMO DCS.

The IMO DCS was adopted in 2016 and was viewed as an acceptable compromise between IMO Member States which are interested to collect reliable information about fuel consumption (and calculate CO_2 emissions) in order to adjust future IMO work and those Member States that wish to collect some more detailed information about transport work and fuel efficiency of ships. The necessary support for this IMO compromise was given with the understanding that the DCS should be simple for the ships and primarily be based on fuel consumption, and most importantly, data relating to fuel consumption under the IMO system will remain anonymous. The purpose of the IMO DCS is to inform future policymaking, rather than to penalise or reward individual ships or ship owners.

The EU MRV Regulation was adopted in 2015, and in addition to the submission of data by ships on fuel consumption, some international shipping stakeholders believe that includes controversial provisions for the transport work, using different metrics to those currently agreed by IMO. Moreover, the verification and certification method that has been developed by the EU seems to be complex. The greatest concern about the EU MRV regulation is that the EC will annually publish commercially sensitive information, along with ship name and company identifiers. This is with the intention of facilitating comparison of the supposed operational efficiency of individual ships. In general, the EU regulation contains many of the elements which most IMO Member States chose to reject when adopting the global IMO DCS. From this fact, we can clearly understand the major competition concerns and possible reactions of shipowners in public availability of their annual emission report.

3.6 Initial IMO Strategy on Reduction of GHG Emissions from Ships

The Marine Environment Protection Committee (MEPC) addresses environmental issues under IMO's remit. The Initial IMO Strategy on reduction of GHG emissions from ships was adopted with resolution MEPC-304 on MEPC 72 (13/04/2018) and encounters for the first time a reduction in total GHG emissions from international shipping, which it says should peak as soon as possible and to reduce the total annual GHG emissions by at least 50% by 2050 compared to 2008 (IMO Greenhouse Gas Emissions 2018).

Levels of ambition also include reviewing with the aim to strengthen the ships' energy efficiency design requirements with the percentage improvement for each phase to be determined for each ship type as appropriate and reduction of CO_2 emissions per transport work, as an average across international shipping by at least 40% by 2030, pursuing efforts towards 70% by 2050 compared to 2008 (Sofiev et al. 2018).

Representing a framework for further action, the Initial Strategy sets the vision for international shipping the levels of ambition to reduce GHG emissions and guiding principles, including possible further measures (short, mid, long term) with possible timelines and their impacts on Member States. The strategy also identifies barriers and supportive measures, including capacity building, technical cooperation, research and development (R&D).

During MEPC 72, the UN Secretary-General highlighted the important contribution of IMO towards achieving the UN's Sustainable Development Goals (SDGs) and he welcomed the action of IMO to adopt the Initial IMO Strategy on reduction of GHG emissions from ships as a major step forward in global action to combat climate change. He also called all nations to contribute and adopt the ambitious Initial Strategy of IMO and support the modernisation of the shipping sector in a manner consistent with the ambitions of the Paris Agreement. As we can realise, this is a historic milestone in the Organization's continuous contribution to global efforts to limit and reduce GHG emissions. This strategy should send a strong signal to the shipping community as a whole to innovative energy-efficient technologies and stimulate investment in the development of low and zero carbon fuels.

Some Member States probably preferred to see the adoption of even more aggressive targets. We should consider that a 50% total cut by 2050 can realistically be achieved only with the development and widespread use of zero CO_2 fuels by a large proportion of the fleet, so the total achievement is not an easy case (Sofiev et al. 2018).

4 Ancillary Benefits of Climate Policies

The policy measures towards climate and environmental protection within the shipping sector and the development of the existing regulation of various institutions are in a manner of continuous contribution to global efforts to limit and reduce GHG emissions, with the ultimate goal of climate change mitigation that comes from shipping sector. Climate change is already negatively impacting our health and if permitted to continue unabated, it will exacerbate direct and indirect health impacts to varying degrees across populations. Reduction of annual premature mortality and morbidity in populations worldwide is one of the objectives of IMO global compliance with 2020 marine fuel sulphur standards. There are significant benefits from consumption of cleaner marine fuels, especially in trading routes and ports close to densely populated areas.

The distributional impacts, associated with air pollution health burdens and the promising benefits of new cleaner marine fuel standards after 2020, should be considered and metered, while shipping activity will continue to produce harmful air emissions and greenhouse gases. This is the first meaningful fuel control regulation since international fleet converted to diesel and petroleum byproducts. It is expected that the energy needs in global trade will increase along with air emissions from shipping (IMO 2019). Low sulphur marine fuels still account annually for ~250k deaths and ~6.4M childhood asthma cases, so additional reductions beyond 2020 standards may prove beneficial (Third IMO GHG Study 2014).

The use of cleaner fuels in marine sector and the reduction of sulphur-based emissions may offer collateral health and climate benefits that merit quantification. For example, 2020 compliant marine fuels may enable or be accompanied with additional $PM_{2.5}$ emissions reductions, such as organic carbon and black carbon particles. Moreover, many control technologies for harmful particulates and ozone precursor emissions perform better under low-sulphur combustion conditions. International policymaking efforts jointly pursuing air pollution health benefits and climate targets may increase the urgency for continued progress to control and mitigate GHG.

4.1 Health Impacts of Climate Change and Health Co-benefits of Mitigation Measures

Accessibility to energy has been fundamental for human development and progress, but the combustion of fossil fuels contributes to climate change, resulting in direct and indirect health impacts. While the attribution of these impacts on human health is challenging, researchers utilise more sophisticated scientific methods and long-term datasets, which are able to quantify and attribute in a better and more accurate way specific health burdens to climate change (Hales et al. 2014; Ebi et al. 2017). The Intergovernmental Panel on Climate Change (IPCC) classifies the

Table 2 An overview of health impacts of climate change (IPCC 2014)

Classification	Potential Impacts—increased morbidity and mortality from:
Direct	Increased exposure to extreme weather conditions; hurricanes, storms, floods; heatwaves, UV radiation
Ecosystem-mediated	Increased exposure to vector-borne and other infectious diseases; food and water-borne infections; air pollution and lung diseases
Human institution-mediated	Poor nutrition; occupational health; mental health; violence and conflict

health impacts of climate change into three categories: direct impacts, ecosystem-mediated (indirect) impacts, and human institution-mediated impacts (see Table 2) (IPCC 2014).

The interrelationship between air quality and climate change is a very complex case, with many air pollutants produced from the combustion of fossil fuels concurrently with greenhouse gas emissions (Fiore et al. 2015). Projections of bad air quality, as a result of climate change, point out increasing premature deaths due to ozone and especially Particulate Matter in coming years (Silva et al. 2017). As a consequence to these events, there are estimations for substantial external economic costs attributable to climate change and air pollution, which point out that global annual Gross Domestic Product (GDP) could be impacted by up to 3.3% by 2060, while labour productivity constitutes one area that will be most significantly impacted. Additional analysis by the Organization for Economic Cooperation and Development (OECD) estimates that the economic consequences of outdoor air pollution will result in healthcare costs of US\$176 billion and 3.7 billion lost working days annually by 2060 (OECD 2016; Workman et al. 2018a).

Realising the size of current and projected health impacts, researchers highlight the potential health co-benefits that result from ambitious mitigation efforts. The term "co-benefits" refers to multiple benefits in different fields resulting from specific actions, strategies or policies. Co-beneficial approaches to climate change mitigation are those that also promote positive outcomes in other areas, such as concerns relating to the environment (e.g. air quality management, health, agriculture, forestry, and biodiversity), energy (e.g. renewable energy, alternative fuels, and energy efficiency) and economics (e.g. long-term economic sustainability, industrial competitiveness, income distribution).

To determine the potential health co-benefits from domestic and global action, more new complex modelling techniques have been created and utilised by researchers and organisations. The findings are consistent; despite the heterogeneity of study methods, prospective health co-benefits studies consistently conclude that the implementation of ambitious mitigation measures can reap significant health benefits for local populations, and partially, if not completely, offset resulting implementation costs. A strong effect of health co-benefits is their immediacy and specifically, health benefits associated with reduced air pollution can materialise

Table 3 Examples of potential health co-benefits from mitigation activities relating to the energy and transport sectors, including the anticipated time lag for the realisation of health co-benefits (Remais et al. 2014)

Mitigation activity	Potential health co-benefits	Anticipated time lags
Reductions in fossil fuel use	Reductions in sudden cardiac death risk; acute respiratory infections; chronic obstructive pulmonary disease exacerbations	Days to weeks; weeks and months; weeks and months
Improvements in fuel economy; incentivise electric vehicle use; tighten vehicle emission standards		
Increases in accessibility to active modes of transport, including walking and cycling	Reductions in type 2 diabetes; depression; breast and colon cancer incidence	Years for all potential health co-benefits identified

promptly after mitigation measures are implemented (see Table 3) (Workman et al. 2018a; Remais et al. 2014).

EU has a defined policy development process for climate change and supporting governance structures in place to develop evidence-based integrated policies with opportunities for input from diverse stakeholders. Specifically, impact assessments developed for climate change mitigation policies are explicit in their consideration of health and other impacts is a good example of procedures and tools that can support the incorporation of multiple considerations into the development of a cross-sectoral policy issue.

Despite a robust policy development process, health co-benefits ultimately play a limited role in the development of climate change mitigation policies. In spite of the EU's commitment to the equal consideration of economic, social and environmental impacts, the realpolitik considers economic costs and energy supply security issues as particularly influential in final climate change mitigation policies. In reality, the Commission's role in this issue requires balancing the provision of cost-effective and evidence-based policy options with politically palatable policy choices for the Member States with their own national interests and diverse stakeholder groups to assuage (Workman et al. 2018b).

4.2 Case for Equity Between Paris Climate Agreement's Co-benefits and Adaptation

Whilst significant co-benefits have been associated with energy and transportation, adaptation offers ancillary benefits for emission reduction through land and forest conservation, which merit to be described as co-benefits because they are enhanced with biodiversity management, nutrient recycling and water purification as part of the indicators.

Although adaptation policy goals do not always have measurable indicators compared to mitigation, its impacts extend beyond human development issues (e.g. land area loss, people displacement, ecosystem loss or change, economic value loss, infrastructure loss, cultural heritage loss, etc.) when viewed from the UN Sustainable Development Goals (SDGs) perspective as outlined in Article 8 of the Paris Climate Agreement (Dovie 2019). Mitigation co-benefits clearly aligns to:

1. SDG 7 on affordable and clean energy
2. SDG 9 on industry, innovation and infrastructure
3. SGD 12 on responsible consumption and production
4. SDG 13, yet intersect with adaptation on the climate action.

The SDGs are a call for action by all countries (poor, rich or with middle-income) to promote prosperity while protecting the planet. They recognise that ending poverty must go hand in hand with strategies that build economic growth and address a range of social needs including education, health, social protection, and job opportunities, while tackling climate change and environmental protection.

Emphasising mitigation (e.g. renewable energy, energy efficiency, sustainable transportation, cleaner fuels) should not diminish adaptation but rather enhance it (e.g. forest protection, land use changes, infrastructure and green building design) which is comparable to co-benefits (e.g. green infrastructure, distributed energy, water and energy conservation, low-input agriculture) (Dovie 2019).

We can discern that there is a need for new forms of multilevel governance of the climate policy schemes, including financing mechanisms and response measures, for enhanced adaptation to effectively protect the integrity of emission reduction, hence the Nationally Determined Contributions that Paris Agreement requests from each country to clarify and communicate their post-2020 climate actions. Nowadays, we can utilise further expansions and compilations of potential co-benefits and we are able to suggest the categorisation as depicted in Table 4 (Mayrhofer and Gupta 2016).

In order to have sufficiently positive impacts, climate policies need to look beyond climate impacts. There are significant negative impacts and limited time available to address the alarming pace of observed global warming. The social and economic co-benefits of climate change mitigation offer an important opportunity to mobilise a strategic and interest-oriented approach to support effective and timely climate actions. Interest-oriented co-benefits of climate change mitigation represent positive net effects of policies and actions beyond those directly related to climate change and global warming processes (such as greenhouse gas emission reduction) that pertain to the following five key attributes (Table 5) (Mayrhofer and Gupta 2016).

Table 4 Co-benefits categorisation of climate change policy

Category	Co-benefit
Climate-related	• Reduce GHG emissions
	• Enhance resilience to climate change
Economic	• Enhance energy security
	• Trigger private investment
	• Improve economic performance
	• Generate employment
	• Stimulate technological change
	• Contribute to fiscal sustainability
Environmental	• Protect environmental resources
	• Protect biodiversity
	• Support ecosystem services
	• Improve soil quality
	• Reduce air pollution
Social	• Enhance energy access
	• Reduce poverty incidence and inequality
	• Contribute to food and water security
	• Improve health
	• Reduce stressors
Political and institutional	• Contribute to political stability
	• Improve democratic quality of governance
	• Contribute to interregional collaboration

Table 5 Key attributes of co-benefits of climate change mitigation

Interest oriented	Benefit can be defined in view of specific interests/interest groups
Identifiable	Benefit can be distinctly described, delimited from other factors, measured, and evaluated
Timely	Benefit unfolds in a time frame crucial for the addressed interest group (usually less than 10 years)
Attributable	Benefit can be connected to a specific intervention and allocated to a specific interest group and reconstructed by members of this group
Opportunity oriented	Benefit can be defined through a resulting opportunity or profit, and not merely through avoided burdens, risks, or losses

4.3 Guidelines for Mobilising the Interest-Oriented Co-benefits of Climate Change Mitigation

The global transformation towards green technologies, renewable resource energy or cleaner fuels, seems to be irreversible in the long term given its many advantages and additionally competitive outlook. In contrast, current investments in heavy fossil fuel-based energy scheme are still present and consists a serious threat for the climate of our planet. For this reason, IMO decided to apply a new regulation for the maritime sector to control and set lower sulphur limit content of marine fuels.

The interest-oriented co-benefits of climate change mitigation act as important players towards enhanced transformation and additionally promote long-lasting political deadlocks in order to prevent environmentally harmful path dependencies. We can mobilise these co-benefits by expanding the view of traditional climate policy evaluation by specifically addressing the net effects of climate policy measures and actions. Also, explicit strategic use of the multiple-benefits approach to climate policy must be promoted.

While at present there is no standard practice for climate change attribution for health outcomes, from the literature, our empirical study, and various case studies, we can conclude that a proportion of the current burden of climate-sensitive health outcomes can be attributed to climate change. Extreme weather effects, undoubtedly increasing the probability to observe more deaths, during heatwaves or floods, which are attributable to climate change and estimate the exact proportion using different approaches.

A conservative and defensible approach would be by attributing deaths above a threshold, related to the degree to which climate change increased ambient temperature over recent decades. Also, sensitivity analyses and assumptions of the linearity between mortality and temperature could be used to provide an uncertainty range around the estimated impact (Ebi et al. 2017).

As climate change unfolds, climate-sensitive health outcomes will continue to emerge. We must urgently gain a better understanding of the distribution of climate change burden on human health by achieving more knowledge about the factors that contribute and affect our health due to climate. Greater knowledge sharing between different science sectors, reliable long-run datasets, refinement of analytic techniques for detection and attribution, will all be important and help policymakers to adjust climate change policy and achieve multiple targeted benefits.

5 An Overview of Wider Impacts

The realisation of the potential multiple impacts of climate change to our planet and our civilisation will lead to a strong engagement towards mitigation and adaptation actions. Possible behavioural changes, including sensitivity to environment, decreased air pollution, recycling, employment of renewable resources and sustainable agriculture practices, are some of the actions that can be developed. An overview of the wider key impacts associated with these actions is provided below:

- Significant health benefits through decreased air pollution have associated multiple economic benefits by reduction in healthcare costs and increase of the size of the workforce as more working-age people are in good health.
- Reduced air pollution and reduced noise as a consequence of alternatively fueled vehicles (ships, trucks, buses, cars) provide health and wellbeing benefits.

However, one of the most significant potential wider benefits comes from a reduced demand for fossil fuels thus increasing energy security.

- Energy efficiency improvements in constructions (vehicles, ships, buildings) provide reduced exposure to cold or hot living environments and increased income due to lower energy bills. Energy efficiency can provide significant health and wellbeing, energy security by reduced fuel dependence and affordable living benefits.
- Measures towards sustainable agriculture and environmentally friendly farming practices protect the environment and natural resources. Reduced nitrogen runoff or less fertilisers or pesticides use is a key outcome and has benefits for water quality, biodiversity and human health.
- Education and behaviour change are closely linked and should complement any technical measures.

However, there are barriers that prevent the implementation of these actions or minimise the wider benefits that can be gained. Some actions are needed to overcome these barriers (Hampshire et al. 2017):

- Clear messages that translate targets into local actions, accompanied by comprehensive and consistent performance monitoring across policies and sectors. By effective communication across various levels of government and relevant stakeholders the most effective policies can be applied and people engagement can be increased.
- Lack of "political appetite and willingness" may occur due to restricted time (i.e. four-year) governmental changes. Thus the precariousness of political actions, combined with the potential costs (economic and/or social) of climate change policies, can lead to a lack of long-term thinking and probably inaction. This barrier can be overcome through reliable political will, awareness raising among the public and global funding for implementation of climate change actions.
- Climate change is a complex global issue, which is hard to understand by individuals who do not actually comprehend the impact that they can exhibit and are reluctant to change their beliefs which are rooted in experiences, knowledge and tradition. This barrier can be overcome by educating communities about all the benefits they spring up from climate change policies. For some, the health of their children or the quality of life is a priority, whilst for others this may be house prices or noise reduction.
- Recognition of the barriers and specific conditions for each geographic area, as these affect the magnitude of wider impacts that can be experienced, will contribute to maximised efficiency. To overcome these barriers, targeted actions are required, climate change policy and action needs to be embedded into wider governmental strategies, as a way of bringing together community, environmental and economic goals.

References

Directorate-General for Mobility and Transport (2011) White paper on transport. Luxembourg: Publications Office of the European Union

Dovie D-BK (2019) Case for equity between Paris Climate agreement's co-benefits and adaptation. Sci Total Environ 656:732–739

Ebi K, Ogden N, Semenza J, Woodward A (2017) Detecting and attributing health burdens to climate change. Environ Health Perspect 125(8):085004

EU Directive 87 (2003) Establishing a scheme for greenhouse gas emission allowance trading within the Community and amending Council Directive 96/61/EC. Official Journal of the European Union L275/32

EU Regulation 601 (2012) On the monitoring and reporting of greenhouse gas emissions pursuant to Directive 2003/87/EC of the European Parliament and of the Council. Official Journal of the European Union L181/30

EU Regulation 757 (2015) On the monitoring, reporting and verification of carbon dioxide emissions from maritime transport, and amending Directive 2009/16/EC. Official Journal of the European Union L123/55

Fiore A, Naik V, Leibensperger E (2015) Air quality and climate connections. Air Waste Manag Assoc 65:645–685

Hales S, Kovats S, Lloyd S, Campbell-Lendrum D. (2014) Quantitative risk assessment of the effects of climate change on selected causes of death, 2030s and 2050s. World Health Organization (WHO), Geneva Switzerland

Hampshire K, Pridmore A, Claxton R (2017) Wider impacts of climate change mitigation and adaptation actions in Jersey. States of Jersey Department of the Environment, Oxford Centre for Innovation, Oxford, UK

IMO (n.d.) [Online]. Available www.imo.org

IMO DCS (Data Collection System) (2019). DNVGL.com [Online]. https://www.dnvgl.com/maritime/imo-dcs/index.html. Assessed 22 February 2019

IMO Greenhouse Gas Emissions (2018) Greenhouse Gas Emissions, 2018 [Online]. http://www.imo.org/en/OurWork/Environment/PollutionPrevention/AirPollution/Pages/GHG-Emissions.aspx. Accessed Feb 2019

IPCC (2014) Human health: impacts, adaptation and co-benefits. In: Climate change 2014: contribution of working group II to the 5th assessment report of the IPCC. Cambridge University Press, Cambridge, UK

Mayrhofer J, Gupta J (2016) The science and politics of co-benefits in climate policy. Environ Sci Policy 57:22–30

MEPC 72 (2019) Marine Environment Protection Committee (MEPC), 72nd session, 9–13 April 2018 [Online]. http://www.imo.org/en/mediacentre/meetingsummaries/mepc/pages/mepc-72nd-session.aspx. Assessed 22 February 2019

OECD (2016) The economic consequences of outdoor air pollution. Organisation for Economic Cooperation and Development (OECD Publishing), Paris, France

Remais J, Hess J, Ebi K, Markandya A, Balbus J, Wilkinson P, Haines A, Chalabi Z (2014) Estimating the health effects of greenhouse gas mitigation strategies: addressing parametric, model, and valuation challenges. Environ Health Perspect 122:447–455

Silva R, West J, Lamarque J-F, Shindell D, Collins W, Faluvegi G, Folberth G, Horowitz L (2017) Future global mortality from changes in air pollution attributable to climate. Nat Clim Chang 7:647–651

Sofiev M, Winebrake J, Johansson L, Carr E, Prank M, Soares J (2018) Cleaner fuels for ships provide public health benefits with climate tradeoffs. Nat Commun 9, Article 406

Third IMO GHG Study (2014) Third IMO GHG Study. International Maritime Organization (IMO), London, UK

www.climaterealityproject.org (2019) [Online]. https://www.climaterealityproject.org/blog/wait-why-climate-change-bad-thing. Assessed March 2019

Workman A, Blashki G, Bowen K, Karoly D, Wiseman J (2018a) The political economy of health co-benefits: embedding health in the climate change agenda. Int J Environ Res Public Health 15:674–692

Workman A, Blashki G, Bowen K, Karoly D, Wiseman J (2018b) Health co-benefits and the development of climate change mitigation policies in the European Union. Clim Pol J 19:585–597. https://doi.org/10.1080/14693062.2018.1544541

Part IV
Climate Actions in Urban Areas and Their Ancillary Benefits

Co-benefits of Climate Change Mitigation and Pollution Reduction in China's Urban Areas

Liu Jie, Pan Jiahua, Liu Ziwei, and Jiao Shanshan

1 Introduction

During the "twelfth five-year" development period (2010–2015), China significantly strengthened its ecological and environmental governance to improve environmental quality and actively promoted international cooperation on climate change, becoming an important participant, contributor, and leader of global ecological civilization construction. During the "thirteenth five-year" development period (2016–2020), China will further enhance the quality and efficiency of development and better meet the growing needs of the people in the aspects of economy, politics, culture, society, and ecology. Adhering to keeping harmony between man and nature calls for China to develop a low-carbon and green economy. Especially, addressing two major challenges of global warming and regional air pollution control for the construction of "Beautiful China" will help to create a good environment for production and living for the people and contribute to global ecological security.

In the Paris Agreement, China's Intended Nationally Determined Contributions (INDC) for greenhouse gas emissions reduction clearly has set a target of 40–45% decrease in carbon intensity by 2020 compared with 2005 and has committed to achieving a carbon emissions peak by 2030 and reducing the carbon intensity by 60–65% compared with 2005. In July 2018, China formulated the "Three Year Action Plan for Blue Sky Protection," which proposed that after three years of efforts, the total amount of major atmospheric pollutants will be drastically reduced, and greenhouse gas emissions and $PM_{2.5}$ concentration will be synergistically

---

L. Jie (✉) · L. Ziwei · J. Shanshan
International Business School, Shaanxi Normal University, Xi'an, China
e-mail: liujiesx@snnu.edu.cn; jiaoshanshan@snnu.edu.cn

P. Jiahua
Institute for Urban and Environmental Studies, Beijing, China

© Springer Nature Switzerland AG 2020
W. Buchholz et al. (eds.), *Ancillary Benefits of Climate Policy*, Springer Climate,
https://doi.org/10.1007/978-3-030-30978-7_16

decreased. More specifically, by 2020, the total amount of sulfur dioxide and nitrogen oxides emissions will be reduced by more than 15% compared with 2015, and the concentration of $PM_{2.5}$ in prefecture-level cities will be more than 18% lower than that of 2015. Days with good air quality of prefecture-level cities will reach 80%, and days with heavy and severe pollution will be decreased by more than 25% compared with the year of 2015.

Since CO_2 and atmospheric pollutants (like, e.g., $PM_{2.5}$ and PM_{10}) are mainly derived from the combustion of fossil energy carriers such as coal, oil, and natural gas, CO_2 and air pollutants emissions show characteristics of homology (Zheng et al. 2015). It is feasible to curb CO_2 emissions while simultaneously reducing atmospheric pollutants such as $PM_{2.5}$. Therefore, quantitative research on the co-benefits of reducing CO_2 emissions and air pollution can provide targeted policy recommendations for haze and carbon reduction in China's urban areas.

The rest of the chapter is organized as follows. Section 2 compares definitions of co-benefits of different sources. In Sect. 3 we conduct a literature review for understanding feasible approaches to quantifying co-benefits of climate mitigation and pollution reduction. In Sect. 4 data and methodologies are introduced, respectively. Section 5 depicts the results of magnitude and influencing factors of co-benefits of CO_2 abatement and $PM_{2.5}$ reduction. Section 6 concludes with a brief summary and some policy recommendations.

2 Definition of Co-benefits

There has been no consensus on the definition of co-benefits, and some international agencies have given similar definitions but with differentiated emphases. The concept of synergistic benefit was first introduced in the Third Assessment Report by the Intergovernmental Panel on Climate Change, and it was defined as the socioeconomic benefits of policies and actions to mitigate greenhouse gas emissions (IPCC 2001). The Organization for Economic Co-operation and Development pointed out that in addition to the direct climate impact benefits, climate change mitigation policies will generate a broader range of benefits. For example, policies aimed at deploying energy technologies or increasing energy efficiency are likely to improve local or indoor air quality, thereby reducing human health risks (Bollen et al. 2009). The US Environmental Protection Agency's Comprehensive Environmental Strategy Handbook states that synergistic benefits should include all positive benefits from local measures to reduce atmospheric pollutants and related greenhouse gases, such as energy conservation, economic benefits, and air quality improvement (Hu et al. 2004). The Asian Development Bank stated that co-benefits could be seen as ancillary benefits generated from and going beyond greenhouse gas emissions reduction (Yang 2013), and the co-benefits of greenhouse gas emissions mitigation also consist of development issues from a local perspective.

A part of the literature equates synergistic effects with co-benefits. The European Environment Agency holds the perspective that synergistic effects refer to other social benefits generated from greenhouse gas reduction measures (Zheng et al. 2015). China's Research Center for Environmental Protection Policy advocates that, on the one hand, synergistic effects refer not only to the declining emissions of other local air pollutants in the process of controlling greenhouse gas emissions, but on the other hand, also the decreasing greenhouse gas emissions while controlling local pollutants and promoting ecological construction (Li et al. 2012). It can be seen that definitions of co-benefits in the literature are limited to the various synergies brought about by climate change mitigation policies. Few studies give adequate emphasis on the interactive mechanisms between climate change mitigation policies and air pollution control measures; among the exceptions are, e.g., Löschel and Rübbelke (2009) and Bollen (2015). The important role of co-benefits for developing countries deserves even further attention, although they were also considered in earlier studies (see, e.g., Aunan et al. 2007; Rive and Rübbelke 2010; Zheng et al. 2011; Löschel et al. 2018).

3 Literature Review

Since the 1970s, a large number of studies have investigated the synergies of climate change mitigation and air pollution control. According to the classification of research methods, the existing literature ranges from the use of qualitative analysis methods, atmospheric environment models, econometric models, comprehensive evaluation models, and other quantitative analysis methods to carry out synergistic analyses at the regional, urban, and industrial levels.

In terms of qualitative discussions, Yang et al. (2013) summarize the research advance in co-benefits of greenhouse gas emissions reduction policies for mitigating environmental pollution and improving human health and social welfare. Zhao and Yuan (2014) formulate and implement the "problem definition-policy formulation-policy coordination" analysis framework to discuss the cross-departmental coordination of air pollution control policies in the Beijing–Tianjin–Hebei region in China, and the authors concluded that the imbalanced economic development levels, incomplete joint mechanism of regional air pollution prevention and control, and information asymmetry are factors constraining the efficiency of policy coordination for air pollution control. Puppim de Oliveira et al. (2013) review several co-benefits studies in cities across Asia and find that embedding climate change mitigation at the local level can help catalyze co-benefits in urban development.

Co-benefits of climate change mitigation and pollution reduction have also been revealed by using atmospheric environment models. Based on the ambient air quality model (CMAQ) and the mesoscale meteorological model (MM5), Li et al. (2008) study the concentrations distribution and transportation status of O_3 and PM_{10} in the near-surface of the Yangtze River Delta in January and July in 2001, respectively. It is believed that the ambient air quality and pollution severity in the Yangtze River Delta region are significantly affected by air pollution transmission and chemical conversion, and solar radiation and wind direction are closely related to the degree of air pollution. Research by the Institute of Atmospheric Physics of

Chinese Academy of Sciences pointed out that global warming led to a reduction in Arctic sea ice, which significantly increased the pollution of smog in eastern China, linking greenhouse gas emissions to atmospheric pollution (Wang et al. 2015). Zou et al. (2017) reveal that the decrease of Arctic Sea ice and the increase of snowfall in Eurasia caused by global warming further changed the structure of regional atmospheric circulation, which in turn aggravated the severe air pollution in winter in China in recent years. In winter, the air pollutants discharged by industry and motor vehicles cannot be effectively removed through horizontal diffusion or vertical mixing. The continuous accumulation of air pollutants finally led to the occurrence of severe haze pollution in China in January 2013. Cai et al. (2017) also demonstrate that global warming caused by greenhouse gas emissions will increase the frequency and duration of haze pollution in future winters.

In the process of applying econometric models and integrated assessment models to study synergies of climate mitigation and pollution reduction, Yan (2017) confirms the significant synergistic relationship between carbon emissions and haze pollution based on the panel threshold regression model of 29 provinces and autonomous regions in China, and conducted an empirical study on the threshold effect of the urbanization level on the synergistic relationship between carbon emissions and haze pollution. Li (2018) estimates the correlation between carbon emissions and haze pollution in 30 provincial capital cities in China based on the varying coefficients model with cross-sectional fixed effects, and drew the conclusion that there was a significant and extremely high correlation between carbon emissions and haze pollution in 30 provincial capital cities of China. Rafaj et al. (2013) study the effects of strict climate mitigation strategies on air quality and human health and ecosystems based on the Greenhouse Gas and Air Pollution Interaction and Synergy Model (GAINS) and found that global greenhouse gas emission reduction policies could reduce the emission levels of traditional air pollutants (SO_2, NO_X, and $PM_{2.5}$) on the global scale as well as in the European Union, China, India, and the USA. Based on the DICE model, Xie et al. (2018) study the common benefit of carbon dioxide emissions reduction for reducing nitrogen oxide emissions. The NO_X emissions jointly emitted with carbon dioxide emissions in a non-policy scenario would exceed 600 million tons/year. However, in the emission reduction policy scenario, reducing carbon dioxide emissions could simultaneously reduce NO_X emissions by at least 15%. The stronger the climate mitigation policy, the more obvious the combined benefits of NO_X reduction. Anenberg et al. (2012) use an integrated assessment model to simulate the impact of greenhouse gas mitigation measures on the outdoor concentration of $PM_{2.5}$ and ozone and used the epidemiologically derived concentration response function to calculate $PM_{2.5}$ and ozone-related premature death. Implementing the methane and black carbon emissions control measures could reduce the global population weighted average $PM_{2.5}$ and ozone concentrations and decrease the global death toll.

For other analytical methods, Qin (2012) analyzed the impact of greenhouse gas emissions reduction on atmospheric pollutant emissions based on different mitigation policy scenarios and studied the impact of air pollution control policies on greenhouse gas emissions as well, proving the existence of co-benefits between the two. The authors finally put forward policy recommendations for future greenhouse gas emission reduction and atmospheric pollutant control in Shenzhen City. Li (2010) studied the ancillary effects of pollution reduction in Panzhihua City and Xiangtan City based on the evaluation method of classified pollution reduction. Different pollutant emission reduction technologies and measures have both positive and negative effects on greenhouse gas emissions reduction, but there are more significant positive effects on the whole.

To sum up, there are two main research directions for synergistic emissions reduction between greenhouse gases and atmospheric pollutants: (1) Regional emissions reduction in atmospheric pollutants leads to synergistic emissions reduction of greenhouse gases; (2) Greenhouse gas emissions reduction leads to the synergistic reduction of regional atmospheric pollutants. Compared with the long-term benefits of GHG emissions reduction, it is easier to achieve the benefits of reducing air pollutant emissions in the short term, such as health benefits and additional ecological benefits generated by pollutant emissions reduction. All or part of the costs of implementing GHG emissions reduction policies will be offset by the ancillary benefits, making the second type of research direction more concerned. There is a two-way correlation between climate change mitigation and air pollution control. The co-benefits assessment needs to incorporate the two into an integrated analytical framework to quantitatively analyze the inner interaction mechanism. Based on the above reasons, this chapter proposes the following hypotheses and conducts an empirical analysis, respectively: First, there is a synergistic trend between low carbon development and green development in China's urban areas. Second, China's urban carbon dioxide emissions reduction and air pollution control have significant synergies. Third, the factors affecting the interaction mechanism between climate change mitigation and air pollution control are different.

4 Methodology and Data

4.1 Variables Definition and Data Description

4.1.1 Endogenous Variables

In order to figure out the co-benefits between climate change mitigation and atmospheric pollutants control, we select carbon dioxide emissions (tons/year) and atmospheric $PM_{2.5}$ concentration (micrograms/cubic meter) to represent China's urban carbon dioxide emissions and atmospheric pollutants concentration.

Since there are uncertainties in China's existing urban carbon emissions data, this chapter calculates CO_2 emissions according to formula (1), based on the urban energy consumption data of 30 provincial capital cities during 2000–2015.

$$c = \sum_{i=1}^{16} e_i \times \lambda_i \tag{1}$$

where c denotes the calculated amount of CO_2 emissions, e_i is the amount of the i-th basic energy consumption, λ_i represents the conversion coefficient corresponding to the i-th basic energy consumption, and i indicates the type of basic energy consumption. Based on the data from 2000 to 2014, the polynomial extrapolation method is used to calculate and fill the missing values of carbon emissions of 2015 for some provincial capital cities.

Among the components of atmospheric pollutants, $PM_{2.5}$ (fine particles with particle size below 2.5 microns) and PM_{10} (inhalable particles with particle size below 10 microns) are the main sources of atmospheric pollutants. Compared with PM_{10}, $PM_{2.5}$ has a smaller particle size, a higher degree of activity, and it is easier to attach toxic and harmful substances, which are more harmful to human health. Therefore, the concentration of $PM_{2.5}$ has become an important indicator of the degree of atmospheric pollution. The statistics of $PM_{2.5}$ for 30 provincial capital cities are gathered from the China statistical yearbook and from the International Earth Science Information Network Center of Columbia University.

4.1.2 Exogenous Variables

The independent variables selected in this chapter include population size, economic development level, industrial structure, energy structure, urbanization level, technological progress, environmental regulation, urban traffic, and climatic factors.

1. Population size (*popu*): The IPCC Fifth Assessment Report (IPCC AR5) discloses that population growth and economic development are the two most fundamental driving factors for the increase of greenhouse gas emissions. Human activities are also the main sources for rapid urban expansion and energy consumption rise. In order to analyze the impact of urban population on carbon emissions and air pollution, this chapter selects the total amount of urban population as a representative indicator of population size.
2. Economic development level (*pgdp*): Urban economic production activities need to consume all kinds of energy, resulting in carbon emissions and air pollution. In order to measure the impact of economic development on urban carbon emissions and air pollution, this chapter selects GDP per capita as the index of the economic development level. After selecting 2000 as the base year, the nominal GDP per capita was adjusted using the GDP deflator to derive the real GDP per capita of each provincial capital city.
3. Industrial structure (*indus*): Industrial restructuring will cause changes in the amount and structure of energy consumption, thus affecting carbon emissions. China's industrial development, which is still in the late stage of industrialization, has been accompanied by high energy consumption and high pollution. Economic development of China has shown significant regional characteristics, and different levels of industrialization have diverse impacts on urban carbon emissions and air pollution. This chapter calculates and adopts the proportion

of urban secondary industry output to GDP to represent the level of urban industrialization.

4. Energy structure (*ener*): Coal is a black energy source in terms of primary energy, and coal combustion has caused a large amount of carbon emissions and severe environmental pollution. Studies have shown that coal-fired heating in northern China is an important cause of the formation and aggravation of smog pollution (Chen et al. 2017). Many cities in northern China control air pollution by limiting coal combustion or deploying "Coal-to-Gas" policy in the winter of 2017–2018. The proportion of urban coal consumption to the total energy consumption is selected as an indicator of the energy structure to analyze the impact of energy structure changes on the co-benefits of carbon emissions reduction and air pollution control.

5. Urbanization level (*urban*): An increase in the urbanization level implies a rise of the overall level of production and consumption of urban residents. In the process of urbanization, the continuous urban expansion, the increase of construction dust and motor vehicles have led to the increase in energy consumption and carbon emissions in China's urban areas. This chapter intends to use the proportion of permanent urban residents to the total population as a measure of the urbanization level.

6. Technological progress (*tech*): Green technology innovation is an effective means to achieve carbon emissions and atmospheric pollutants reduction. In order to explore the impact of the technological innovation level on the co-benefits between carbon emissions reduction and air pollution control in China's urban areas, the market turnover of technology is adopted to represent the level of technological innovation in each provincial capital city.

7. Environmental regulation (*policy*): Government environmental regulation is a useful means to effectively reduce CO_2 and atmospheric pollutant emissions, including energy conservation and emission reduction policies, environmental tax instruments, and traffic restriction measures. In order to analyze the impact of environmental regulation intensity on the synergies of urban carbon emissions reduction and air pollution control, this chapter introduces total investment in air pollution control as a proxy of environmental regulation intensity. The more investment in air pollution control, the lower the amount of atmospheric pollutants discharged.

8. Urban traffic (*trans*): Vehicle emissions are important sources of $PM_{2.5}$. The impact of changes in the number of motor vehicles on urban air pollution varies significantly with different levels of economic development. Urban traffic is expressed by the year-end area of urban paved roads.

9. Climate condition (*prep*): Urban climatic conditions contribute to distinctive levels of air pollution. For example, during the winter half year in China, haze pollution in Beijing and Tianjin with relatively lower air humidity is much more severe than that in the Pearl River Delta region. This chapter intends to utilize annual urban precipitation to analyze the impact of climate factors on urban atmospheric pollutant emissions.

4.2 Methodologies

4.2.1 Decoupling Index of Carbon Dioxide Emissions and $PM_{2.5}$

The decoupling index of carbon dioxide emissions indicates the trend of CO_2 decoupling from urban economic growth, which reflects the level of low carbon development. The decoupling index of $PM_{2.5}$ indicates the trend of $PM_{2.5}$ emissions decoupling from urban economic growth, which reflects the level of green development. The decoupling indexes of CO_2 and $PM_{2.5}$ are calculated for the 30 provincial capital cities in China according to formulas (2) and (3), respectively.

$$e_{CO_2} = \frac{\triangle co_2}{\triangle gdp} \cdot \frac{gdp}{co_2} \tag{2}$$

$$e_{PM_{2.5}} = \frac{\triangle pm_{2.5}}{\triangle gdp} \cdot \frac{gdp}{pm_{2.5}} \tag{3}$$

4.2.2 Simultaneous Equations Model

The co-benefits are defined as a two-way causality between climate change mitigation and atmospheric pollutants control. Therefore, the two endogenous variables need to be integrated into an analytical framework so as to scientifically assess the synergistic benefits between CO_2 and $PM_{2.5}$ emissions control and to fully analyze the interaction mechanisms between the two endogenous variables. Since the simultaneous equations model is a multi-equation system, two or more endogenous variables can be included in an analytical framework while estimating model parameters. Based on the establishment of the simultaneous equations of carbon emissions and $PM_{2.5}$ (see Eqs. 4 and 5), this chapter reveals the two-way causality between the two and uses ordinary least squares (OLS), two-stage least squares (TSLS), and three-stage least squares (3SLS) methods to estimate and compare model parameter results.

According to the variables selected in Table 1, the simultaneous equations are listed as follows:

$$\ln carb_{it} = \alpha_0 + \alpha_1 \ln pm_{it} + \alpha_j \sum_{j=2}^{8} \ln X_{itj} + \mu_{it} \tag{4}$$

$$\ln pm_{it} = \beta_0 + \beta_1 \ln carb_{it} + \alpha_k \sum_{k=2}^{8} \ln X'_{itk} + \varepsilon_{it} \tag{5}$$

where $\ln carb_{it}$ and $\ln pm_{it}$ are logarithms of urban CO_2 and $PM_{2.5}$ emissions; $\ln X_{itj}$ denote control variables affecting urban CO_2 emissions, including urban population size, economic development level, industrial structure, energy structure, urbaniza-

Table 1 Variable definition and data source

Variable	Meaning	Indicator	Unit	Data source
Carbon emissions	Total CO_2 emissions	*carb*	tons/year	Wind database
Air pollution	$PM_{2.5}$ concentration	*pm*	$\mu g/m^3$	CIESIN; City Statistical Yearbook
GDP per capita	Per capita GDP	*pgdp*	10,000 yuan	City Statistical Yearbook
Industrial structure	Secondary industry output value/GDP	*indus*	%	City Statistical Yearbook
Energy structure	Coal consumption/total energy consumption	*ener*	%	City Statistical Yearbook
Population	Urban total population at year-end	*popu*	10,000/year	City Statistical Yearbook
Urbanization level	Urban population/total population at year-end	*urban*	%	City Statistical Yearbook
Technological progress	Market turnover of technology	*tech*	%	City Statistical Yearbook
Environmental regulation	Total investment in pollution control	*policy*	10,000 yuan	City Statistical Yearbook
Urban traffic	Area of city paved roads at year-end	*trans*	10,000 m^2	City Statistical Yearbook
Precipitation	Annual precipitation	*prep*	mm	City Statistical Yearbook

tion level, technological progress, and environmental regulation policies; and $\ln X'_{itk}$ are control variables affecting urban $PM_{2.5}$ emissions, including urban population size, economic development level, energy structure, urban traffic, technological progress, environmental regulation policies, and urban precipitation. α_1 refers to the elasticity of urban CO_2 emissions to $PM_{2.5}$, and β_1 corresponds to the elasticity of urban $PM_{2.5}$ emissions to CO_2 emissions.

5 Results Analysis

5.1 Synergistic Analysis of Low-Carbon Development and Green Development

The decoupling elasticity is widely used to reflect the fact that economic growth and material consumption are possibly not synchronized (Sun and Li 2011; Ma et al. 2019). The criterion for evaluating low-carbon development in an economy refers to the negative growth of greenhouse gas emissions (CO_2 emissions) while sustaining positive economic growth. The nature of low-carbon development is that greenhouse gas emissions are constantly decoupled from economic growth. Similarly, the criterion for evaluating green development in an economy refers to the negative growth of atmospheric pollutants ($PM_{2.5}$ emissions) while keeping economic growth. The essence of green development is that air pollution is gradually decoupled from economic growth.

According to different values of decoupling elasticities (e for abbreviation), the low-carbon development levels and green development levels of China's provincial capital cities can be divided into different categories, as illustrated in Table 2.

The low-carbon development of the 30 provincial capital cities in China from 2001 to 2015 (Fig. 1) shows that the relationship between CO_2 emissions and economic growth has changed from tight coupling to decoupling. The CO_2 decoupling elasticities during 2001 to 2006 mainly belonged to relative decoupling or negative decoupling. For example, negative decoupling cities include Beijing, Shanghai, Hefei, Fuzhou, and Jinan in the east, Hohhot and Zhengzhou in the middle, and Lanzhou and Urumqi in the west. From 2007 to 2012, the decoupling elasticities have gradually transitioned to the values of relative decoupling and weak decoupling, with a strong decoupling phenomenon in a few cities. Most of the negative decoupling cities before 2006 have moved into the categories of weak decoupling, and the rest moved promptly into the type of strong decoupling. After 2013, the four types of the CO_2 decoupling index tended to be more balanced, and the number of cities with strong decoupling has doubled compared with the period of 2007–2012. Cities with high decoupling elasticities include Beijing, Shanghai, Lanzhou, and Taiyuan (all belonging to strong decoupling). Economically less-developed areas show relatively lower decoupling values, including three cities in the Xiang–Gan–E region (i.e., Changsha, Nanchang, and Wuhan), two cities in the

Table 2 Types of decoupling with respect to low-carbon and green development levels

Type of decoupling	Decoupling elasticity	Low-carbon development levels	Green development levels
Negative decoupling	$e > 1$	High-carbon development	Grey development characterized by destruction of ecological balance, massive consumption of energy and resources, and damage to human health
Relative decoupling	$0 < e < 1$	Relatively high-carbon development	Relatively grey development
Weak decoupling	$-1 < e < 0$	Low-level low-carbon development	Low-level green development
Strong decoupling	$e < -1$	Low-carbon development	Green development with a balanced economy characterized by maintaining environmental quality and protecting resources and energy

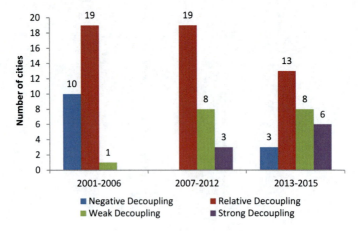

Fig. 1 Low-carbon development trend of China's provincial capitals

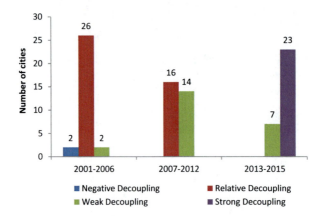

Fig. 2 Green development trend of China's provincial capitals

Shan–Gan–Ning region (i.e., Xi'an and Yinchuan), and three cities in the Qian–Gui–Dian region (i.e., Guiyang, Nanning, and Kunming).

According to the trend of green development of the 30 provincial capital cities from 2001 to 2015 (Fig. 2), the decoupling of $PM_{2.5}$ from economic growth has been much faster than that of low-carbon development. For instance, from 2001 to 2006, the $PM_{2.5}$ decoupling values are mainly dominated by the type of relative decoupling, and four cities are in the type of negative decoupling (i.e., Changchun and Nanning) and weak decoupling (i.e., Yinchuan and Urumqi), respectively. The relative decoupling and weak decoupling have become the two prominent types of green development from 2007 to 2012. During 2013 to 2015, all of the provincial capital cities have stepped into weak decoupling and strong decoupling, and the number of cities with strong decoupling exceeds 2/3 of the total. The evolution of the decoupling process of $PM_{2.5}$ during 2001 to 2015 demonstrates that the persistent efforts to reduce atmospheric pollutants emissions significantly contribute to a nonparallel trend between air pollution and economic development in China's urban areas.

Based on the averaged decoupling elasticities of CO_2 and $PM_{2.5}$ from 2014 to 2015, the 30 provincial capital cities of China could be further classified to reveal regional differences in synergies of low-carbon development and green development (Table 3). The cities with $PM_{2.5}$ decoupling elasticities less than -1 are divided into strong–weak decoupling (low carbon and low-level green development), weak–weak decoupling (low-level low-carbon and low-level green development), relative–weak decoupling (relative high-carbon and low-level green development), and negative–weak decoupling (high-carbon and low-level green development), respectively. Similarly, the cities with $PM_{2.5}$ decoupling elasticities more than -1 are categorized into strong–strong decoupling (low-carbon and green development), weak–strong decoupling (low-level low-carbon and green development), relative–strong decoupling (relative high-carbon and green development), and negative–strong decoupling (high-carbon and green development), respectively.

The results of Sect. 5.1 show that low-carbon development and green development in China's urban areas depict a synergistic trend changing from tight coupling to decoupling, and the decoupling of $PM_{2.5}$ shows higher magnitudes than that of CO_2 emissions in a few capital cities of China, and more and more cities have transitioned from the mode of high-carbon and grey economy to the mode of low-carbon and green economy.

5.2 Analysis of Co-benefits of China's Urban CO_2 Mitigation and Pollution Reduction

5.2.1 Regional Differences in Co-benefits of CO_2 and $PM_{2.5}$ Emissions Reduction

Based on model (3) cross-elasticities of carbon dioxide emissions and $PM_{2.5}$ can be estimated and plotted to describe the urban differences in the synergies of climate change mitigation and air pollution reduction (Fig. 3). It is obvious that there are significant synergies between urban carbon dioxide emissions and $PM_{2.5}$ concentration in China. The elasticities of $PM_{2.5}$ concentration to carbon emissions are in the range of 0.0658–0.2531. Except for Haikou, Kunming, Xining, and Urumqi, the elasticity values of other cities show statistical significance at the 0.05 significance level; the elasticities of carbon emissions to $PM_{2.5}$ are in the range of -0.0241 to 0.7219. Except for the three cities of Harbin, Kunming, and Xining, all the other coefficients are statistically significant at the 0.05 significance level.

According to the regionally differentiated co-benefits, China's 30 provincial capital cities can be divided into three groups. In the first group, 13 cities with high co-benefits are verified, including Shanghai, Beijing, Tianjin, Zhengzhou, Jinan,

Table 3 Regional differences in synergies of low-carbon development and green development

Elasticity of $PM_{2.5}$	Elasticity of CO_2	Provincial cities	Development levels
$-1 < e_{PM_{2.5}} < 0$	$e_{CO_2} < -1$	Beijing	Low carbon and low-level green development
	$-1 < e_{CO_2} < 0$	Zhengzhou, Jinan	Low-level low-carbon and low-level green development
	$0 < e_{CO_2} < 1$	Changchun, Chongqing	Relative high-carbon and low-level green development
	$e_{CO_2} > 1$	Yinchuan, Shenyang	High-carbon and low-level green development
$e_{PM_{2.5}} < -1$	$e_{CO_2} < -1$	Shanghai, Lanzhou, Taiyuan, Fuzhou, Hohhot	Low-carbon and green development
	$-1 < e_{CO_2} < 0$	Hefei, Tianjin, Guangzhou, Hangzhou, Xining, Harbin	Low-level low-carbon and green development
	$0 < e_{CO_2} < 1$	Guiyang, Haikou, Wuhan, Nanjing, Nanchang, Chengdu, Xi'an, Nanning, Kunming, Shijiazhuang, Urumqi	Relative high-carbon and green development
	$e_{CO_2} > 1$	Changsha	High-carbon and green development

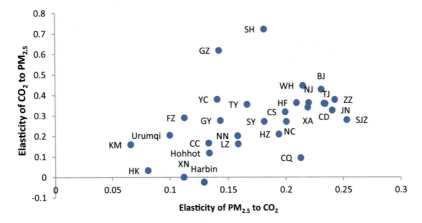

Fig. 3 Regional differences in co-benefits of CO_2 and $PM_{2.5}$ emissions reduction

Guangzhou, Wuhan, Chengdu, Changsha, Hefei, Nanjing, Xi'an, and Shijiazhuang. In the third group, there are seven cities with low co-benefits, including Haikou and Xining, Harbin, Kunming, Hohhot, Urumqi, Changchun, and Chongqing. The rest of the cities belong to the second group, including Fuzhou, Guiyang, Lanzhou, Nanning, Hangzhou, Shenyang, Nanchang, Yinchuan, and Taiyuan. Most of the cities showing high co-benefits are provincial capital cities with a high level of economic development and large population size in the eastern part of China. By contrast, the provincial capital cities with a relatively less-developed economy and smaller population size in western China are accompanied by a low level of co-benefits. Co-benefits in Shanghai are the highest among the 30 provincial capitals, as a one percent reduction in carbon emissions will contribute to a 0.1811 percent drop in $PM_{2.5}$ concentration, and a one percent drop in $PM_{2.5}$ concentration will result in a 0.7219 percent reduction in carbon emissions. Co-benefits of Haikou are the lowest of the 30 provincial capitals. A one percent reduction in carbon emissions will only cause a 0.0809 percent drop in $PM_{2.5}$ concentration, and a one percent reduction in $PM_{2.5}$ concentration will lead to a 0.0330 percent reduction in carbon emissions.

5.2.2 Temporal Variation of Co-benefits of CO_2 and $PM_{2.5}$ Emissions Reduction

Based on the time-varying cross-elasticities of carbon dioxide emissions and $PM_{2.5}$ concentration in China's urban areas, temporal variations of the co-benefits between carbon dioxide emissions and atmospheric pollutants reduction are analyzed (Fig. 4). The impact of carbon dioxide emissions on atmospheric pollutants unfolds with time. From 2000 to 2012, the elasticities of $PM_{2.5}$ concentration with respect to carbon dioxide emissions have been maintained in a stable range of 0.3596–0.3722

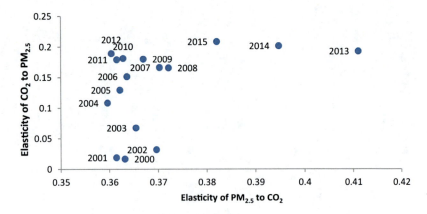

Fig. 4 Temporal variation of co-benefits of CO_2 and $PM_{2.5}$ emissions reduction

and increased to a higher range of 0.3821–0.4110 during 2013–2015. The highest elasticity of 0.4110 is detected in 2013, which means that a one percent increase in carbon emissions will cause an increase in $PM_{2.5}$ concentration by 0.4110 percent, and vice versa. Moreover, the impact of $PM_{2.5}$ on carbon emissions is continuously intensified over time. The elasticities of carbon emissions to $PM_{2.5}$ concentration in 2000–2015 have been on the upward trend, increasing from the lowest value of 0.0168 in 2000 to the highest value of 0.2086 in 2015. In 2015, a one percent increase in $PM_{2.5}$ concentration contributed to a 0.2086 percent increase in carbon emissions, and vice versa.

The above analysis illustrates that there have been significant co-benefits between carbon dioxide emissions and $PM_{2.5}$ concentrations in the 30 provincial capital cities in China from 2000 to 2015, and the synergies of the two show obvious regional and temporal variations. Section 5.3 will further empirically investigate the main influencing factors affecting the co-benefits between China's urban carbon dioxide mitigation and atmospheric pollutants control.

5.3 Influencing Factors of Co-benefits of CO_2 and $PM_{2.5}$ Emissions Reduction

Carbon dioxide emissions and $PM_{2.5}$ concentration are integrated into an equation system, and the parameters of the simultaneous equations are estimated by ordinary least squares (OLS), two-stage least squares (TSLS), and three-stage least squares (3SLS), respectively. Table 4 shows the impact of urban $PM_{2.5}$ concentration changes on carbon dioxide emissions and the corresponding main influencing factors. Table 5 depicts the impact of urban carbon dioxide emissions changes on $PM_{2.5}$ concentration and the critical influencing factors. The results derived from the

Co-benefits of Climate Change Mitigation and Pollution Reduction in China's... 295

Table 4 Influencing factors of co-benefits (endogenous variables: CO_2 emissions)

Variables	Coefficients		
	Model 1 (OLS)	Model 2 (TSLS)	Model 3 (3SLS)
lnpm	0.5946***(0.0477)	0.6378***(0.1615)	0.7358***(0.1590)
lnpopu	0.1870***(0.0551)	0.1736**(0.0732)	0.0988(0.0713)
lnpgdp	0.3327***(0.0486)	0.3187***(0.0696)	0.3078***(0.0689)
lnindus	0.6898***(0.1222)	0.6664***(0.1482)	0.7285***(0.1464)
lnener	0.0348(0.0778)	0.0176(0.0990)	−0.0650(0.0970)
lnurban	0.6320***(0.0954)	0.6350***(0.0961)	0.4751***(0.0907)
lntech	−0.0093(0.0215)	−0.0109(0.0222)	−0.0011(0.0220)
lnpolicy	−0.0379*(0.0217)	−0.0366*(0.0222)	−0.0283(0.0219)
R-squared	0.6738	0.6733	0.6648
SSR	92.35	92.52	94.90

Note: *represents p-value <0.1, **represents p-value <0.05, and ***represents p-value <0.01

Table 5 Influencing factors of co-benefits (endogenous variables: $PM_{2.5}$)

Variables	Coefficients		
	Model 1 (OLS)	Model 2 (TSLS)	Model 3 (3SLS)
lncarb	0.3853***(0.0319)	0.3052***(0.0880)	0.3046***(0.0871)
lnpopu	0.2201***(0.0567)	0.2170***(0.0572)	0.1634***(0.0550)
lnpgdp	0.0460(0.0518)	0.0697(0.0575)	0.0277(0.0559)
lnener	0.2741***(0.0618)	0.3093***(0.0719)	0.3249***(0.0711)
lntrans	0.1473***(0.0544)	0.1801***(0.0642)	0.2203***(0.0628)
lntech	−0.0240(0.0183)	−0.0210(0.0187)	−0.0156(0.0184)
lnpolicy	−0.0300*(0.0176)	−0.0280*(0.0179)	−0.0220(0.0176)
lnprep	−0.1062***(0.0321)	−0.1027***(0.0325)	−0.0597**(0.0302)
R-squared	0.5928	0.5873	0.5853
SSR	64.54	65.41	65.74

Note: *represents p-value <0.1, **represents p-value <0.05, and ***represents p-value <0.01

three parameter estimation methods demonstrate that estimates of model parameters are robust.

The elasticity coefficient of the explanatory variable ($PM_{2.5}$) and its statistical significance indicate that there are significant synergies between urban $PM_{2.5}$ and CO_2 emissions. A one percent increase in $PM_{2.5}$ concentration can lead to an increase in CO_2 emissions by 0.6378 to 0.7358 percent. Among the control variables listed in Table 4, population size, economic development level, industrial structure, and urbanization level have greatly promoted CO_2 emissions in China's urban areas. In the case of model 2 with the TSLS estimation method, compared with other control variables, the industrial structure has the highest elasticity of 0.6664 indicating that a one percent increase in the proportion of the industrial structure will lead to an increase of 0.6664 percent in carbon emissions. Besides, urbanization level, economic development level and population size, with the elasticity coefficients of 0.6350, 0.3187, and 0.1736 respectively, are influencing factors that remarkably promote carbon dioxide emissions. However, it is worth noting that technological progress and environmental regulation policy can play a role in curbing carbon emissions. The two-stage least squares estimation for the elasticity of environmental policy gives a significant value of -0.0366 at the 0.1 statistical significance level, indicating that a one percent increase in investment for urban pollutants control will evidently reduce carbon dioxide emissions by 0.0366 percent. It can be concluded that reducing urban $PM_{2.5}$ emissions can simultaneously decrease carbon emissions, and atmospheric pollutants control investment aimed at reducing $PM_{2.5}$ concentration plays a role in curbing carbon emissions; the main contributors to China's urban CO_2 emissions include the second industry-dominated industrial structure, the rapidly increasing urbanization level, the growing urban population, and the level of economic development.

The elasticity coefficient of the explanatory variable (CO_2) and its statistical significance indicate that there are significant synergies between urban CO_2 emissions and $PM_{2.5}$ concentrations (Table 5). A one percent increase in CO_2 emissions can result in an increase in $PM_{2.5}$ concentration by approximately 0.31 percent. Among the control variables listed in Table 5, population size, energy structure, and urban traffic have significantly increased the urban $PM_{2.5}$ concentration. In the case of model 2 with the TSLS estimation method, the elasticity coefficient of energy structure reaches 0.3093, which is the highest among all the control variables. A one percent increase in coal consumption will lead to an increase in $PM_{2.5}$ concentration by 0.3093 percent. In addition, population size and urban traffic, with coefficients of 0.2170 and 0.1801, also contribute positively to urban $PM_{2.5}$ concentration. On the contrary, control variables including technological progress, environmental regulation, and precipitation contribute negatively to $PM_{2.5}$ concentration. The elasticity coefficient of urban precipitation in model 2 is -0.1027 going beyond the aggregate effects of technological progress and environmental regulation policy on atmospheric pollutants. The analysis above shows that reducing urban CO_2 emissions can significantly reduce $PM_{2.5}$ concentration simultaneously; urban coal consumption, a rapidly increasing number of urban motor vehicles, and urban population are the main contributors to urban atmospheric pollutant emissions in China; the investment in air pollution control aimed at reducing $PM_{2.5}$ concentration has played a role in constraining atmospheric pollutant emissions, but it is far less effective than urban precipitation.

6 Conclusions and Discussions

Developing a low-carbon and green economy is an unavoidable path for China's ecological civilization, and it is also an inevitable choice for China to achieve its Intended Nationally Determined Contributions by 2030 set out in the Paris Agreement. Therefore, it is necessary to address the two major challenges of global warming and regional air pollution control. Based on the empirical analysis of the co-benefits of carbon dioxide emissions mitigation and air pollution control in 30 provincial capital cities in China, this study proposes targeted policy recommendations on haze and carbon reduction in China's urban areas.

There is a clear synergistic trend between low-carbon development and green development in China's urban areas. From 2001 to 2015, the carbon dioxide emissions and $PM_{2.5}$ concentrations of the 30 provincial capital cities in China all showed a common trend of gradually decoupling from economic growth. In particular, air pollution control has been continuously strengthened in recent years, and green development is more rapid than the low-carbon development. Economic development in China's urban areas has gradually transitioned from the "high carbon-gray development" model to the "low carbon-green development" model.

China's urban carbon dioxide emissions and air pollution control have significant co-benefits, and show spatial and temporal characteristics. Cities with higher co-benefits are distributed in the eastern regions with high economic development levels and large population size, while cities with low co-benefits are distributed in the western regions with relatively low economic development levels and small population size. From 2000 to 2015, the elasticity of $PM_{2.5}$ concentration to carbon dioxide emissions fluctuates little, but the elasticity of carbon dioxide emissions to $PM_{2.5}$ concentration has been on an upward trend, increasing from the lowest value of 0.0168 in 2000 to the highest value of 0.2086 in 2015.

Parameter estimation based on a simultaneous equations model shows that the co-benefits of climate change mitigation and air pollution control are significantly influenced by other control variables in China. Industrial structure, level of urbanization, level of economic development, and the size of the urban population are factors that significantly promote carbon dioxide emissions, and air pollution control investment aimed at reducing $PM_{2.5}$ concentration has played a role in curbing carbon dioxide emissions. Energy structure, population size, and urban traffic are contributing factors that significantly increase $PM_{2.5}$ concentration in China's provincial cities. Air pollution control investment has played a role in restraining pollutant emissions, but it is far less effective than urban rainfall.

In the process of industrialization and urbanization, China has achieved the goal of coordinating carbon dioxide emissions reduction and air pollution control, but it is necessary to further fully explore the co-benefits between carbon dioxide emissions reduction and air pollution control. First, it is crucial to optimize the urban industrial structure and promote green industrial development. Second, it is urgent to accelerate the process of urban energy structure adjustment, build a clean, low-carbon and efficient energy system, and continue to control total coal consumption in some key cities. Third, actively adjusting the structure of the urban traffic system, developing a green transportation system, and strengthening pollution control of mobile sources are important prerequisites for urban pollutants reduction. Fourth, the implementation efficiency of environmental regulation policies should be substantially improved to reduce emissions of air pollutants, and to establish and improve regional cooperation and coordination mechanisms for preventing and controlling air pollution among different urban regions.

Appendix

Decoupling characteristics of CO_2 and $PM_{2.5}$ emissions of 30 provincial capital cities in China

City	Variable	2001–2006	2007–2012	2013–2015
Beijing	CO_2	Negative	Strong	Strong
	$PM_{2.5}$	Relative	Relative	Weak
Tianjin	CO_2	Relative	Weak	Weak
	$PM_{2.5}$	Relative	Relative	Strong
Shijiazhuang	CO_2	Relative	Relative	Relative
	$PM_{2.5}$	Relative	Relative	Strong
Taiyuan	CO_2	Negative	Strong	Strong
	$PM_{2.5}$	Relative	Weak	Strong
Hohhot	CO_2	Negative	Strong	Strong
	$PM_{2.5}$	Relative	Relative	Strong
Shenyang	CO_2	Relative	Relative	Negative
	$PM_{2.5}$	Relative	Relative	Weak
Changchun	CO_2	Relative	Relative	Relative
	$PM_{2.5}$	Negative	Relative	Weak
Harbin	CO_2	Relative	Relative	Weak
	$PM_{2.5}$	Relative	Relative	Strong
Shanghai	CO_2	Negative	Weak	Strong
	$PM_{2.5}$	Relative	Weak	Strong
Nanjing	CO_2	Relative	Relative	Relative
	$PM_{2.5}$	Relative	Weak	Strong
Hangzhou	CO_2	Relative	Relative	Weak
	$PM_{2.5}$	Relative	Weak	Strong
Hefei	CO_2	Negative	Weak	Weak
	$PM_{2.5}$	Relative	Relative	Strong
Fuzhou	CO_2	Negative	Weak	Strong
	$PM_{2.5}$	Relative	Weak	Strong
Nanchang	CO_2	Relative	Relative	Relative
	$PM_{2.5}$	Relative	Weak	Strong
Jinan	CO_2	Negative	Weak	Weak
	$PM_{2.5}$	Relative	Weak	Weak
Zhengzhou	CO_2	Negative	Weak	Weak
	$PM_{2.5}$	Relative	Relative	Weak
Wuhan	CO_2	Relative	Relative	Relative
	$PM_{2.5}$	Relative	Relative	Strong
Changsha	CO_2	Relative	Relative	Negative
	$PM_{2.5}$	Relative	Weak	Strong
Guangzhou	CO_2	Relative	Relative	Weak
	$PM_{2.5}$	Relative	Weak	Strong

(continued)

City	Variable	2001–2006	2007–2012	2013–2015
Nanning	CO_2	Relative	Relative	Relative
	$PM_{2.5}$	Negative	Relative	Strong
Haikou	CO_2	Relative	Relative	Relative
	$PM_{2.5}$	Relative	Weak	Strong
Chongqing	CO_2	Relative	Relative	Relative
	$PM_{2.5}$	Relative	Relative	Weak
Chengdu	CO_2	Relative	Relative	Relative
	$PM_{2.5}$	Relative	Weak	Strong
Guiyang	CO_2	Relative	Relative	Relative
	$PM_{2.5}$	Relative	Weak	Strong
Kunming	CO_2	Relative	Relative	Relative
	$PM_{2.5}$	Relative	Relative	Strong
Xi'an	CO_2	Relative	Relative	Relative
	$PM_{2.5}$	Relative	Weak	Strong
Lanzhou	CO_2	Negative	Weak	Strong
	$PM_{2.5}$	Relative	Weak	Strong
Xining	CO_2	Relative	Weak	Weak
	$PM_{2.5}$	Relative	Relative	Strong
Yinchuan	CO_2	Weak	Relative	Negative
	$PM_{2.5}$	Weak	Relative	Weak
Urumqi	CO_2	Negative	Relative	Relative
	$PM_{2.5}$	Weak	Relative	Strong

Note: $e < -1$ = strong decoupling, $-1 < e < 0$ = weak decoupling, $0 < e < 1$ = relative decoupling, > 1 = negative decoupling

References

Anenberg SC et al (2012) Global air quality and health co-benefits of mitigating near-term climate change through methane and black carbon emission controls. Environ Health Perspect 120(6):831–839

Aunan K, Berntsen T, O'Connor D, Persson TH, Vennemo H, Zhai F (2007) Benefits and costs to China of a climate policy. Environ Dev Econ 12(3):471–497

Bollen J (2015) The value of air pollution co-benefits of climate policies: analysis with a global sector-trade CGE model called WorldScan. Technol Forecast Soc Chang 90:178–191

Bollen J et al (2009) Co-benefits of climate change mitigation policies: literature review and new results. OECD Publishing, Paris

Cai WJ et al (2017) Weather conditions conducive to Beijing severe haze more frequent under climate change. Nat Clim Chang 7(4):257–262

Chen Q, Sun FK, Xu YX (2017) Haze caused by heating in winter? Evidence from urban panels in North China. Nankai Econ Res 4:27–42 (in Chinese)

Hu T, Tian CX, Li LP (2004) Policy impact of synergy on climate change in China. Environ Protect 9:56–58 (in Chinese)

IPCC (2001) Climate change 2001: the scientific basis. In: Houghton JT, Ding Y, Griggs DJ et al (eds) Contribution of working group I to the third assessment report of the intergovernmental panel on climate change. Cambridge University Press, Cambridge, UK, 881p

Li L et al (2008) Simulation of regional pollution characteristics of atmospheric O_3 and PM_{10} in the Yangtze River Delta. Environ Sci (in Chinese) 29(1):237–245

Li LP, Zhou GM, Ji HY (2010) Evaluation of the synergistic effect of pollution reduction: a case study of Panzhihua City. China Popul Resour Environ 20(S2):91–95 (in Chinese)

Li LP et al (2012) Study on the synergistic effect of the emission reduction measures on greenhouse gas emission reduction in Xiangtan city. Environ Sustain Develop (in Chinese) 37(01):36–40

Li ZP (2018) Study on the relationship between urban carbon emissions and smog pollution in China and its influencing factors. Thesis for Master's degree. Shaanxi Normal University (in Chinese)

Löschel A, Rübbelke DT (2009) Impure public goods and technological interdependencies. J Econ Stud 36(6):596–615

Löschel A, Pei J, Sturm B, Wang R, Buchholz W, Zhao Z (2018) The demand for global and local environmental protection–experimental evidence from climate change mitigation in Beijing. CESifo working paper

Ma MD, Cai W, Cai WG, Dong L (2019) Whether carbon intensity in the commercial building sector decouples from economic development in the service industry? Empirical evidence from the top five urban agglomerations in China. J Clean Prod 222:193–205

Puppim de Oliveira JA, Doll CNH, Kurniawan TA et al (2013) Promoting win-win situations in climate change mitigation, local environmental quality and development in Asian cities through co-benefits. J Clean Prod 58:1–6

Qin XL (2012) Research on the co-benefits of greenhouse gas and air pollution control. South China University of Technology, dissertation, Guangdong

Rafaj P et al (2013) Co-benefits of post-2012 global climate mitigation policies. Mitig Adapt Strateg Glob Chang 18(6):801–824

Rive N, Rübbelke DT (2010) International environmental policy and poverty alleviation. Rev World Econ 146(3):515–543

Sun YH, Li ZM (2011) Research on the relationship between economic development and carbon emissions in China's provinces and regions. China Popul Resour Environ (in Chinese) 21(05):87–92

Wang HJ, Chen HP, Liu JP (2015) Arctic sea ice decline intensified haze pollution in eastern China. Atmos Oceanic Sci Lett 8(1):1–9

Xie X, Weng YW, Cai WJ (2018) Co-benefits of CO2 mitigation for NO_X emission reduction: a research based on the DICE model. Sustainability 10(4):1109

Yan YX et al (2017) Analysis of synergy between carbon emissions and haze pollution. J Environ Econ (in Chinese) 2(02):52–63

Yang X, Teng F, Wang GH (2013) Synergistic benefits of greenhouse gas reduction. Ecol Econ (in Chinese) 08:45–50

Zhao XF, Yuan ZW (2014) Study on coordination of air pollution control policy among Regional Governments in Beijing, Tianjin and Hebei. China Adm (in Chinese) 11(3):18–23

Zheng X, Zhang L, Yu Y, Lin S (2011) On the nexus of SO_2 and CO_2 emissions in China: the ancillary benefits of CO_2 emission reductions. Reg Environ Chang 11(4):883–891

Zheng JJ et al (2015) Synergistic effect of greenhouse gas emission reduction and air pollution control: a review. Ecol Econ (in Chinese) 31(11):133–137

Zou YF et al (2017) Arctic sea ice, Eurasia snow, and extreme winter haze in China. Sci Adv 3(3):e1602751

Climate Co-benefits in Rapidly Urbanizing Emerging Economies: Scientific and Policy Imperatives

Mahendra Sethi

1 Introduction: The Emerging Economies

Emerging economies are broadly defined as nations in the process of rapid growth and industrialization. Often, these nations are transitioning to an open market economy with a growing working age population (Fleury and Houssay-Holzschuch 2012; Kuepper 2019). The term itself was coined in the 1980s, by Antoine van Agtmael, as a more positive alternative to the then popular term "less economically developed country" or LEDC. An emerging market economy describes a nation's economy that is progressing toward becoming more advanced, usually by means of rapid growth and industrialization. The latest Emerging Markets Index (MSCI 2019) reveals that several developing nations in Asia, Africa, Eastern Europe, and Latin America are now evolving as emerging economies, most importantly it includes 21 countries, namely Brazil, Chile, China, Colombia, Czech Republic, Egypt, Hungary, India, Indonesia, Korea, Malaysia, Mexico, Morocco, Peru, Philippines, Poland, Russia, South Africa, Taiwan, Thailand, and Turkey.

As evident, emerging economies vary considerably on account of geographical size, natural resource base, socioeconomic parameters, etc.; yet these exhibit common features like their transitional status, young and growing population, developing infrastructure, and increasing foreign investments. Transitional economy assumes that emerging markets are often in the process of moving from a closed economy to an open market economy. While most hope the transition would bear favorable policies, there is also a heightened political and monetary policy risk involved. Young and growing population are capable of spurring strong long-term

---

M. Sethi (✉)
Technical University Berlin, Berlin, Germany

Mercator Research Institute on Global Commons and Climate Change (MCC), Berlin, Germany
e-mail: m.sethi@campus.tu-berlin.de

© Springer Nature Switzerland AG 2020
W. Buchholz et al. (eds.), *Ancillary Benefits of Climate Policy*, Springer Climate,
https://doi.org/10.1007/978-3-030-30978-7_17

growth rates by replenishing aging workers and consuming goods, but it can also lead to an increased risk of political instability. Developing infrastructure infers that emerging markets are often in the early stages of building infrastructure. While this means there is often pent-up demand for government spending, it can also mean higher costs and less efficiency for businesses. Lastly, emerging markets see strong and increasing foreign direct investment, which can be a good sign of anticipated economic growth ahead. However, too much capital can quickly lead to an overheated market ripe for a correction. In addition to the above risks and uncertainties, these countries also exhibit numerous physical, infrastructural, and societal transformations because of rapid and sporadic urbanization on the ground, making an uphill task for their settlements to develop sustainably. These economies are thus stressed with mounting population in cities and constantly meeting with growing needs of public goods and services. The situation gets complicated by overcrowding, housing problems, degradation of the local environment, insanitary conditions, air and noise pollution, and overlooking of best practices. Thus, it becomes imperative to investigate how cities and urbanization auger with their original role of economic development in the crisis of global warming.

While research in the causes and impact of climate change has witnessed widespread growth since the 1980s and 1990s, cities have only recently started taking interest and active role in climate action (for details on chronological evolution, refer Sethi 2017). There is a strong and complex interplay between economic development, greenhouse gas (GHG) emissions, and urbanization pathways of countries. Evidence from 200 plus countries during 1960–2010 shows that GHG emission contributions and energy consumption are shifting toward fast urbanizing regions of the developing world (Sethi and Puppim de Oliveira 2015). Yet these cities have little resilience to the impact of global warming, heat waves, floods, storm surges, landslides, heat islands, and so on. They also host a large population of poor people living in the most environmentally vulnerable locations (Satterthwaite 2009; IPCC 2014a). It is further observed that environmental policies are least prioritized in cities as issues related to provision of basic services, shelter, and jobs take center stage. Nevertheless, simultaneously responding to local environmental problems and changing climate, the rapidly urbanizing regions in emerging economies are in the right position to effectively assess multiple challenges and the trade-offs associated with high economic growth, sustainable development, mitigating excessive emissions, and adapting to global warming. Thus, a sample study of few emerging economies becomes necessary to understand the challenges faced by developing countries.

As evident, there are no major precedents or best practices as urban challenges for emerging economies are significantly different from those of developed countries. While cities in industrialized nations have undoubtedly initiated several climate projects at the local level, there is a lot of debate about appropriate consideration of co-benefits. Early evidence from Cities for Climate Protection Campaign (Kousky and Schneider 2003) shows eight municipal co-benefits in transport, energy, waste, clean air, local economy, etc., these were barely accounted for, showed greater uncertainty in measurement suggesting that municipalities were interested to focus more on moving beyond the business-as-usual approach or demonstration of climate projects than actually quantifying the ancillary benefits. Thus, cities in developed countries do not appear to seriously pursue co-benefits as a scientifically driven policy instrument. In addition, there are marked differences in cities of industrialized nations and those in developing context, in terms of legislative powers, political economy, functional autonomy, technical capabilities, and mitigation targets. Thus, the direct transfer of lessons (that might work in a developed country) may not be the most appropriate solution for a problem in a developing country. In fact, the prime question in emerging economies appears to be "how can we improve the standards of living of our population without putting too much GHG in the atmosphere and simultaneously adapt to the changing climate?" Could this gap be bridged with ongoing climate action paradigm that focuses on mitigation or adaptation initiatives alone? Are the resulting benefits adequate to bring about a positive change or there is a need to search for peripheral, ancillary, and mutual benefits arising from early climate experiments in emerging economies? Secondly, how these measures stand up to agenda on global climate change, sustainable development, and urbanization? For instance, in principle the Paris Agreement, Urban Sustainable Development Goals (SDGs), New Urban Agenda and Sendai Framework, all emphasize greater collaboration between diverse communities, multilevel institutions, and development sectors to co-generate positive impacts. All the above questions were explored as a part of the panel discussion convened by the author at the Cities and Climate Change Science Conference (2018).[1] The ensuing research investigates the role of co-benefits in rapidly urbanizing emerging economies (Sect. 2), assessment of co-benefits in urban areas (Sect. 3), case studies from Brazil, China, India, and Turkey (Sect. 4), followed by conclusions (Sect. 5) and recommendations (Sect. 6).

---

[1]The Cities and Climate Change Science Conference (https://citiesipcc.org/) was held during March 5–7, 2018, in Edmonton (Canada) bringing together representatives from the academia, scientific bodies, and agencies; concerned member states of the United Nations; city and regional governments; and urban and climate change practitioners. The main aim was to stimulate scientific research around cities and climate change to provide input to be assessed by the three Working Group Reports and three Special Reports (SR) of the sixth assessment cycle (AR6) and to establish the foundation for the SR on cities and climate change that will be undertaken during the seventh assessment cycle (AR7). The conference was co-organized by several organizations, including C40, Cities Alliance, ICLEI, Future Earth, SDSN, UCLG, UN-Habitat, UN Environment, and WCRP. The session convened by the author was titled, "Co-Benefits for Emerging Economies: Practical Experiences and Policy Imperatives." The complete proceedings of the conference could be accessed at: http://www.urbangateway.org/system/files/documents/urbangateway/cities_ipcc_proceedings_final_for_email-s.pdf

2 The Role of Climate Co-benefits in Rapidly Urbanizing Emerging Economies

In the last two decades, co-benefits or ancillary benefits has become an area of growing interest in climate policy and action. The United States Environment Protection Agency (USEPA) considers co-benefits as all of those positive outcomes associated with multiple, simultaneous emissions reductions (USEPA 2005). IPCC, in its fourth assessment cycle, referred it as the non-climate benefits of GHG mitigation policies that are explicitly incorporated into the initial creation of mitigation policies. Co-benefits reflects that most policies designed to address GHG mitigation also have other, often at least equally important, rationales involved at the inception of these policies (e.g., related to objectives of development, sustainability, and equity). The intended positive benefits of a policy and decision-making distinguish it from unintended positive side effects (IPCC 2007). Thus, co-benefits approach is regarded to be a "win-win strategy aimed at capturing both development and climate benefits in a policy or measure" (IGES 2010). In fact, it is now increasingly being used to study, analyze, and promote local level action. IPCC's recent assessment elaborates that a government policy or a measure intended to achieve one objective often affects other objectives, either positively or negatively. For example, mitigation policies can influence local air quality. When the effects are positive they are called "co-benefits," also referred to as "ancillary benefits" and can be measured in monetary or non-monetary units (IPCC 2014b). In a recent review of case studies of climate co-benefits globally, Doll and Puppim de Oliveira 2017 underpin that climate co-benefits are generated if, (1) One outcome of an intervention is a benefit in terms of climate change mitigation or adaptation and (2) Benefits of certain development interventions in tackling climate change as compared to other viable development alternatives (including doing nothing or business as usual). Meanwhile, co-benefits are also generated by climate action while purposely mitigating climate change and leading to development benefits at the local or regional level, such as reduction in air pollution, creation of jobs, or energy cost savings.

The global scientific and policy discourse clearly identifies the typologies of climate co-benefits approach. As per IPCC's review of evidence of actions with co-benefits include: (1) improved energy efficiency and cleaner energy sources, leading to reduced emissions of health damaging, climate-altering air pollutants; (2) reduced energy and water consumption in urban areas through greening cities and recycling water; (3) sustainable agriculture and forestry; and (4) protection of ecosystems for carbon storage and other ecosystem services (IPCC 2014b). Meanwhile, the Paris Agreement recognizes the social, economic, and environmental values of voluntary mitigation actions and their co-benefits for adaptation, health, and sustainable development (UNFCCC 2015). It entrusted a Technology Executive Committee to prepare a summary for policy makers, with information on specific policies, practices, and actions representing best practices and with the potential to be scalable and replicable.

Cities located in rapidly developing countries until recently have shown limited empirical research on climate action and co-benefits; that too primarily concentrated on the adaptation agenda. There are several research questions arguing this bias, most importantly that issues in developing cities are devolved from their national narratives on climate change. It is actually an intellectual and policy vacuum that developing countries identify themselves as hardly contributing to the global GHG burden, but rather facing higher climate risks and vulnerabilities. However, the local situation in these countries is gradually but surely changing. In the process, they have to simultaneously deal with local development priorities and environmental challenges of migration, slums, sanitation, transport, environmental pollution, waste, parks, and biodiversity, wherein the approach of urban co-benefits provides a good opportunity to meet diverging objectives.

One of the most significant international initiative in this area was led by the United Nations University with the support of Ministry of Environment Japan. It considers urban climate co-benefits approach in the implementation of initiatives (policies, projects, etc.) that simultaneously act in reducing the contribution to man-made global climate change while solving local environmental problems in cities, and in turn potentially having other positive developmental impacts, such as improvements in citizens' health, energy security, and income generation (UNU-IAS 2013). Drawing from research experiences in developing Asian cities, several studies (Geng et al. 2013; Jiang et al. 2013; Kurniawan et al. 2013; Puppim de Oliveira et al. 2013) underscore that though many actions meant to combat climate change, inadvertently have other local benefits, the co-benefits approach seeks to purposefully multiply and mainstream climate co-benefits into the development process. This approach has been further developed into a conceptual framework (Fig. 1) to understand and analyze potential co-benefits at the intersection of multiple development and environmental objectives, GHG emission sectors across different spatial scales (Sethi and Puppim de Oliveira 2018), and in doing so how these could be utilized for innovations, affirmative action, and reforms in the urban realm. Nevertheless, the effect of co-benefits and adverse side effects from climate policies on the overall social systems has not yet been thoroughly examined. A major area of concern is on account of systematic/methodical and reasonably accurate estimation of co-benefits in urban areas.

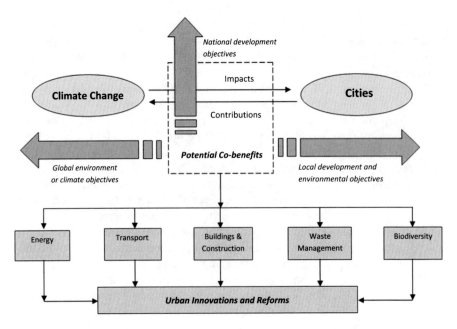

Fig. 1 A conceptual framework to understand and analyze potential co-benefits at the intersection of multiple development and environmental objectives, GHG emission sectors across different spatial scales

3 Assessment of Co-benefits in Urban Areas

The impacts of pursuing climate actions have been studied through traditional cost–benefit approaches (IPCC 2014a). Internationally, mechanisms based on measurement, reporting, and verification (MRV) have been accepted to guide climate action. These take into account multi-criteria analysis of risks, socioeconomic and monetary costs and benefits. Measuring mitigation benefits within a particular sector is relatively easy based on activity data and level of technology. The assessment typically follows Kaya Identity that takes into account emission factors, energy intensity, carbon intensity, etc. (Kaya 1990). Accordingly, UNFCCC has standard IPAT-based methodologies (Chertow 2000) for various GHG producing processes adhering to principles of adopting transparent, consistent, comparable, complete, and accurate (TCCCA) methods that, while selecting an activity–technology mix could be used to measure carbon saved, as a measure of co-benefits. However, as one moves from process-based activities (that typically follow life cycle analysis or input–output analysis) to system based ones as in case of transactions in complex urban systems (which involve parallel activities generating GHG emissions from different functional sectors), it becomes a complicated procedure to measure their throughput, the co-benefits even more.

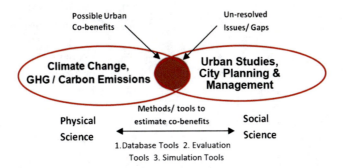

Fig. 2 Conceptual understanding of research methods to estimate co-benefits in the inter-disciplines of Climate Change and Cities

Any assessment tool or methodology has a trade-off between technical sophistication and its general application (UNU-IAS 2013). Precise quantification requires a method or tool that will be difficult to use for all policy makers from differing levels of technical expertise. This would limit its applicability and consequently its use. More quantification would be costly and time consuming to apply too. A tool may not be viable in the context of many developing countries or in case of small initiatives with limited resources. Applying a co-benefit tool in urban areas faces a similar challenge. To simplify this rather analytical- and calculation-based process, and assess its possible impacts and ramifications while selecting a policy alternative, there are several models or assessment tools being used worldwide that estimate actual or relative co-benefits and facilitate decision-making, refer Fig. 2 for clarity.

Based on this theory, Sethi (2018) presents a first of its kind global database of 44 urban-co-benefit assessment tools, 27 in mitigation, and 17 in adaptation. The features of assessment tools are exhaustively reviewed by accessing those, studying their manuals, and product literature made openly accessible by their proponents on the Internet. The findings underscore that (1) Estimation of co-benefits is impaired by inconsistencies in the definition of a city and accounting a city's GHG emissions and (2) There is apparently no single methodology to account for urban level risks and vulnerabilities from multiple and diverse natural processes like floods, heat waves, sea-level rise, etc. Nevertheless, there are extensive normative benefits that are local, development-oriented, and promoting socioeconomic welfare that ought to be accounted for.

The co-benefit assessment tools are broadly classified into three categories: (1) Informative/database tools, (2) Evaluation tools, and (3) Simulation tools (Sethi 2018). On the one hand, tools dealing with mitigation co-benefits are more GHG quantification-based and abundant in practice, co-benefit tools in adaptation are comparatively scarce, deal with non-quantifiable data, and rather end up recreating a normative though crucial decision-making process. While researchers and software developers have begun to create meticulous co-benefit assessment tools, their practical use by city managers to definitively predict and decide planning alternative for future, to offset emissions or contribute to a resilient society is still not apparent. In addition, the pace of growth in emerging economies leaves little time to fully test these tools.

Yet diverse urban perspectives can have new and more contextual takes on this matter. As more and more developing countries are contributing to the climate change discourse, emerging economies like Brazil, China, and India form the frontier of innovations and reforms. Recent evidence affirms that it is possible to have the same economic state but lesser emissions by managing urbanization in a controlled manner (Sethi and Puppim de Oliveira 2015). Thus, there lie several opportunities for emerging economies to harp on co-benefit methods/initiatives for achieving multiple objectives. In particular, their first-order cities have high level of emissions, and thus greater stakes to combat climate change. The case studies in this research represent such cities in Brazil, China, India, and Turkey, demonstrating how identification and assessment of co-benefits are being experimented within their specific contexts.

4 Case Studies from Emerging Economies

4.1 Brazil

The country inherently demonstrates special circumstances in addressing climate change. Brazil utilizes biofuels in transportation and sources more than 70% electricity from hydropower (IEA 2017), nevertheless its largest source of emissions is from land use change in the Amazon area. Despite a major economic turndown leading to a fall in energy use, Brazil's emissions rose by 8.9% in 2016, largely driven by a hike in illegitimate deforestation (Maisonnave 2017). Brazil's Intended Nationally Determined Contribution (INDC), submitted to the UN during the Paris Agreement, targets a 37% cut in GHGs by 2025 compared to 2005 levels (UNFCCC 2015). Some assessments underrate this target as lack of ambition, making it crucial to explore greater synergy between different emissions sectors and functions on the one hand and within urban–rural linkages on the other.

Rio de Janeiro and Sao Paulo are principal cities of Brazil and thus taken as a sample to study experimentation of co-benefits approach with diverse methods. It is noteworthy how through good governance, convergence of multiple challenges, certain development actions have converged to bring ancillary benefits. For example, the program "Rio Green Capital," a strategic plan for land use aims to consolidate 2000 hectares of land that has already been reforested, in the region of Marapendi, Chico Mendes e Prainha. In addition, Rio also benefits from an urban afforestation program covering 170,000 m^2 of parks and squares. Replanting trees on Rio's steep slopes has multiple benefits: it protects the soil preventing landslides, it also lowers the daily peak temperature, increases air quality, helps avoid the spread of wildfires (especially in inhabited areas), and protects the fauna and flora. The reforestation project aimed in 150 hectares helped 150 communities while creating employment for around 800 people (Kahn 2015).

In transportation, a series of initiatives such as the expansion of the metro lines and Bus Rapid Transit (BRT) lines was unleashed in Rio de Janerio (UNU-IAS 2013). Shifting modal share from private to public vehicles will be testing, as it requires changing people's habits, but it could yield multiple benefits beyond GHG emission reductions, such as improved air quality, less road congestion, and more connected, accessible, and livable cities. For example, the adverse health impacts of air pollution in Sao Paulo have been estimated at US$208 million per year, including respiratory and cardiovascular diseases and related mortality (Miraglia et al. 2005). In addition, the city of Rio has been encouraging bicycles, by designating over 450 km of special lanes in the city, to offer a healthy alternative to automobiles. While supporting this infrastructure, the city has further set up a bike-sharing service with over 200,000 users.

Another urban co-benefit is being realized through concerted policy actions in systematic municipal waste management. This primarily involved organized composting by creating community gardens and engaging people to reduce garbage production (Puppim de Oliveira 2009). In addition to significantly reducing the municipal waste being collected, treated and disposed to a landfill site, it further led to urban gardening, production of domestic vegetables and fruits. Similar opportunities within water–energy–land nexus are observed in municipal sewage management in Brazilian cities (Stepping 2016) that unlocks co-benefits in generating electricity. In both the cases, it is seen that the key is to understand what are the immediate needs and problems of a community and how local governments and experts can answer those while simultaneously aiming for climate change action. For instance, providing people with non-renewable source of energy and at the same time reducing waste-related pollution.

Since the co-benefits approach creates supplementary advantages to monetary or commercial ones in traditional urban projects, in all likelihood they create a larger social good. In doing so, poor people impacted most by the climate change are the ones who benefit most from this approach because they are provided with goods and services they need at lower and sustainable costs to the local governments. For example, public lighting is essential to make urban areas safer and more enjoyable at night, for traffic safety, and to illuminate monuments, buildings, and landscapes. Energy-efficient technologies such as LEDs make it possible to deliver these benefits saving up to 30% on energy use; while energy savings with traffic lights can be as high as 90%. Since 2000, more than 2.5 million public lighting units have been replaced by more efficient ones in more than 1300 Brazilian settlements, at a cost of about R$500 million (Kahn 2015). The case of Brazilian cities shows that co-benefits can mitigate GHG emissions or reduce the climate impacts, but are not the silver bullet. Recently, the government has explored the possibility of using a national emissions trading scheme (World Bank 2019), but the attempts are mostly tentative and do not rely on qualified methods of evaluation. These would have to be integrated with local priorities, development needs, and willingness of a society to bring transformations from the status quo.

4.2 China

With high industrialization in China for the last three decades, mounting GHGs in urban areas is a huge challenge, with local air pollution causing increasing expenditures on public health. In its INDC, the People's Republic of China (PRC) announced four principal climate goals: (1) To achieve the peaking of carbon dioxide emissions around 2030, making best efforts to peak early; (2) To lower carbon dioxide emissions per unit of GDP by 60–65% from the 2005 level by 2030; (3) To increase the share of non-fossil fuels in primary energy to around 20% by 2030; and (4) To increase the forest stock volume by around 4.5 billion cubic meters from 2005 levels by 2030 (PRC 2015). As evident, the second and third objectives are crucial to realize co-benefits from energy efficiency and rooftop solar generation in urban areas.

If one compares major cities in China like Guangzhou, Beijing, Shanghai, Chongqing, Wuhan, Xian, etc.; these clearly have different economic structures, emissions disparities, and development priorities (Liu et al. 2012). For instance, how to support manufacturing was imperative in relatively less developed cities. In such variable conditions, only an integrated approach bearing mutual benefits could help solve this challenge. The widespread use of coal in thermal plants, industrial processes and space heating as well as activities like housing construction and transportation are some of the most polluting functions that local governments should also focus on (Kishimoto et al. 2017; Dwortzan 2017; Hove and Enoe 2015). In a 2015 national survey, 76% of respondents said that air pollution is a "big problem" and 35% of respondents said it is a "very big problem" (Wike and Parker 2015). Lung cancer mortality rates in Hebei, one of China's most polluted provinces nearly tripled from 1973–1975 to 2010–2011. Research also shows a strong relationship between air pollution and cardiovascular disease in China (Rohde and Muller 2015; Zhao et al. 2017). In September 2013, the Chinese government announced the Action Plan on Prevention and Control of Air Pollution calling for a 10% reduction in PM10 concentrations by 2017 in cities across China, with more stringent targets in three key regions (Beijing–Tianjin–Hebei, Yangtze River Delta, and Pearl River Delta) through delineating "10 tasks" for cleaning the air (Zhang 2013).

In the recent past, many Chinese provinces and localities have further committed to climate goals as well. As of June 2016, 23 provinces and cities committed to peaking CO_2 emissions before 2030 as part of China's Alliance of Pioneer Peaking Cities. For example, in December 2017, the city of Wuhan issued a Carbon Emissions Action Plan that included a commitment to peak CO_2 emissions by 2022 (Fong 2016; Energy Foundation 2015). China's 13th Five-Year Plan (2016–2020) gives priority to fighting air pollution. Measures to control coal burning have been a top priority. Stricter vehicle fuel efficiency and emissions standards have also been adopted (PRC 2016). During 2017, strict policies with respect to coal burning, industrial activities, and traffic were announced for the Beijing–Tianjin–Hebei area. Most measures to fight local air pollution in China also help fight climate change. Policies that promote natural gas as an alternative to coal help cut both local air pollution (by 90% or more, depending on the pollutant) and carbon emissions by roughly 50% (SIPA 2018). Similarly, the policies that promote energy efficiency also reduce both local air pollution and carbon emissions. Policies promoting industrial energy efficiency are especially important, as are policies to improve the energy efficiency of Chinese buildings. China's fuel efficiency standards for vehicles also cut both local air pollution and carbon emissions as well as reduce China's reliance on imported oil (Dwortzan 2017). China's policies to promote electric vehicles provide local air pollution benefits, since electric vehicles do not have tailpipe emissions. There is a debate among experts about the extent to which electric vehicles help mitigate carbon emissions in China, since those vehicles increase power demand from China's coal-heavy electric grid, although the significant benefits accrued because of clean air in cities cannot be undermined.

Along with following its IPCC agenda, China is promoting the national carbon market. In addition, it is trying to encourage the private sector to forward their climate change agendas. Several empirical assessment studies (Geng et al. 2013; Creutzig and He 2009; Liu et al. 2013; Ramaswami et al. 2017) have tried to predict the future impacts and potential benefits against the probable economic loss. In a more comprehensive study, Dong et al. (2017) set up seven different co-benefit scenarios to consider all possible impacts. The use of simulation tools clearly demonstrates that while implementing climate co-benefit actions in cities, we can actually gain GDP and minimize losses to the environment. The Chinese findings suggest that cities that are next to each other or having similar context, priorities, and challenges should coordinate with each other to further climate change agenda and have more positive impact.

4.3 India

India's constitution adopted in 1950 was one of the first in the world to safeguard the national environment, forests, and wildlife. The National Environment Policy approved in 2006 aim to integrate environment and development and the National Action Plan on Climate Change (NAPCC 2008) mandates a multi-sectoral approach requiring states to produce their own climate action plans. In course to the Paris Agreement, the country pledged for a 33–35% reduction in emissions associated with each unit of economic output by 2030, compared to 2005 levels (PIB 2015). India's GHG emissions in 2015 stood at 3571 m tonnes of CO_2 equivalent (Potsdam Institute for Climate Impact Research 2019). About two-third of the national emissions can be attributed to urban areas (Sethi and Mohapatra 2013). India's per capita emissions stood at 2.7tCO2e in 2015, around a seventh of the US figure and less than half the world average of 7.0tCO2e. One in every eight deaths in India is due to air pollution, according to a recent report in the Lancet Planetary Health, while India is home to 13 of the world's 20 most polluted cities (World Health Organization 2018). A majority of the emissions are from industries and transportation activities in these cities. Thus, there is a complex relationship between energy supplies, air-pollution, and potential human health benefits.

There are different perspectives and lessons emerging from different Indian cities. In a rapidly growing scenario, urban infrastructure is a huge part in financial investment be it in energy, water, transportation, buildings, waste, or biodiversity that invariably has several co-benefits. A case in point is co-benefits in transportation, if one were to study how the sector is essentially about people and the different ways to make them commute safely and conveniently while causing lower environmental impacts. This innate approach of interrelated functional assessment needs to be considered in all urban infrastructural domains. One has to identify specific problems in each city and their concerted needs instead of merely branding and expecting these cities to become low-carbon or smart. Thus, life-cycle assessment of urban projects becomes important to evaluate co-benefits potential. But then how does one internalize life-cycle issues? This alludes that climate change action needs to permeate all areas of decision-making at the city scale—from planning to implementation. The above also indicates that the government should not always consider urban projects on the basis of the lowest cost bid but consider potential co-benefits being offered too.

Application in fast-paced cities, the co-benefits approach requires consideration of several adaptation and local development goals using research tools: GIS mapping of urban data, topological information on pressing environmental concerns like urban heat waves and floods. This should address the question of how to avoid people moving to flood-prone areas just because land is cheap there. This information then needs to be transferred to the ground using flood zonation maps for preemptive planning and disaster response. It was observed in the devastating floods of Chennai (2015) that thousands of people were stranded on terraces and food had to be air dropped for a week (BBC 2017). The city office was itself inundated to act swiftly. Later, it mapped all the vulnerable infrastructure such as schools and hospitals where people aggregate, in addition to critical water supply and sanitation infrastructure. Thus, mapping of public infrastructure and amenities is key to realize potential co-benefits in avoiding disasters, monetary losses, and proactively realizing economic goods. Secondly, to attain local development priorities, comparative assessment of co-benefits in existing urban services becomes imperative, say private vs. public transport to rationalize what is the additional number of people that should shift toward the metro? For example, Delhi metro transports more than two million people per day, yet citizens remain largely unaware how much do they contribute in minimizing carbon emissions and local pollutants by using the metro.

The INDC noted that around 70% of its population depends on traditional biomass energy, which is inefficient and causes high levels of indoor air pollution. India is promoting the use of biomass to generate power, expanding the access of LPG stoves and electricity in all households, which it regards as cleaner and more efficient. Against the target of 10GW of such bioenergy by 2022 India already reached 9GW in 2018 (MoEFCC 2018), which is bound to realize multiple health benefits. In addition, there are many urban-centric programmes like promotion of electric rickshaws as para-transit, full electric bus fleet in cities by 2030, and support for more efficient electric appliances that collectively bear ancillary social and health benefits like jobs, clean air, human productivity, health, and comfort. Thus, early experiences of co-benefits in India shows how one could expand learnings from urban cases through the process of actively mainstreaming climate change in city planning and management. With immense variability in cities, the Indian case further demonstrates that each city is unique in development situation, infrastructure, climate situation, its GHG emissions, vulnerabilities, and thus mandates greater systematic co-benefits assessment and implementation.

4.4 Turkey

The country is an emerging economy and an energy importer (70% of its total energy needs), with growing emission from fossil fuels. Energy consumption is the main source of Turkey's emissions: 72.5% of total GHG emissions and 85.2% of CO_2 emissions are attributed to the energy sector (Sahin 2016). The national energy policy has made coal the preferred fuel for expanding Turkey's energy production capacity and therefore the energy–economy nexus is strongly dependent on fossil fuels creating a constant rise in emissions. An economic analysis shows that economic growth between 2003 and 2009 became more energy and pollution intensive compared to 1995–2002, and high carbon economic activities related to construction, such as real estate as well as transportation were among the leading sectors (Ahmet 2015), rendering pollution in major cities like Istanbul, Ankara, Adana, Amasya, and Manisa (Ahval 2018). While the demand for sustainable energy funding is increasing, the government's National Climate Change Strategy, 2010–2020 has a vision of the country fully integrating climate change-related objectives into its development policies, disseminating energy efficiency, increasing the use of clean and renewable energy resources, actively participating in the efforts for tackling climate change within its "special circumstances," and providing its citizens with a high quality of life and welfare with low-carbon intensity (Ministry of Environment and Urbanization 2010). The National Climate Change Action Plan (NCCAP) further lays down cross-sectorial mitigation measures from short-, medium-, and long-term goals under eight topics (energy, industry, forestry, agriculture, buildings, transportation, and waste and climate change adaptation) including provisions for cross-cutting issues for data collection, reporting, monitoring, and verification (Ministry of Environment and Urbanization 2011). The necessary transformation is expected to bring co-benefits such as the continuity of ecosystems and biodiversity, preservation of public health, increase in qualified and clean jobs, reduction on foreign source dependency particularly for energy (Day et al. 2016; Sahin 2016).

Integrated Urban Development Strategy and Action Plan (2010–2023), commonly known as KENTGES (Ministry of Public Works and Settlement 2010), was adopted in 2010 for the period of 2010–2023. The plan mentions studies and actions to be performed at central and local levels regarding transport, housing and land supply, disasters, natural and cultural assets, climate change, life quality, social policies, and participation. In the Plan, the main principles and values regarding climate change and sustainable development are: (1) Paying attention to ecological balance in natural resource use; (2) Ensuring healthy, safe, and quality environment, free from natural and technological disasters and risks; (3) Improving use of a sustainable transportation systems as well as use of renewable energy resources; (4) Paying attention to environmental, natural, and ecological equality; and (5) Encouraging methods to decrease impacts of consumption patterns on natural and cultural environment in settlements. KENTGES is a cross-cutting strategy document in terms of integrated urbanization, low carbon, and sustainable development applicable to large and small cities alike. Recently, the Strategic Plan of Ministry of Environment and Urbanization (2015) has been developed and adopted in 2018. It targets to support local authorities in providing environmentally sustainable urban areas and livelihoods; promoting and supporting mechanisms to establish smart cities; increasing energy efficiency standards of buildings. The policies noted above have strong urban traction and need to be assessed for the progress on co-benefit projects at the local level.

The national government, post 2005 political reforms announced a support mechanism for local municipalities. This has been an important driver for improved environmental conditions in urban areas and some cities are working on landfill site projects, particularly on co-generation of energy. Scientific management and disposal of waste has always been an important environmental issue. It involves huge cost and new landfills are always challenging sites to find when cities expand. Undoubtedly, such projects have substantive socioeconomic co-benefits for the community. Thus at local level, cities have gradually started to engage in climate change actions. One can observe small-scale projects in transportation and energy production sectors that also have a component of climate mitigation. Nevertheless, the most important issue is their financial sustainability. The municipalities still do not have adequate financial capacity to deal with large development projects and there is insufficient interest from the community too. This essentially leaves city governments on their own. It is seen that the progress is largely based on motivation of an individual decision maker (mayor) or some co-sponsorship by certain private funder.

In line with the goal of reducing energy demands and carbon emissions of buildings and popularizing environment friendly sustainable buildings that use renewable energy, an Energy Efficiency Strategy Paper was released in 2012. It targeted that at least one-fourth of building stock in 2010 may be converted to sustainable buildings till 2023 (Ministry of Environment and Urbanization 2014). It is reported that efforts are being carried on for analyzing the potentials in collective housing projects for utilizing sustainable energy resources, cogeneration or micro-generation, central and localized heating and cooling systems, and heating pump systems, and for encouraging such practices (OECD 2017). Furthermore, studies are being carried on regarding construction materials, building inspection and issuing energy identity card for buildings. In 2011, energy identity card practice was initiated, and so far more than 34,000 buildings have been awarded, over 90% of which are new buildings. The monitoring of new buildings shows that the energy consumption reduces more than 40% as compared to the old ones.

The support mechanism under the urban realm has encouraged the private sector and initiated projects that harness climate co-benefits. Notable co-benefits of GHG reduction are energy consumption in the water sector (treatment, conveyance, and distribution), clean air in health and transportation sector, lesser demand in expanding landfill sites, or commissioning new ones. There are two crucial points worth considering while implementing co-benefits approach at local scales: (1) The economy of cities is almost never in a good shape and requires financial inputs, and (2) Does the approach bring about some local autonomy or has some kind of incentive or reward for municipalities? It would be imperative to utilize co-benefits tools to evaluate and predict costs and benefits in the future. Moreover, the next obvious step to upscale and expand this approach would be allocating compulsory climate budgets and giving bonuses to good climate practices among cities. The Voluntary Carbon Market in Turkey represents a very small percentage within the World Carbon Market (Ministry of Environment and Urbanization 2015), and further research needs to answer this practical need. Also, it is pertinent to have standard procedures and guidelines to identify and implement co-benefits in cities. If the local governments can monetize this approach, it would be relatively easy to convince funding agencies and local companies to invest in climate co-benefit projects too.

5 Conclusions

It is a known fact that efforts to pursue just climate mitigation or adaptation have not lead to significant outcomes on the ground. Cities in emerging economies till recently had limited research on climate response, largely preoccupied with the adaptation agenda. As many of these economies harbor fast growing cities, they would have to simultaneously respond to the mounting urban development pressures, local environmental issues, and the test of global climate change. Thus, cities in emerging economies are on the frontline to effectively assess multiple challenges and their trade-offs associated with rapid growth, sustainable development, mitigating excessive emissions, and adapting to climate change. At the same time, there are several gaps in estimating and applying co-benefits in urban settlements. The country cases analyzed in this chapter exhibit a varying degree of conceptual, methodological, empirical, and policy governance gaps, supporting normative gaps in co-benefits theory (Sethi 2018). This section discusses key research and technical aspects to actualize these on the ground that eventually leads to recommendations for policy application.

Case studies from emerging economies demonstrate multiple and diverse kinds of urban co-benefits opportunities in the interphase of transport, waste management, clean environment, sewerage, and energy domains. Turkey shows a lead in accruing co-benefits in municipal waste management sector. Likewise, Chinese cities show more attempts of capturing ancillary benefits between transport, air pollution, and health sectors. Depending upon individual cities, Brazilian and Indian cities have shown harnessing of co-benefits in both transport and waste sectors. Thus with fresh evidence, this research corroborates the existing literature, underpinning that emerging economies and other developing nations have a huge potential to adopt climate mitigation (IPCC 2014b). The above does not mean that these emerging economies are not experimenting with some other co-benefits in their cities. Limited examples from each case country can be attributed to the adopted case study approach that largely selects the most visible and reported co-benefits projects. Nevertheless, it could be concluded with a reasonable degree of confidence that there still lie many other co-benefits significantly untapped in challenging sectors like buildings and construction, land planning, land use, and biodiversity, as none of the emerging economy cities seem to show any focused approach and initiatives in capturing these. This is in spite the fact that most of the new infrastructure, industries, housing, etc. that is going to be created in this century would be in the urban settlements of these countries (IPCC 2014a). Such co-benefits could only be realized if systematically integrated with urban planning, building codes, use of energy-saving construction practices and electric appliances in upcoming greenfield developments. As such, there are several emerging issues that could be incorporated and integrated into the research agenda on cities and climate change that include promotion of clean energy generation (rooftop solar) and use, public transport typically mass or light transit (as applicable), non-motorized transport, electric vehicles, energy-efficient buildings or green habitats and waste-to-energy technologies. In addition, it is vital to prioritize those climate policies that bring co-benefits, as they improve public health (Doll and Puppim de Oliveira 2016) and can most benefit the poorest, such as those in informal settlements (Ahmad and Puppim de Oliveira 2016). Thus, a major area of concern is to localize climate responsible planning, mainstreaming climate agendas in urban plans, smart city plans, and building capacities of municipal agencies to implement these.

Cases from four emerging economies investigated in this research show limited application of scientific assessment tools. Most decisions are based on normative understanding, preliminary databases with little comparative evaluation of co-benefits or the use of simulation tools to project long-term impacts. This is most prominently evident in the evaluation of transportation, GHGs, air pollution, and health co-benefits in Chinese cities. Lack of mid- and long-term temperature scenarios at the city level is an impediment to appropriate adaptation planning and co-benefits thereof. A substantive understanding of urban climate action indicates institutional and financial weakness of local governments, and the lack of focused leadership and policy pressures from upper levels of governance as well as within the community. While no one in the public sector questions the relevance of co-benefits, their awareness and willingness to pursue them seriously is limited, chiefly because many potential co-benefits involve multifunctional and multilevel cooperation between different government bodies and sector portfolios. This could pose further challenges to political and financial decision-making with their already constrained capacities. In addition to the challenges of working within the complex urban systems, there are several externalities and jurisdictional issues while considering mitigation co-benefits. For instance, how can cities be held accountable as they use hinterland for carbon sinks, to source energy from thermal plants in rural areas, or import raw materials from outside for raising housing and infrastructure. Yet, there are immense possibilities for leveraging external factors and involving broader stakeholders for larger economic good, like urban agriculture can be used as a co-benefits policy response for informal settlements located in urban fringes. In order to systematically internalize such external factors, there is an urgent need to have integrated and scientific methods and tools that could exhaustively evaluate and precisely simulate energy and carbon offsets, measure co-benefits (monetary and non-monetary) of a city within and across different geographical and administrative sub-national units.

6 Recommendations

A fundamental question for science, policy, and practice community is how to maximize ancillary benefits at the intersection of global climate change, national environmental goals, and local development. Focusing on local, regional, or national climate initiatives in an isolated manner leads to blind spots and runs into the danger of overlooking challenges arising from the multi-scalar nature of climate change. This mandates a concrete strategy to address multiple challenges pertaining to how to design, redesign low-carbon and climate resilient cities and manage the existing ones. This research verifies prevailing knowledge that there is hardly any single policy tool that helps to evaluate and compare benefits of greenfield and brownfield development (Sethi 2018). Nevertheless, there is an urgent need to identify and revise planning norms and standards that lock in more emissions in local transport, energy sector, commercial and residential buildings, etc. Bringing the science, policy, and practice communities together can help reduce research and policy gaps in this area and ensure a timely realization of co-benefits in climate mitigation and adaptation. Accordingly, it becomes essential to recommend what local needs, data architecture, and application platforms are crucial to address the priority research and policy gaps.

6.1 Addressing Knowledge Needs

As this research reveals, there is insufficient knowledge on climate co-benefits in scientific, policy, and practice communities, especially in developing countries. For instance, there is an urgent need to develop future climate and energy scenarios until 2080 at the city level. Currently available projections are at broad regional scale, which have limitations in their applicability at local/city scale. In addition, there is a need to conduct studies and develop methodologies for comparative assessment of co-benefits of various technologies used—across different sectors. There should be national-level stakeholder workshops organized on co-benefits to enhance awareness and cooperation of these diverse groups along with the citizenry. There is a major requirement of having information dossiers for politicians, public representatives, administrators/bureaucrats, civil society, etc. The scientific community and academia need to be provided with research funds to do more co-benefits studies and demonstration projects. Joint labs could be set up in established urban and engineering institutions with support from multilateral agencies, local or state governments and private donors where researchers could learn, collaborate, and co-generate action-based climate co-benefits projects. These labs can further instill climate co-benefits in environmental and development policies of national and sub-national governments. In addition, cities ought to be devolved special funds and technical assistance (through UN agencies and national governments) to prepare local-level climate co-benefit plans. Most importantly, practitioners working on master plans should be trained to use planning tools on co-benefits. Such tools could further be introduced in academic curriculum of climate and environmental studies, urban planning, and management schools.

6.2 Data Architecture, Systems, and Modelling

As case studies from China, India, and Turkey demonstrate the effective application of co-benefits in rapidly urbanizing emerging economies hinges on enhanced use of scientific data and assessments. This is in tune with Creutzig et al. (2019) that outline three routes for upscaling urban data science for global climate solutions: (1) Mainstreaming and harmonizing data collection in cities worldwide; (2) Exploiting big data and machine learning to scale solutions while maintaining privacy; and (3) Applying computational techniques and data science methods to analyze published qualitative information for the systematization and understanding of first-order climate effects and solutions. This necessitates for having an integrated tool that could methodically, precisely, and transparently evaluate and simulate energy/carbon offsets, monetary and non-monetary co-benefits. The tool would require building of a robust database of cities, multi-sectoral modelling and its methodology should consider the following aspects:

1. As a baseline, it takes into account risks, vulnerabilities, and hazards on cities and also contributions of city-based GHG emissions in various mitigation sectors.
2. Utilizes available national/regional datasets and indicators. Downscales national scenarios on urbanization, GHG emissions, climate variability, etc. as regional level inputs, and at the same time upscale local urban data, user, and their behavioral data collected through surveys at locality to city scale.
3. The tool should also be able to quantify co-benefits based on simulation of different factors particularly user choices, spatial planning, density, type of built

form, and their energy needs, city structure, land use mix, transportation modes, and electric appliance choices.

4. The tool ought to perform an evaluation of scenarios, apply spatial-temporal and real-time analysis for simulations and periodic monitoring for policy guidance, decision-making, regulations, and mid-term course corrections.

6.3 Fostering of a Science-Policy-Practice Partnership-Based Platform

Collective experience from new evidence of case studies examined in this research aligns with the limited but growing body of knowledge on ancillary benefits. It augers with the argument that urban climate co-benefits can be initially achieved with simple technologies already available, there is no need for "rocket science" (Puppim de Oliveira 2013). The research findings support current discourse with more credible evidence from emerging economies now, that favors bottom-up initiatives through public policies and international mechanisms in order to scale-up results and disseminate innovations (Doll and Puppim de Oliveira 2017; ESCAP, UNU, UNEP and IGES 2016). Thus, established and functional channels of coordination and cooperation should be utilized in the vertical dimension between the international, national, regional, and local spheres, and horizontally in networks between cities in the national and transnational context, both within and across sectors. In order to widen the application of co-benefits in policies and practice, there has to be knowledge-based advisory platform that guides practitioners and decision makers about which tools and policy alternatives would serve what specific purpose. As discussed above, setting up of urban climate labs in large cities would be crucial to bridge the knowledge gaps between multilevel governments, the private sector, international agencies, professionals, academia, and the people. Some pertinent examples being the setting up of Climate and Energy Lab in New Delhi and the Sino-German Climate Partnership in renewable energies in three Chinese cities Dunhuang, Xintai, and Jiaxing. In addition, cooperation and comparative research between similar cities followed by handholding of smaller cities would help mainstream climate action on ground.

6.4 Realizing Systemic Linkages, Synergies, and Trade-offs Between Cities and Climate Change

The co-benefits approach is perhaps the only assessment framework that proactively considers systemic linkages, synergies, and trade-offs between cities and climate change. While there is a debate about how economic cost–benefit analysis seems to miss some of the ancillary benefits, it is important to highlight that it is a necessary evil. Decision makers in cities of developing countries usually prioritize a strong financial or economic benefit in climate change policies. This investigation underscores that future research should focus on developing new and robust methodologies, inclusive of benchmarks, unit values, etc., to translate the non-economic benefits of climate change policies into monetary terms, which may, in turn, raise the awareness of decision makers on real value of co-benefits policies (as emphasized by Rashidia et al. 2017). There are also arguments on how projects on climate action and co-benefits should be executed—in project mode setting up new organizations/SPV or should they be housed in existing institutions. It is felt that for an effective climate policy environment where wider co-benefits are likely to occur, institutional reform and organizational change should be in place, the scale and intensity of it depending on the local context. Thus, the true success of climate policy in future, especially in emerging economies invariably relies on decentralization of governance systems, empowerment of local bodies, and greater use of co-benefits assessment tools, supported by greater scientific, technical, and financial assistance.

Acknowledgments Earlier versions of this article were presented in the WMO and UNEP sponsored Cities and Climate Change Science Conference, Edmonton and research meeting at the Institute for Global Environmental Strategies, Tokyo (2018). The author acknowledges the Alexander von Humboldt Foundation and the Federal Ministry for Education and Research (Germany) for the research fellowship. The views presented are independent without any influence or conflict of interest.

References

Ahmad S, Puppim de Oliveira JA (2016) Determinants of urban mobility in India: lessons for promoting sustainable and inclusive urban transportation in developing countries. Transp Policy 50:106–114. https://doi.org/10.1016/j.tranpol.2016.04.014

Ahmet AA (2015) On the sustainability of the economic growth path of Turkey: 1995–2009. Renew Sustain Energ Rev 52:1731–1741

Ahval (2018) Only six Turkish cities have clean air, report says. Ahvalnews.com. Accessed 10 May 2019

BBC (2017) Viewpoint: Why poor planning leads to floods in Chennai and Houston. https://www.bbc.com/news/world-asia-india-41144776 (6 September 2017). Accessed 21 February 2019

Chertow MR (2000) The IPAT equation and its variants. J Ind Ecol 4:13–29. https://doi.org/10.1162/10881980052541927

Creutzig F, He D (2009) Climate change mitigation and co-benefits of feasible transport demand policies in Beijing. Transp Res Part D: Transp Environ 14(2):120–131

Creutzig F, Lohrey S, Bai X, Baklanov A, Dawson R, Dhakal S et al (2019) Upscaling urban data science for global climate solutions. Global Sustain 2:e2

Day T, Gonzales S, Röschel L (2016) Co-benefits of climate action: assessing Turkey's climate pledge. Climate Action Network Europe, Ankara

Doll C, Puppim de Oliveira JA (2016) Governance and networks for health co-benefits of climate change mitigation: lessons from two Indian cities. Environ Int 97:146–154, 7.49, 254, 67

Doll C, Puppim de Oliveira JA (2017) Urbanization and climate co-benefits: implementation of win-win interventions in cities. In: Doll C, Puppim de Oliveira JA (eds) Urbanization and climate co-benefits: implementation of win-win interventions in cities. Routledge, Oxford

Dong H, Dai H, Geng Y, Fujita T, Liu Z, Xie Y et al (2017) Exploring impact of carbon tax on China's CO_2 reductions and provincial disparities. Renew Sust Energ Rev 77:596–603

Dwortzan M (2017) Tackling air pollution in China. MIT News (May 2017). http://news.mit.edu/2017/tackling-air-pollution-in-china-0517. Accessed 15 March 2019

Energy Foundation (2015) Mega-city Wuhan issues carbon peaking plan with emissions cap. http://www.efchina.org/News-en/EF-China-News-en/news-efchina-20180210-en. Accessed 5 March 2019

ESCAP, UNEP, UNU and IGES (2016) Transformations for sustainable development: promoting environmental sustainability in Asia and the Pacific. United Nations, Bangkok. Available from http://www.unescap.org/publications/transformations-sustainable-development-promoting-env ironmental-sustainability-asia-and

Fleury A, Houssay-Holzschuch M (2012) For a social geography of emerging countries. Introduction to the themed issue. *EchoGéo* 21

Fong WK (2016) 23 Chinese cities commit to peak carbon emissions by 2030. World Resources Institute (June 8, 2016). http://www.wri.org/blog/2016/06/23-chinese-cities-commit-peak-carbon-emissions-2030. Accessed 12 March 2019

Geng Y, Ma Z, Xue B, Ren W, Liu Z, Fujita T (2013) Co-benefit evaluation for urban public transportation sector–a case of Shenyang, China. J Clean Prod 58:82–91

Hove A, Enoe M (2015) Climate change, air quality and the economy: integrating policy for China's economic and environmental prosperity. Chicago: Paulson Institute (June 2015). http://www.paulsoninstitute.org/. http://www.paulsoninstitute.org/wp-content/uploads/2016/07/Climate-Change-Air-Quality-and-the-Economy-Integrating-Policy-for-China%E2%80%99s-Economic-and-Environmental-Prosperity.pdf . Accessed 18 March 2019

IEA-International Energy Agency (2017) World Energy Outlook 2017. IEA/OECD, Paris

IGES (2010) Fact Sheet No. 1 What are co-benefits? The Institute for Global Environmental Strategies. http://pub.iges.or.jp/modules/envirolib/upload/3378/attach/acp_factsheet_1_what_co-benefits.pdf. Accessed 5 March 2015

IPCC (2007) Intergovernmental panel on climate change. In: Metz B, Davidson O, Swart R, Pan J (eds) Fourth assessment report. Climate change mitigation. Cambridge University Press, Cambridge

IPCC (2014a) Intergovernmental panel on climate change. Seto KC, Dhakal S, Bigio A, Blanco H, Delgado GC, Dewar D, Huang L, Inaba A, Kansal A, Lwasa S, McMahon JE, Müller DB, Murakami J, Nagendra H, Ramaswami A (2014) Human settlements, infrastructure and spatial planning. In: Edenhofer O, Pichs-Madruga R, Sokona Y, Farahani E, Kadner S, Seyboth K, Adler A, Baum I, Brunner S, Eickemeier P, Kriemann B, Savolainen J, Schlömer S, von Stechow C, Zwickel T, Minx JC (eds) Climate change 2014: mitigation of climate change. Contribution of working group III to the fifth assessment report of the intergovernmental panel on climate change. Cambridge: Cambridge University Press

IPCC (2014b) Intergovernmental panel on climate change. Climate change 2014: synthesis report. In: Core Writing Team, Pachauri RK, Meyer LA (eds), Contribution of working groups I, II and III to the fifth assessment report of the intergovernmental panel on climate change. Geneva: IPCC

Jiang P, Yihui C, Geng Y, Dong W, Xu B, Xue B, Li W (2013) Analysis of the co-benefits of climate change mitigation and air pollution reduction in China. J Clean Prod 58(1):130–137

Kahn S (2015) The contribution of low-carbon cities to Brazil's Greenhouse gas emissions reduction goals. U.S. Center – Stockholm Environment Institute, Seattle

Kaya Y (1990) Impact of carbon dioxide emission control on GNP growth: interpretation of proposed scenarios. Response Strategies Working Group, Paris

Kishimoto PN, Karplus VJ, Zhong M, Saikawa E, Zhang X, Zhang X (2017) The impact of coordinated policies on air pollution emissions from road transportation in China. https://www.sciencedirect.com/science/article/pii/S1361920916305041?via=ihub. Accessed 18 March 2019

Kousky C, Schneider SH (2003) Global climate policy: will cities lead the way? Clim Pol 3(4):359–372

Kuepper J (2019) What are emerging markets? Finding and investing in emerging markets. Int Invest. https://www.thebalance.com/what-are-emerging-markets-1978974. Accessed 11 March 2019

Kurniawan TA, Puppim de Oliveira JA, Premakumara DGJ, Nagaishi M (2013) City-to-city level cooperation for generating urban co-benefits: the case of technological cooperation in the waste sector between Surabaya (Indonesia) and Kitakyushu (Japan). J Clean Prod 58(1):43–50

Liu Z, Liang S, Geng Y, Xue B, Xi F, Pan Y et al (2012) Features, trajectories and driving forces for energy-related GHG emissions from Chinese mega cites: the case of Beijing, Tianjin, Shanghai and Chongqing. Energy 37(1):245–254

Liu F, Klimont Z, Zhang Q, Cofala J, Zhao L, Huo H et al (2013) Integrating mitigation of air pollutants and greenhouse gases in Chinese cities: development of GAINS-City model for Beijing. J Clean Prod 58:25–33

Maisonnave F (2017) Brazil's carbon emissions rose 8.9% in 2016, despite recession. https://www.climatechangenews.com/2017/10/27/brazils-carbon-emissions-rose-8-9-2016-despite-recession/. Accessed 5 February 2019

Ministry of Environment and Urbanization (2010) National climate change strategy 2010–2020. Republic of Turkey Ministry of Environment and Urbanization, Ankara

Ministry of Environment and Urbanization (2011) National climate change action plan 2011–23. Republic of Turkey Ministry of Environment and Urbanization, Ankara

Ministry of Environment and Urbanization (2014) Turkey HABITAT-III National Report, December 2014. Ankara: Republic of Turkey Ministry of Environment and Urbanization. http://unhabitat.org/wp-content/uploads/2014/07/Turkey-national-report.pdf. Accessed 18 January 2019

Ministry of Environment and Urbanization (2015) Strategic plan of ministry of environment and urbanization. Republic of Turkey Ministry of Environment and Urbanization, Ankara

Ministry of Public Works and Settlement (2010). KENTGES: integrated urban development strategy and action plan, 4 November 2010. Ankara: Republic of Turkey Ministry of Public Works and Settlement. www.kentges.gov.tr. Accessed 20 June 2016

Miraglia SGEK, Saldiva PHN, Böhm GM (2005) An evaluation of air pollution health impacts and costs in São Paulo, Brazil. Environ Manag 35(5):667–676. https://doi.org/10.1007/s00267-004-0042-9

MoEFCC (2018) India: Second Biennial Update Report to the United Nations framework convention on climate change. Ministry of Environment, Forest and Climate Change, Government of India

MSCI (2019) MSCI emerging markets index. Retrieved from https://www.msci.com/index-tools#EM

NAPCC (2008) National action plan on climate change. Govt. of India, Ministry of Environment & Forests, New Delhi

OECD (2017) The State of National Urban Policy in Turkey. https://www.oecd.org/regional/regional-policy/national-urban-policy-Turkey.pdf. Accessed 2 January 2019

PIB (2015) India's intended nationally determined contribution. Government of India Ministry of environment, forest and climate change. New Delhi: Press Information Bureau

Potsdam Institute for Climate Impact Research (2019) Climate & Weather Data. https://www.pik-potsdam.de/services/climate-weather-potsdam. Accessed 16 December 2018

PRC (2015) Enhanced actions on climate change: China's intended nationally determined contributions (June 2015). Beijing: People's Republic of China. http://www4.unfccc.int/ndcregistry/PublishedDocuments/China%20First/China%27s%20First%20NDC%20Submission.pdf. Accessed 25 March 2019

PRC (2016) 13th Five year plan for economic and social development of the People's Republic of China. Beijing: NDRC. http://en.ndrc.gov.cn/newsrelease/201612/P020161207645765233498.pdf. Accessed 28 February 2019

Puppim de Oliveira JA (2009) The implementation of climate change related policies at the subnational level: an analysis of three countries. Habitat Int 33(3):253–259

Puppim de Oliveira JA (2013) Learning how to align climate, environmental and development objectives: lessons from the implementation of climate cobenefits initiatives in urban Asia. J Clean Prod 58(1):7–14

Puppim de Oliveira JA, Doll CNH, Kurniawan TA, Geng Y, Kapshe M, Huisingh D (2013) Promoting win–win situations in climate change mitigation, local environmental quality and development in Asian cities through co-benefits. J Clean Prod 58(1):1–6

Ramaswami A, Tong K, Fang A, Lal RM, Nagpure AS, Li Y et al (2017) Urban cross-sector actions for carbon mitigation with local health co-benefits in China. Nat Clim Chang 7(10):736

Rashidia K, Stadelmannb M, Patta A (2017) Valuing co-benefits to make low-carbon investments in cities bankable: the case of waste and transportation projects. Sustain Cities Soc 34(2017):69–78

Rohde RA, Muller RA (2015) Air pollution in China: mapping of concentrations and sources. PLOS One. http://journals.plos.org/plosone/article?id=10.1371/journal.pone.0135749. Accessed 25 March 2019

Sahin U (2016) Warming a frozen policy: Challenges to Turkey's climate politics after Paris. Turkish Policy Quarterly (September 23). http://turkishpolicy.com/article/818/warming-a-frozen-policy-challenges-to-turkeys-climate-politics-after-paris. Accessed 2 February 2019

Satterthwaite D (2009) The implications of population growth and urbanization for climate change. Environ Urban 21(2):545–567

Sethi M (2017) Climate change and urban settlements – a spatial perspective of carbon footprint and beyond. Routledge, Taylor & Francis, London & NY

Sethi M (2018) Co-benefits assessment tools and research gaps. In: Sethi M, Puppim de Oliveira JA (eds) Mainstreaming climate co-benefits in Indian cities. Springer Nature, Singapore. https://doi.org/10.1007/978-981-10-5816-5_2

Sethi M, Mohapatra S (2013) Governance framework to mitigate climate change: challenges in urbanising India. In: Governance approaches to mitigation of and adaptation to climate change in Asia. Palgrave Macmillan, London, pp 200–230

Sethi M, Puppim de Oliveira J (2015) From global 'North–South' to local 'urban–rural': a shifting paradigm in climate governance? Urban Clim 14(4):529–543

Sethi M, Puppim de Oliveira JA (2018) Cities and climate co-benefits. In: Sethi M, Puppim de Oliveira JA (eds) Mainstreaming climate co-benefits in Indian cities. Springer Nature, Singapore. https://doi.org/10.1007/978-981-10-5816-5_1

SIPA (2018) Guide to Chinese climate policy: urban air pollution. New York: Sipa Center on Global Energy Policy, Columbia University in The City of New York. https://chineseclimatepolicy.energypolicy.columbia.edu/en/urban-air-pollution. Accessed 12 May 2019

Stepping KMK (2016) Urban sewage in Brazil: drivers of and obstacles to wastewater treatment and reuse: governing the water-energy-food nexus series. Deutsches Institut für Entwicklungspolitik gGmbH

UNFCCC (2015) Intended nationally determined contribution. https://www4.unfccc.int/sites/ndcstaging/PublishedDocuments/Brazil%20First/BRAZIL%20iNDC%20english%20FINAL.pdf. Accessed 21 January 2019

United States Environmental Protection Agency (2005) Integrated Environmental Strategies (IES) Program. Seoul, Korea: International Conference on Atmosphere Protection. CGE Training Workshop on Mitigation Assessments

UNU-IAS (2013) Urban development with climate co-benefits: aligning climate, environmental and other development goals in cities. Yokohama: United Nations University Institute of Advanced Studies

Wike R, Parker B (2015) Corruption, pollution, inequality are top concerns in China. Washington, DC: Pew Research Center. http://www.pewglobal.org/2015/09/24/corruption-pollution-inequality-are-top-concerns-in-china/. Accessed 22 March 2019

World Bank (2019) Carbon pricing dashboard. https://carbonpricingdashboard.worldbank.org/map_data. Accessed 25 March 2019

World Health Organization (2018) WHO global ambient air quality database (update 2018). https://www.who.int/airpollution/data/cities/en/. Accessed 5 February 2019

Zhang Q (2013) National action plan on prevention and control of air pollution. China's State Council. http://www.sustainabletransport.org/wp-content/uploads/2017/08/National-Action-Plan-of-Air-Pollution-Control.pdf. Accessed 5 March 2019

Zhao L, Liang HR, Chen FY, Chen Z, Guan WJ, Li JH (2017) Association between air pollution and cardiovascular mortality in China: a systematic review and meta-analysis. Oncotarget 8(39):66438–66448. https://doi.org/10.18632/oncotarget.20090

Protocol of an Interdisciplinary and Multidimensional Assessment of Pollution Reduction Measures in Urban Areas: MobilAir Project

Sandrine Mathy, Hélène Bouscasse, Sonia Chardonnel, Aïna Chalabaëv, Stephan Gabet, Carole Treibich, and Rémy Slama

1 Introduction

A large literature explores the impact, in the economic evaluation of climate policies, of integrating air pollution reduction as a co-benefit of these policies. This literature shows that considering pollution reductions induced by climate policies leads to an extremely significant reduction in the cost of reducing greenhouse gas emissions (Ekins 1996; Nemet et al. 2010; Vandyck et al. 2018). It also increases the willingness to pay on the part of the population to reduce GHG emissions (Longo et al. 2012). The co-benefits of climate policies encourage an increase in the ambition of national policies to reduce GHG emissions (Zenghelis 2017). This is why these pollution reduction policies are perceived as a political lever to obtain short-term benefits relevant for the implementation of climate policies whose specific benefits will only be felt in the long term (Aunan et al. 2003; Altemeyer-Bartscher et al.

S. Mathy (✉) · C. Treibich
Univ. Grenoble Alpes, CNRS, INRA, Grenoble INP, GAEL, Grenoble, France

Institute of Engineering, Univ. Grenoble Alpes, Grenoble, France
e-mail: sandrine.mathy@univ-grenoble-alpes.fr

H. Bouscasse
CESAER, Agrosup Dijon, INRA, Bourgogne Franche-Comté Univ., Dijon, France

S. Chardonnel
PACTE, Laboratoire de sciences sociales, Grenoble, France

A. Chalabaëv
SENS, Univ. Grenoble Alpes, Grenoble, France

S. Gabet · R. Slama
Team of Environmental Epidemiology Applied to Reproduction and Respiratory Health, Institute for Advanced Biosciences (IAB), Inserm, CNRS, and Grenoble-Alpes University, U1209
Grenoble, France

© Springer Nature Switzerland AG 2020
W. Buchholz et al. (eds.), *Ancillary Benefits of Climate Policy*, Springer Climate,
https://doi.org/10.1007/978-3-030-30978-7_18

2014). These pollution reduction policies are all the more relevant in the case of urban areas (Krupnick et al. 2000; Jack and Kinney 2010; Harlan and Ruddell 2011). Indeed, in many countries, they face an alarming health impact from air pollution, and they are central actors in implementing actions to reduce greenhouse gas emissions with decisive sectoral issues, whether for the energy consumption of residential and tertiary buildings or the organization of mobility in the catchment area between home, leisure and professional activities. This is why the MobilAir project, whose protocol we present here, focuses on policies to reduce pollution and greenhouse gas emissions in an urban area. Nevertheless, we consider here that the agglomeration is setting up a programme to reduce particulate air pollution in order to reduce its impacts in the short term. Air quality is thus the primary benefit of the MobilAir project while greenhouse gas emission reduction can be considered as secondary benefits.

Reducing air pollution and combating climate change are two inextricably linked issues. In many cities, the main sectors contributing to these two major environmental problems are transport and heating. Reducing fuel consumption in road transport and the share of individual cars in mobility are the main common challenge facing the transport sector. In the buildings and residential sector, the coherence of the actions to be taken to fight on both fronts is more complex. Indeed, in this sector, the major problem in combating global warming is to reduce energy consumption and promote the penetration of non-carbon energies, including wood heating that is considered as carbon neutral. On the other hand, wood heating, if it is not efficient, is a major source of air pollution.

Outdoor air pollution kills more than three million people across the world every year and causes health problems from asthma to heart disease for many more (OECD 2014; Lelieveld et al. 2015; WHO 2016). The cost of the health impact of air pollution in OECD countries (including deaths and illness) was about USD 1.7 trillion in 2010 (OECD 2014). Nevertheless, despite these alarming figures, there are few examples of the implementation of programmes at urban levels that have significantly reduced the impacts of air pollution.

A first explanation lies in the difficulty that the measures implemented have in structurally changing individual behaviour towards sustainable practices. It seems necessary to make progress in understanding the determinants of individual behaviour. It is clear that developing infrastructures for road alternative mobility, e.g., bike lane is not enough to strongly increase bike use. Mobility behaviours are complex and are not determined only by cost of transport, time spent in transportation, and transport offer. Individual drivers related to altruism, perceptions, and social norms, as well as habits, also play a role. It is thus important to improve the comprehension of mobility behaviours. Identifying if subjects being offered adapted alternative (cleaner) transportation modes can adopt these modes, and if not, understanding which obstacles exist, would be very innovative.

A second explanation is that policies to reduce pollution are often dimensioned without explicit consideration of a targeted health impact. At the best, they rely on an ex ante environmental evaluation ignoring any health consequence. Starting from a target health impact to appropriately dimension urban policy measures would be a logical and important change of approach. It raises scientific challenges, implying to develop a reverse approach consisting of starting from a target formulated in terms of improvement of health (e.g., a reduction by 20% in PM-related mortality) and subsequently identifying policy measures allowing to reach such a health target. Such an approach would be of great relevance to decision-making.

Finally, when these air pollution reduction programmes are implemented, they are not subject to a systematic multi-dimensional assessment approach. Cities in Europe and elsewhere have undertaken measures to limit air pollution emission, especially from transportation and heating sources. In France, in 2012, "ZAPA" (meaning "priority action area for air", an ancient acronym for low emission zones—LEZ) had been planned, but abandoned. A reason put forward was that such measures were deemed socially unequal (with the more deprived people being disproportionally touched by traffic restriction measures). However, very few, if any, rigorous evaluations of any social inequality in cost (and also benefit) of low emission zones have to our knowledge been conducted. In cities where ambitious programs were implemented, environmental evaluations have documented decreases in PM_{10} by as much as 50% in Tokyo between 2001 and 2010 (Hara et al. 2013), and between 5% and 13% in Germany (Cyrys et al. 2014; Fensterer et al. 2014). ADEME's review of LEZ (ADEME is the French Environment and Energy Management Agency; ADEME 2017) shows that only in rare cases was the environmental evaluation supplemented with a Health Impact Assessment, or HIA (Clancy et al. 2002; Cesaroni et al. 2012). Estimating the cost of the measures taken by the local authorities would also be very important, and would allow conducting cost–benefit analyses of such policies. In the USA, analyses of benefits and costs of the Clean Air Act law are planned by the law and indicate that the benefits exceed costs by a factor of 30–90. No such figure is available in France. Making such figures available in the French and European context, where atmospheric pollution standards are much higher than in the USA (regulatory limit of 25 $\mu g/m^3$ for $PM_{2.5}$ yearly concentration in Europe, compared to 10 in the USA), would be very relevant for citizens and decision-makers.

The MobilAir project, therefore, aims to contribute to the significant reduction of air pollution and greenhouse gas emissions in urban areas. To that end, it relies on an interdisciplinary methodology.

2 MobilAir Project

2.1 Objectives

MobilAir overarching aim is to contribute to air quality improvement in urban areas. To do so, two major scientific issues will be investigated: the characterization of the exposure level of the population to atmospheric pollution and of its health impact (WP1), and a better comprehension of levers and obstacles to air quality improvement, particularly concerning mobility from behaviours to urban planning

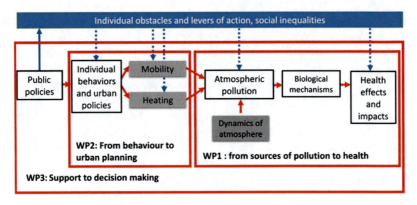

Fig. 1 Structure of the MobilAir project

(WP2). These two WPs will provide inputs to build an interdisciplinary modelling tool to provide support to decision-making towards healthier cities (WP3) (Fig. 1).

To meet these scientific challenges, the MobilAir project relies on a set of disciplinary skills existing on a specific aspect related to air pollution in the laboratories of the Grenoble university campus. Seven research laboratories working in seven different disciplines make up the MobilAir consortium:

- In sociology, geography, and planning, the PACTE laboratory is implied in research related to barriers to behavioural change, mobility, and urban planning.
- In behavioural psychology, the SENS laboratory studies the motivation for physical and sports activities, or active mobility.
- In environmental economics, the GAEL laboratory is involved in transport and health economics, consumption behaviour, and cost–benefit analyses of environmental policies.
- In biology and health, the IAB evaluates the health impact of early exposures and mechanisms for the action of pollutants.
- In air quality, the IGE analyses the particulate matter toxicity.
- The LEGI develops atmospheric dispersion models.
- The INRIA has built a land use and transport interaction model.
- The observatory of air quality in the Rhone Alps region ATMO-AURA develops and uses a numerical model to assess particulate matter concentrations based on emission scenarios.

2.2 Grenoble Urban Area as Study Field

Grenoble is the main city located inside the Alps. About 500,000 inhabitants live in the urban area called Grenoble Alpes Métropole, which is composed of 49 municipalities. Grenoble is known in France for its long-standing commitment to the fight against climate change and to programmes to reduce air pollution.

In July 2005, the Grenoble Alpes Métropole was the first urban area in France to sign a local climate plan. By 2014, it sets a "3 × 14" target, namely a reduction of at least 14% in greenhouse gas emissions compared to 2005, a 14% reduction in per capita energy consumption compared to 2005 and an increase in the share of renewable energy to 14% of the agglomeration's total energy consumption. The mitigation objective in 2020 is set to −35%, and in 2030 to −50% compared to 2005.

In 2015, emission reduction observed go beyond targets, with GHG emission level 23% lower than in 2005. This is the result of the reduction in the use of fossil fuels in favour of renewable energies and electricity, but also, to a large extent, of a significant reduction in energy consumption in the industry as a result of gains in energy efficiency but also and above all of a decrease in industrial activity. Even if all sectors are down, this decrease remains less marked in the residential, tertiary, and transportation sectors, which are major sources of energy consumption and also the main sources of atmospheric pollution.

Grenoble is one of the French cities with high air pollution exposure (mean yearly population $PM_{2.5}$ exposure, 18 µg/m^3 (Morelli et al. 2016; Fig. 2). The vast majority of residents (96%) are exposed to an average $PM_{2.5}$ level higher than the WHO guideline (10 µg/m^3). These high levels are partly explained by the basin configuration of Grenoble, the low winds, specific winter meteorological conditions (low mixing height) and frequent use of old (inefficient) wood heating

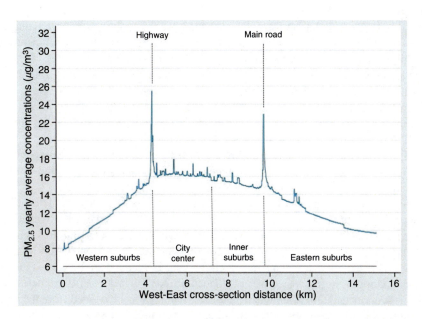

Fig. 2 PM2.5 levels in a West-East cross section passing through the city centre and going through the suburban cities of (i) Grenoble conurbation (yearly average concentrations during the period 2015–2017, in µg/m^3). From Morelli et al. (2019)

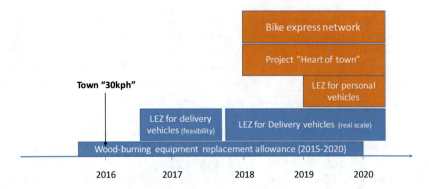

Fig. 3 Time-scale of the main policy measures aiming at reducing air pollution currently planned in Grenoble urban area. Town 30 kph: Restriction of maximal speed to 30 km/h for all vehicles

stoves. One partner of the MobilAir consortium evaluated the health impacts of air pollution exposure in the year 2012, concluding that at this time, $PM_{2.5}$ exposure was responsible of an estimated 3–10% of non-accidental mortality cases and lung cancers (Morelli et al. 2016).

Grenoble Metropole has undertaken several measures (Fig. 3) to improve air quality through the Ministry of the Environment's program "Breathable cities within 5 years". The plan targets the transportation of goods and mobility and non-efficient individual wood heating.

To reduce particulate emissions from wood heating, an air–wood premium has been in place since 2015. It aims to encourage the replacement of inefficient and highly polluting installations with the most efficient stoves on the market. The level of the premium was set at 800€, recently its level has been doubled to 1600€ and the most modest households benefit from a higher premium of 2000€.

In transport, in 2016, the speed limit was reduced to 30 km/h on most roads within the urban area except the main roads. To accelerate the shift in the current vehicle fleet composition towards a more sustainable and less polluting vehicle fleet, vehicles are classified into different categories (Crit'Air sticker) based on their particulate matters emissions. The higher level of sticker, the higher the level of pollution.

Since 2017, a "commercial LEZ" only for goods transport vehicles (light commercial vehicles and heavy goods vehicles), is implemented in the extended city centre of Grenoble. Only the unclassified (the less efficient) goods transport vehicles are banned from driving in the LEZ between 6 am and 7 pm Monday to Friday. In June 2019, the LEZ for good transports has been extended to 10 new volunteer municipalities. A progressive ban is planned, in 2019 diesel Euro III vehicles will be first concerned until 2025 when all diesel vehicles will be banned from the LEZ. Consideration is being given to extending the low-emission area to all vehicles to also cover passenger cars. Similarly, the "city centre" project aims to reduce the

Protocol of an Interdisciplinary and Multidimensional Assessment of Pollution... 331

amount of road space dedicated to individual motorized transport to give pedestrians more space, and to increase the network of cycle paths.

Thus, the territory of the Grenoble urban area appears to be particularly well adapted by its particularly high exposure to pollution problems, by its determination in the fight against climate change and the reduction of air pollution, by the richness and disciplinary diversity of research on issues related to pollution and global warming. For all these reasons, Grenoble can be a pilot city to test the implementation of the MobilAir protocol.

3 Better Understanding of Individual Drivers in Mobility Behaviour

Despite the many negative externalities of the private car (health and well-being impacts related to pollution, noise, increased sedentarization, contribution to climate change, accidents, etc.), public policies have most often failed to significantly change individual behaviour in the transport sector. The use of private cars remains the norm. Quantitative surveys show that a large number of people are aware of the need to reduce car use and agree with the idea of changing their mobility habits by considering the possibility of switching from car to another mode of transport (BCG and Ipsos 2017; Kaufmann et al. 2010). However, beyond this stated intention, these same people face multiple obstacles to effectively and sustainably change their travel practices.

Research distinguishes obstacles that act at different levels (Sallis et al. 2006): in terms of the urban environment, we find, for example, the low access to public transport network (in terms of distance from dwelling to transport network, frequency of urban public transport, difficult implementation of multimodality), the lack of secure cycle paths, or the lack of walkability (i.e. extent to which an area is walkable); economically, the cost of cycling equipment or the cost of public transport can be a barrier, especially for the most disadvantaged populations; at the personal level, the well-anchored habits, the lack of knowledge about pollution exposure for each mode of transport or the health benefits of active modes, or the lack of confidence in people's ability to change their own behaviour in a sustainable manner, can also explain the difficulty of reducing personal car use, as well as family constraints (e.g. accompanying children) or the sequence of activities (work, leisure, shopping).

Among the behavioural theories, the Theory of Planned behaviour (TPB; Ajzen 1991) has been successfully used to explain mode choice (see, e.g. Bamberg et al. 2003). This theory posits that behaviour, in order to be effective, must first be decided/planned and that it is necessary to act on three types of factors: judgments about the desirability of the behaviour and its consequences (attitudes); the influence and opinion of others on the behaviour (subjective norms) and beliefs about the subject's ability to succeed in the behaviour (self-efficacy). According to the TPB, these three factors influence intentions which, in turn, influence actual behaviour. To study situations characterized by high behavioural costs like a mode shift, the TPB is considered as more powerful than other theories (Gifford et al. 2011) like the norm-activation modal (Schwartz 1977) or the value-belief-norm theory (VBN theory; Stern et al. 1999) used in environmental psychology. Our research, therefore, draws on the TPB.

However, the literature has shown that, even though intentions are the main predictor of behaviour—theoretically but also empirically, see for instance Lanzini and Khan 2017—there is still a gap between intentions and behaviours that needs to be understood (Sheeran 2002). Several avenues of research have been explored by psychologists to bridge this gap. For instance, Verplanken et al. (2008), Chen and Chao (2011) or Bamberg and Schmidt (2003) studied the role of habits, a key variable of the Triandis model (1977). One key area of research is to study how habit strength moderate the intention–behaviour relationship in the TPB (see for instance De Bruijn 2010, for an application on fruit consumption). To this end, we do not only consider habits as the intensity of use of a transport mode but we also measure the automatic nature of modal choice habits as well as test the impact of disrupting habits.

Other variables may be used to reinforce the predictive power of the TPB. One originality of our research is to consider the use of alternative transport modes, and in particular actives modes, as a pro-environmental behaviour but also as a physical activity. We therefore draw on this strand of literature, in particular by using intrinsic and extrinsic self-determined motivations[1] (Deci and Ryan 2002). Both are supposed to influence the strength of the relationship between intention and behaviour.

Therefore, MobilAir research questions on mobility behaviours deal with:

- The formation of intentions for transport mode choice: We will assess the extent to which the impact of transport modes on individual and collective health can be a lever for changing transport mode choices, as well as the role of various psychological variable.
- Understanding the determinants of the transition from intentions to effective and sustainable behavioural change: We will deploy a population-based intervention on the territory of the Grenoble urban area aimed at changing the modes of transport of the participants in the intervention.

---

[1]The self-determination theory posits that there are two main types of motivation—intrinsic and extrinsic—and that they are both powerful forces in shaping who individuals are and how they act. Intrinsic motivation is endogenous to the individual and depends on the personal values, interest or sense of morality. Extrinsic motivation is a drive to behave in ways that comes from external incitation and results in external rewards (e.g. evaluations and admiration of others).

3.1 What Role Does the Impact of Transport Modes on Public and Individual Health Play in Our Mobility Choices?

Walking and cycling for transportation provide substantial health benefits from increased physical activity. Globally, more than 30% of all adults are estimated to perform insufficient physical activity (Hallal et al. 2012). A lack of physical activity is associated with all-cause mortality, cardiovascular diseases, type 2 diabetes, cancer, and impaired mental health (Physical Activity Guidelines Advisory Committee 2008). The promotion of walking and cycling for transportation complemented by public transportation, presents a promising strategy to not only address problems of urban traffic strain, environmental pollution and climate change, but also to provide substantial health co-benefits (De Hartog et al. 2010).

In MobilAir, we will implement a protocol to evaluate to what extent the impact of modal choice on public or individual health has a leverage effect on intentions towards active modes of transport or towards public transports.

Modal choice is multifactorial and depends on economic constraints (travel cost), time constraints (travel time), quality of service (frequency of public transport use, comfort), spatial characteristics (density, topography) and individual characteristics (age, gender, attitudes, perceptions).

Until recently, the analysis of mode choice and its behavioural determinants have been partitioned between disciplines. On the one hand, economists explain modal choice with observable explanatory variables (cost, time, service level, age, gender) using discrete choice models. On the other hand, psychologists rely on theories that chart the path between different internal mental states that lead to a decision, as in the TPB previously described. The combination of these two approaches, economic and psychological, has only been possible in recent years thanks to the development of new statistical tools, hybrid choice models (Walker 2001). Hybrid choice models make it possible to model choices and calculate economic quantities based on observable quantities (time, cost, etc.) but also unobservable variables such as attitudes or perceptions. Introducing this type of variable into economic models makes it possible to better understand the process of choice formation and thus to identify public policies that can influence choices towards mobility practices that favour "soft" modes (public transport and active modes) (Bouscasse 2017, p. 108). While these models are beginning to be applied fairly widely worldwide in transport, they still struggle to fully integrate the theories of environmental psychology (Bouscasse 2018).

Discrete choice models explaining mode choice have rarely incorporated observable health variables (see Sottile et al. 2015a, b for exceptions). However, the modal choice may be influenced by public health considerations (atmospheric pollution level) and individual health considerations (cardiovascular diseases related to inadequate physical activity). For instance, are people conscious of the consequences of car pollution or aware that walking or biking can improve their own health more likely to adopt an active mode of transport?

One reason explaining the low integration of environmental psychology theories in discrete choice models is the lack of data encompassing the psychological, sociological and economic dimensions.

Therefore, we propose to implement a phone survey (1300 individuals representative of the Grenoble urban area) collecting data on knowledge and perceptions of pollution combined with a web stated preferences survey (1000 individuals) also including measures of the components of the TPB, as well as habits, altruism and satisfaction when using different transport modes (Ettema et al. 2011). The stated preferences survey (choice experiment) consists in asking the respondent to choose between options (here transport mode) described by objective attributes such as travel mode, travel cost and sanitary risks (see Fig. 4).

The objective of the choice experiment is to simulate situations of modal choice and to build an economic model based on the choices made.

Each interviewee will have to make eight series of choices between different modes of transport (car, public transport, cycling, walking) that exist in real life in the Grenoble metropolitan area. These choices will be personalized (origin, destination, time, cost) for each individual through a trip he or she will have described beforehand. If the distance of the reference trip is more than three kilometers, the walking alternative will not be presented to the respondent. With regard to the bicycle alternative, either the conventional bicycle or the electric bicycle will be presented to the respondent according to the latter's stated preferences.

First, a route calculation tool will be used to describe all the trips that can be made within the Grenoble metropolitan area for the various modes of transportation considered.

Mode of transport	🚌	🚲	🚶	🚗
Travel time	30 min	20 min	35 min	25 min
Cost per trip	1,50 €	-	-	0,50 €
Physical activity By using this mode of transport every day, your risk of developing cardiovascular disease...	is equal to 28%	is equal to 24%	is equal to 20%	is equal to 30%
Air Pollution If 75% of the population adopts this mode of transport, the average risk of developing cardiovascular disease for a person in the agglomeration...	is equal to 29%	is equal to 26%	is equal to 25%	is equal to 30%
What is your choice?	☐	☐	☐	☐

Fig. 4 Example of a choice sheet

Each of the possible modes of travel will be described with a number of attributes, including time, cost and frequency of public transport use and impact on individual and public health of transport mode choices. More precisely, scenarios will include various reductions in relative risk of developing cardiovascular and respiratory diseases.

Since the level of these attributes will vary from one question to another, we will be able, through the discrete choice model, to estimate the weight of each attribute in the modal choice. These models will allow us to evaluate demand elasticities, willingness to pay, time equivalents that can feed into operational evaluation and decision support tools.

The originality of the proposed design lies in two specificities. First, it distinguishes altruistic motivations related to public health (impact of the mode of transport on pollution) from selfish motivations related to individual health (impact of walking or cycling instead of driving on its own physical activity and its own health). Second, the whole design relies on a complete model of environmental psychology to explain individuals' intentions: the TPB.

Our work will also allow us to analyze how norms and altruism may influence intentions to use alternative transport modes and, more specifically, how they may explain the heterogeneity in the weight given to the pollution-related sanitary attribute. The definition of social norms, moral norms and altruism is sometimes floating (Nyborg 2018). Following Nyborg et al. (2016), we retain the definition of a social norm as "a predominant behavioral pattern within a group, supported by a shared understanding of acceptable actions and sustained through social interactions within that group". In the questionnaire, social norms are measured by asking the proportion of persons in their entourage using alternative transport modes, as well as to which extent they are supported and encouraged in their behaviour by their entourage (subjective norm, which is a subset of social norms). Moral norms, "a rule of ethically appropriate behavior, enforced by the individual herself trough inner feelings" (Nyborg 2018), are measured by asking respondents how important it is for them to use alternative transport mode, and how guilty or ashamed they would feel if they did not. Altruism, the fact of "at least partly internalizing the utility of someone else in the individual's own utility function" (Nyborg 2018), is measured by using items of the Big 5 scale (Goldberg 1990).

Overall, this approach will help us to study how intentions, the first predictor of modal choice (Lanzini and Khan 2017), can evolve by highlighting the impact of our modes of transport on individual and collective health, as well as studying precisely the role of psychological variables. However, as mentioned above, there is a gap between intention and actual behaviour. This is why we are also implementing a population-based intervention in MobilAir to identify the levers for active mobility.

3.2 Implementing Lasting Shifts in Transportation Mode Toward Active Modes

Interventions[2] to reduce car use have targeted different levels of barriers identified in research related to determinants of modal choice. More specifically, structural and psychological interventions are distinguished (e.g. Fujii and Kitamura 2003). The first includes so-called "hard" levers for change, by modifying the physical environment (improving the accessibility of public transport, building secure cycle paths) or the economic cost of transport modes (reducing the cost of public transport, tolls at the entrance to cities). The latter include so-called "soft" change levers, which aim to modify beliefs, attitudes or perceptions in order to motivate a voluntary change in mode of transport (Graham-Rowe et al. 2011). However, while many interventions for modal change already exist (for literature reviews, see Chillon et al. 2011; Ogilvie et al. 2007; Scheepers et al. 2008; Yang et al. 2010), several obstacles remain:

1. Most existing interventions target a single level of behaviour change lever (hard or soft); however, interventions that target several levels of factors are needed to promote sustainable behaviour change (Sallis et al. 2006). For example, interventions that include economic incentives (e.g. subsidies for public transit passes) generally only have an effect for the period during which individuals benefit from them, without generating sustainable behavioural change (e.g. Brög et al. 2009). It seems necessary to link these incentives with behaviour change techniques targeting psychological processes, such as habits, attitudes or self-efficacy, in order to promote a change in sustainable transport mode.
2. "Soft" interventions most often seek to promote positive attitudes towards active modes of transport, or to increase the individual's perceived ability to change behaviour, by using informational techniques (information on the availability of alternative transport to the car at the participants' place of residence; information on the health benefits of active travel). However, while attitudes and beliefs significantly predict the intention to change behaviour, intention is not sufficient to implement action, a phenomenon known as the intention–behaviour gap, particularly because past habits and behaviours strongly predict behaviour (Lanzini and Khan 2017). Behaviour change models in health psychology (e.g. health action process approach, Schwarzer 2008) consider that once the individual has decided to modify his/her behaviour, he or she must be able to plan it ("what-where-when-how"?) in order to promote habit change. While some interventions have focused more specifically on the transition from intention to behaviour, these remain rare.
3. Existing interventions have many methodological biases: few randomized controlled trials are conducted, with studies most often conducting pre/post-intervention comparisons without control groups; no systematic evaluation of

---

[2]Behavioural intervention is an action aimed at changing behaviour.

the statistical significance of the results; lack of objective measures, particularly to measure mobility behaviour itself, with studies most often using self-reported behavioural measures; relatively short duration of intervention and monitoring, which do not allow for a possible assessment of sustainable changes.

The MobilAir project aims to address these obstacles by pursuing the following objectives:

1. Quantify to what extent an intervention comparing hard (economic incentives) and soft (psychological incentives) levers can permanently change behaviour towards the adoption of sustainable mobility (walking, cycling, urban public transport). In order to study this objective, the project is based on an innovative interdisciplinary collaboration between economists, geographers, psychologists and epidemiologists. This collaboration will make it possible to: (1) compare the weight of a hard intervention with that of a soft intervention, but also to (2) understand the socio-spatial (e.g. location of activities, urban planning) and socio-demographic (e.g. age, income) factors that promote or hinder the effectiveness of the intervention and (3) identify the psychological mechanisms (attitudes, intention, perceived behavioural control, social norms, etc.) that can explain why the intervention is effective.

2. Test soft levers that have so far been little studied in the literature, based on the most recent knowledge in health and physical activity psychology, developed as part of the health action process approach (Schwarzer 2008), studies on habits (e.g. Gardner 2009), or studies on self-control (Kotabe and Hofmann 2015). This will involve a combination of pre-intentional techniques (e.g. information on the risks of the car and the benefits of alternative transportation modes) to promote the development of the intention to change behaviour, with post-intentional techniques to promote the transition from intention to actual behaviour change (e.g. setting individualized objectives, daily monitoring logbook, identifying barriers and resolution strategies, social support).

3. Test the effectiveness of this intervention using a rigorous methodology, involving the implementation of a randomized controlled trial in which participants will be randomly assigned to one of the intervention groups, or to the control group that will not benefit from the intervention. The effect of the intervention will be analyzed using appropriate inference statistical tests, based on an a priori power analysis that will allow us to limit the type II error (i.e. not detecting an effect due to lack of statistical power). In addition, the project will include subjective measures of psychological constructs coupled with objective measures of behaviour (i.e. GPS to measure daily mobility, accelerometers to measure physical activity). Finally, the measurements will be carried out over a period of 2 years to assess whether the behavioural changes that are taking place are sustainable.

4 Support to Decision-Making Towards Healthier Cities

4.1 Interdisciplinary and Multidimensional Assessment of Pollution Mitigation Measures

The aim of this task is to build an interdisciplinary modelling chain at the urban area scale, to assess actions leading to lower greenhouse gas emissions and lower pollutants emissions.

First, measures actually implemented by the Grenoble urban area will be evaluated based on the combination of the modelling chain, field measurement, and population surveys related to population perception, mobility, wood heating and quality of life. Such measurements at different years can be confounded by differences in meteorological conditions between the compared years, an issue that can be limited by first modelling the influence of meteorology, season and year on PM levels over the whole period through a regression model, and adjusting for this influence.

Then, more theoretical and contrasted scenarios of measures will be assessed. They will include already planned measures (wood heater replacement incentive), hypothesized (e.g. ban of Euro 1-2-3 vehicles in the city centre) or possible measures (taken, e.g. from examples implemented in other cities).

The approach is described in Fig. 5. The left side describes the methodology that will be developed for the assessment of measures already implemented and the right side the modelling approach that will be implemented for the assessment of scenarios.

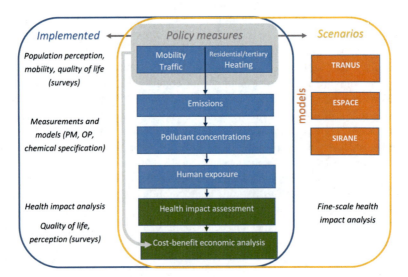

Fig. 5 Overview of the approach for the interdisciplinary and multidimensional assessment of pollution mitigation measures

The dimensions evaluated are the following:

1. Traffic and mobility
2. Wood heating: uses and old stove renewal
3. Greenhouse gas emissions
4. Air quality assessment
5. Health impacts (non-accidental mortality, lung cancer incidence and low birth-weight incidence). As done in the previous study from Morelli et al. (2016), results will be stratified on the European Deprivation Index, a measure of socioeconomic deprivation, available at the IRIS scale, the finest spatial scale for France.
6. Cost–benefit analyses will include (1) *the direct costs of implementation* of measures, (2) *the indirect economic costs/gains* induced by the measures on each category of agent: local authorities, households, economic activities, (3) *external costs*, corresponding to the economic assessment of the decrease of health impacts of pollution and (4) *other external costs* (time spent for mobility, road safety, noise, greenhouse gases emissions, health benefits of the development of active transportation modes).
7. Inequalities related to pollution exposure and to the impact of policies on social inequalities.

4.2 Evaluation of Measures Already Implemented or Planned

The assessment of measures actually implemented during the study period will rely on measurements done by the regional air quality-monitoring network, and dispersion modelling based on actual emission data. The parameters assessed include $PM_{2.5}$ and PM_{10} mass concentration, NO_2, ozone, as well as, for the first and last years of the study period, PM_{10} chemical speciation, including the levoglucosan (a marker of biomass burning) content. The latter parameter, which will only be assessed during the first and last years of the study period (and not continuously throughout the study period), will allow providing an evaluation of changes in the composition of particulate matter, reflecting possible changes in the nature of the main pollution sources, such as a decrease in the role of wood burning facilities.

Then, more theoretical and contrasted scenarios of measures will be assessed. They will include already planned measures (*wood heater replacement* incentive), hypothesized (e.g. ban of Euro 1-2-3 vehicles in the city centre) or possible measures (taken, e.g. from examples implemented in other cities). This approach will rely on the modelling chain described in Fig. 5. TRANUS is an integrated land use and transport model. In addition to allow short-term traffic simulation, TRANUS can simulate the long-term mutual interactions between the location of firms and households and transport offering.

Pollutant emissions are estimated by ESPACE developed by ATMO AuRA, which relies on comprehensive database of activities (vehicles number, ratio of heavy duty vehicles, speed ...). Emission data generated are then used as inputs in the SIRANE model (Soulhac et al. 2011, 2012) developed by ATMO AuRA to estimate the concentration of atmospheric pollutants (PM concentration, nitrogen oxides). This will allow assessing specific prospective evolution scenarios of the configuration of Grenoble area in the 2050 horizon (Fiore et al. 2015; Morton et al. 2008). Other emission drivers (technological evolutions of a fleet of vehicles, energy efficiency of buildings, energy mix of the residential and service sectors, decrease of old diesel vehicles and low-quality wood heating stoves) will also be investigated.

4.3 Identification of Measures Aiming to Reach Given Air Quality and Public Health Targets

Here, we will first investigate the impact of given reductions in $PM_{2.5}$ mass concentration of various amplitudes on health. The considered scenarios are described in Table 1. This approach will be implemented in parallel for Grenoble and Lyon urban areas. For each scenario, the expected change in life expectancy, all-cause mortality, lung cancer incidence as well as the associated economic costs, will be assessed.

Scenarios S1 to S5 are also graphically summarized in Fig. 6.

The estimates of health impact will rely on a health impact assessment approach described by Morelli et al. (2016). They imply in particular to consider the dose-response function between $PM_{2.5}$ concentration and each of the health outcomes considered, which are given in Table 2.

In a further step, we will take a reverse approach starting from the formulation of targeted benefits relevant to public health (a 20% decrease in PM-related mortality or a 2-month improvement in life expectancy for example). These targets will be formulated by the local authorities. These objectives will be then translated into an average change in atmospheric pollution (PM) concentration, using a reverse health impact assessment approach (implemented through iterative forward health impact assessment studies assuming various decreases in PM levels). We will then identify urban measures allowing to reach such a target PM concentration distribution in the urban area using inverse modelling techniques; these measures will be chosen among options concerning both traffic and wood heating based on an available inventory of highly polluting wood burning heaters in the area.

Table 1 Description of the ten hypothetical scenarios of fine particulate matter ($PM_{2.5}$) exposure reduction considered (from Morelli et al. 2019)

Scenario number	Scenario description	Scenario name	$PM_{2.5}$ yearly level reduction
S1	Spatially homogeneous target value in the whole area	"WHO guideline"	Down to WHO yearly guideline (10 µg/m³)
S2		"No anthropogenic $PM_{2.5}$ emissions"	Down to lowest nationwide levels (4.9 µg/m³)[a]
S3		"Quiet neighbourhood"	Down to lowest study area district levels (tenth percentile of exposure)[b]
S4	Homogeneous $PM_{2.5}$ decreases in the whole area	"−1 µg/m³"	Baseline[c] −1 µg/m³
S5		"−2 µg/m³"	Baseline[c] −2 µg/m³
S6	Targeted reduction in $PM_{2.5}$-related mortality in the whole area[d]	"−1/3 of mortality"	Equivalent to decreasing homogeneously and sufficiently the baseline[c] exposure to achieve the indicated health objective[e]
S7		"−1/2 of mortality"	
S8		"−2/3 of mortality"	
S9	2008/50/EU Directive[f] "2020 target"	In the whole study area"	Baseline[c] −15%
S10		Restricted to $PM_{2.5}$ exposure hotspots"	Baseline[c] −15%, only if baseline ≥90th centile of $PM_{2.5}$ levels[g]

[a] Corresponding to the fifth percentile of $PM_{2.5}$ concentration distribution among French rural towns (Pascal et al. 2016)

[b] The tenth percentile of $PM_{2.5}$ exposure by Housing Block Regrouped for Statistical Information (IRIS) in the study area (corresponding to 10.3 and 12.4 µg/m³ in Grenoble and Lyon conurbations, respectively)

[c] Baseline corresponds to the $PM_{2.5}$ exposure average for the 2015–2017 period, taken as a reference in the present study

[d] Mortality reduction targets expressed as a proportion of the non-accidental death cases attributable to $PM_{2.5}$ exposure that can be prevented under the scenario S2: "No anthropogenic $PM_{2.5}$ emissions"

[e] S6: −2.9 and −3.3 µg/m³ in Grenoble and Lyon conurbations, respectively; S7: −4.4 and −5.1 µg/m³; S8: −6.0 and −6.9 µg/m³

[f] Inspired by the 2008/50/EU Directive, which targets relative $PM_{2.5}$ yearly average decreases to obtain by 2020. The decrease value depends on the exposure average for the last three years (2015–2017): −15% in the case of Grenoble and Lyon conurbations

[g] The 90th percentile corresponded to 16.0 and 17.4 µg/m³ in Grenoble and Lyon conurbations, respectively

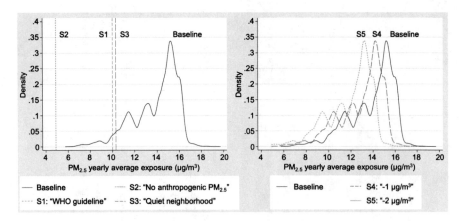

Fig. 6 Expected fine particulate matter (PM$_{2.5}$) exposure levels for the Grenoble conurbation population (yearly average exposure, in μg/m^3) under each PM$_{2.5}$ level reduction scenario: (1) scenarios targeting a spatially homogeneous value in the whole area (S1 to S3) and (2) scenarios decreasing homogeneously PM$_{2.5}$ in the whole study area (S4 and S5)

Table 2 Dose-response functions used to estimate the long-term effects of air pollution exposure to fine particulate matter (PM$_{2.5}$) on health

Health event	Study	Relative risk (95% CI) for a 10 μg/m^3 increase in PM$_{2.5}$ exposure
Non-accidental mortality	World Health Organization (2014)[a]	1.066 (1.040–1.093)
Lung cancer incidence	Hamra et al. (2014)[a]	1.09 (1.04–1.14)
Term low birth weight[b]	Pedersen et al. (2013)	1.392 (1.124–1.769)[c]

[a]Meta-analysis-based relative risk
[b]Occurrence of low birth weight birth cases (<2500 g) among term births (those occurring after the end of the 37th gestational week)
[c]The original odds ratio was reported for a 5 μg/m^3 increase in exposure and was 1.18 (1.06–1.33)

5 Conclusion

Atmospheric pollution in cities is a major challenge for public health in both developed and developing countries. This project aims to show the synergies between short-term public health issues related to pollution and the reduction of greenhouse gas emissions. Adopting a thoroughly interdisciplinary approach, the MobilAir project aims to identify precise measures to reduce significantly atmospheric pollution in cities and their impacts. Drawing on the considerable pluridisciplinary diversity of the Grenoble campus, MobilAir will develop an integrated approach in the urban area of Grenoble, which is a relevant pilot area. MobilAir will seek to develop methods and instruments, which can be copied in other cities in France and in other countries.

References

ADEME (2017) Zones à faibles émissions (Low Emission Zones – LEZ) à travers l'Europe (Les). ADEME. [En ligne]. Disponible sur: http://www.ademe.fr/zones-a-faibles-emissions-low-emission-zones-lez-a-travers-leurope. Consulté le: 09-Sept-2017

Ajzen I (1991) The theory of planned behavior. Org Behav Hum Decis Process 50(2):179–211

Altemeyer-Bartscher M, Markandya A, Rübbelke D (2014) International side-payments to improve global public good provision when transfers are refinanced through a tax on local and global externalities. Int Econ J 28(1):71–93

Aunan K, Mestl HE, Seip HM, Fang J, DO'Connor DC, Vennemo H, Zhai F (2003) Co-benefits of CO2-reducing policies in China-a matter of scale? Int J Global Environ Issues 3(3):287–304

Bamberg S, Schmidt P (2003) Incentives, morality, or habit? predicting students' car use for university routes with the models of ajzen, schwartz, and triandis. Environ Behav 35(2):264–285

Bamberg S, Ajzen I, Schmidt P (2003) Choice of travel mode in the theory of planned behavior: the roles of past behavior, habit, and reasoned action. Basic Appl Soc Psychol 25(3):175–187

BCG & Ipsos (2017) Du lundi au vendredi, les Français passent en moyenne 7h12 à se déplacer selon une étude Ipsos/BCG. Observatoire Européen des Mobilités, 1. Repéré à http://www.unionroutiere.fr/wp-content/uploads/2017/04/2017-04-26-BCG-IPSOS-Observatoire_europeen_des_mobilites_France.pdf

Bouscasse H (2017) Essays on travel mode choice modeling: a discrete choice approach to the interactions between economic and behavioral theories. Doctoral dissertation, Université Louis Lumière-Lyon II

Bouscasse H (2018) Integrated choice and latent variable models: a literature review on mode choice. Working paper

Brög W, Erl E, Ker I, Ryle J, Wall R (2009) Evaluation of voluntary travel behaviour change: experiences from three continents. Transp Policy 16:281–292

Cesaroni G, Boogaard H, Jonkers S, Porta D, Badaloni C, Cattani G, Forastiere F, Hoek G (2012) Health benefits of traffic-related air pollution reduction in different socioeconomic groups: the effect of low-emission zoning in Rome. Occup Environ Med 69(2):133–139

Chen CF, Chao WH (2011) Habitual or reasoned? Using the theory of planned behavior, technology acceptance model, and habit to examine switching intentions toward public transit. Transport Res F: Traffic Psychol Behav 14(2):128–137

Chillon P, Evenson KR, Vaughn A, Ward DS (2011) A systematic review of interventions for promoting active transportation to school. Int J Behav Nutr Phys Act 8(1):10

Clancy L, Goodman P, Sinclair H, Dockery DW (2002) Effect of air-pollution control on death rates in Dublin, Ireland: an intervention study. Lancet 360(9341):1210–1214

Cyrys J, Peters A, Soentgen J, Wichmann H-E (2014) Low emission zones reduce PM10 mass concentrations and diesel soot in German cities. J Air Waste Manag Assoc 64(4):481–487

De Bruijn GJ (2010) Understanding college students' fruit consumption. Integrating habit strength in the theory of planned behaviour. Appetite 54(1):16–22

De Hartog J, Boogaard H, Nijland H, Hoek G (2010) Do the health benefits of cycling outweigh the risks? Environ Health Perspect 118:1109–1116

Deci EL, Ryan RM (2002) Handbook of self-determination research. University Rochester Press, Rochester, NY

Ekins P (1996) The secondary benefits of CO_2 abatement: how much emission reduction do they justify? Ecol Econ 16(1):13–24

Ettema D, Gärling T, Eriksson L, Friman M, Olsson LE, Fujii S (2011) Satisfaction with travel and subjective well-being: development and test of a measurement tool. Transport Res F: Traffic Psychol Behav 14(3):167–175

Fensterer V et al (2014) Evaluation of the impact of low emission zone and heavy traffic ban in Munich (Germany) on the reduction of PM10 in ambient air. Int J Environ Res Public Health 11(5):5094–5112

Fiore AM, Naik V, Leibensperger EM (2015) Air quality and climate connections. J Air Waste Manag Assoc 65(6):645–685

Fujii S, Kitamura R (2003) What does a one-month free bus ticket do to habitual drivers? An experimental analysis of habit and attitude change. Transportation 30(1):81–95

Gardner B (2009) Modelling motivation and habit in stable travel mode contexts. Transport Res F: Traffic Psychol Behav 12(1):68–76

Gifford R, Steg L, Reser JP (2011) Environmental psychology, pages 440–470. Wiley Online Library

Goldberg LR (1990) An alternative "description of personality": the big-five factor structure. J Pers Soc Psychol 59(6):1216

Graham-Rowe E, Skippon S, Gardner B, Abraham C (2011) Can we reduce car use and, if so, how? A review of available evidence. Transp Res A Policy Pract 45(5):401–418

Hallal PC, Andersen LB, Bull FC, Guthold R, Haskell W, Ekelund U (2012) Global physical activity levels: surveillance progress, pitfalls, and prospects. Lancet 380:247–257

Hamra GB, Guha N, Cohen A, Laden F, Raaschou-Nielsen O, Samet JM et al (2014) Outdoor particulate matter exposure and lung cancer: a systematic review and meta-analysis. Environ Health Perspect 122(9):906–911

Hara K, Homma J, Tamura K, Inoue M, Karita K, Yano E (2013) Decreasing trends of suspended particulate matter and PM2. 5 concentrations in Tokyo, 1990–2010. J Air Waste Manag Assoc 63(6):737–748

Harlan SL, Ruddell DM (2011) Climate change and health in cities: impacts of heat and air pollution and potential co-benefits from mitigation and adaptation. Curr Opin Environ Sustain 3(3):126–134

Jack DW, Kinney PL (2010) Health co-benefits of climate mitigation in urban areas. Curr Opin Environ Sustain 2(3):172–177

Kaufmann V, Tabaka K, Louvet N, Guidez J (2010) Et si les Français n'avaient plus seulement une voiture dans la tête ? Évolution de l'image des modes de transport (à partir de l'analyse de 19 Enquêtes Ménages Déplacements), CERTU, Coll. Dossiers

Kotabe HP, Hofmann W (2015) On integrating the components of self-control. Perspect Psychol Sci 10(5):618–638

Krupnick A, Burtraw D, Markandya A (2000) The ancillary benefits and costs of climate change mitigation: a conceptual framework. In: Ancillary benefits and costs of greenhouse gas mitigation, pp 53–93

Lanzini P, Khan SA (2017) Shedding light on the psychological and behavioral determinants of travel mode choice: a meta-analysis. Transport Res F: Traffic Psychol Behav 48:13–27

Lelieveld J, Evans JS, Fnais M, Giannadaki D, Pozzer A (2015) The contribution of outdoor air pollution sources to premature mortality on a global scale. Nature 525(7569):367–371

Longo A, Hoyos D, Markandya A (2012) Willingness to pay for ancillary benefits of climate change mitigation. Environ Resour Econ 51(1):119–140

Morelli X, Rieux C, Cyrys J, Forsberg B, Slama R (2016) Air pollution, health and social deprivation: a fine-scale risk assessment. Environ Res 147(p):59–70

Morelli X, Gabet S, Rieux C, Bouscasse H, Mathy S, Slama R (2019) Which decreases in air pollution should be targeted to bring health and economic benefits and improve environmental justice? Environ Int 129(August):538–550

Morton BJ, Rodríguez DA, Song Y, Cho EJ (2008) Using TRANUS to construct a land use-transportation-emissions model of Charlotte, North Carolina. In: Transportation land use, planning, and air quality, pp 206–218

Nemet GF, Holloway T, Meier P (2010) Implications of incorporating air-quality co-benefits into climate change policymaking. Environ Res Lett 5(1):014007

Nyborg K (2018) Social norms and the environment. Ann Rev Resour Econ 10:405–423

Nyborg K, Anderies JM, Dannenberg A, Lindahl T, Schill C, Schlüter M et al (2016) Social norms as solutions. Science 354(6308):42–43

OECD (2014) The cost of air pollution: health impacts of road transport, Éditions OECD, Paris. https://doi.org/10.1787/9789264210448-en

Ogilvie D, Foster CE, Rothnie H, Cavill N, Hamilton V, Fitzsimons CF et al (2007) Interventions to promote walking: systematic review. Br Med J 334:1204

Pascal M, de Crouy Chanel P, Wagner V, Corso M, Tillier C, Bentayeb M et al (2016) The mortality impacts of fine particles in France. Sci Total Environ 571:416–425

Pedersen M, Giorgis-Allemand L, Bernard C, Aguilera I, Andersen AMN, Ballester F, et al (2013) Ambient air pollution and low birthweight: a European cohort study (ESCAPE). Lancet Respir Med 1(9):695–704

Sallis JF, Cervero RB, Ascher W, Henderson KA, Kraft MK, Kerr J (2006) An ecological approach to creating active living communities. Annu Rev Public Health 27:297–322

Scheepers CE, Wendel-Vos GCW, Den Broeder JM, Van Kempen EEMM, Van Wesemael PJV, Schwarzer R, Lippke S, Ziegelmann JP (2008) Health action process approach: a research agenda at the Freie Universität Berlin to examine and promote health behavior change. Zeitschrift für Gesundheitspsychologie 16(3):157–160

Schwartz SH (1977) Normative influences on altruism. Adv Exp Soc Psychol 10:221–279

Schwarzer R (2008) Modeling health behavior change: how to predict and modify the adoption and maintenance of health behaviors. Appl Psychol 57(1):1–29

Sheeran P (2002) Intention—behavior relations: a conceptual and empirical review. Eur Rev Soc Psychol 12(1):1–36

Sottile E, Cherchi E, Meloni I (2015a) Measuring soft measures within a stated preference survey: the effect of pollution and traffic stress on mode choice. Transp Res Procedia 11:434–451

Sottile E, Meloni I, Cherchi E (2015b) A hybrid discrete choice model to assess the effect of awareness and attitude towards environmentally friendly travel modes. Transp Res Procedia 5:44–55

Soulhac L, Salizzoni P, Cierco F-X, Perkins R (2011) The model SIRANE for atmospheric urban pollutant dispersion; part I, presentation of the model. Atmos Environ 45(39):7379–7395

Soulhac L, Salizzoni P, Mejean P, Didier D, Rios I (2012) The model SIRANE for atmospheric urban pollutant dispersion; PART II, validation of the model on a real case study. Atmos Environ 49:320–337

Stern PC, Dietz T, Abel TD, Guagnano GA, Kalof L (1999) A value-belief-norm theory of support for social movements: the case of environmentalism. Hum Ecol Rev 6(2):81–97

Triandis HC (1977) Interpersonal behavior. Brooks, Cole, Monterey

Vandyck T, Keramidas K, Kitous A, Spadaro JV, Van Dingenen R, Holland M, Saveyn B (2018) Air quality co-benefits for human health and agriculture counterbalance costs to meet Paris Agreement pledges. Nat Commun 9(1):4939

Verplanken B, Walker I, Davis A, Jurasek M (2008) Context change and travel mode choice: combining the habit discontinuity and self-activation hypotheses. J Environ Psychol 28(2):121–127

Walker JL (2001) Extended discrete choice models: integrated framework, flexible error structures, and latent variables. Doctoral dissertation, Massachusetts Institute of Technology

World Health Organization (2016) Ambient air pollution: a global assessment of exposure and burden of disease. World Health Organization

Yang L, Sahlqvist S, McMinn A, Griffin SJ, Ogilvie D (2010) Interventions to promote cycling: systematic review. BMJ 341:c5293

Zenghelis D (2017) Climate policy: equity and national mitigation. Nat Clim Chang 7(1):9

The IMO has addressed ship pollution under the MARPOL convention and required a gradual decrease of air emissions (NO_x, SO_x and Particulate Matters) originating from ship engines. In addition, major energy efficiency improvements for vessels have been proposed through the application of the Energy Efficiency Design Index (EEDI) and Ship Energy Efficiency Management Plan (SEEMP).

The regulation of air pollution by ships was defined in MARPOL (annex VI) and was firstly adopted in 1997 and revised in 2008, including a progressive reduction of SO_x and NO_x and indirectly Particulate Matter (PM) in Emission Control Areas (ECAs) including EU-territories (such as the Baltic Sea, the English Channel and the North Sea). The Marpol Annex VI ECAs are the Baltic Sea area, the North Sea area, the North American area (covering designated coastal areas of the USA and Canada), and the US Caribbean Sea area (around Puerto Rico and the US Virgin Islands) (Directorate-General for Mobility and Transport 2011).

When ECAs were firstly introduced, the limit in the content of sulphur for marine fuels was 1%. Nowadays, under the recent IMO regulations framework, 0.1% is defined as the maximum sulphur content for marine fuel for all ship operations in Emission Control Areas since 1 January 2015. This is the only global regulation that addresses the control of air emissions from ships in a comprehensive manner, while at present the marine fuels sulphur limit is 3.5%. Marpol Annex VI started with a global sulphur cap of 4.5% and in 2012 it was lowered to 3.5%. From 1 January 2020, IMO will enforce a new 0.5% global sulphur cap on marine fuels content, lowering from the present 3.5% limit (see Fig. 1).

The global fuel sulphur cap is part of the IMO's response to global environmental concerns due to harmful air emissions from ships. 2020 as a deadline was confirmed at the 70th session of IMO's Marine Environment Protection Committee (MEPC) held in October 2016.

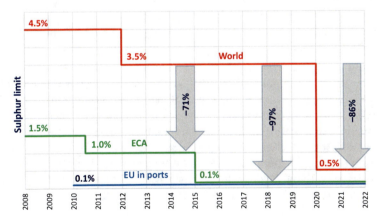

Fig. 1 IMO Marpol Annex VI Sulphur limits timeline

EU has expressed its willingness to widen the enforcement of MARPOL Annex VI sulphur restrictions to all European seas. Also, concerns were expressed through the Strategy for Sustainable Development published on the EU White Paper on Transport Policy (EU Regulation 757 2015) about the impact of maritime transport on air quality. This has led to the establishment of the EU Regulation 2016/802 for sulphur content in marine fuels.

The result of the above directive is that all Member States shall take all necessary actions to ensure that sulphur content of marine fuels used in the areas of their territorial seas, exclusive economic zones and pollution control zones does not exceed:

(a) 3.50% as from 18/06/2014;
(b) 0.50% as from 01/01/2020.

This applies to all vessels of all flags, including those whose journey began outside the European Union. Also during port stays, all vessels calling at EU ports should either use low sulphur fuel (0.1% max) or a shoreside electricity connection if staying longer than two hours. Additionally, passenger ships operating on regular services to or from any EU port shall use marine fuel which not exceeds 1.50% by mass, until 1 January 2020. In European designated ECAs, all ships have to burn fuel with a sulphur content of no more than 0.1% since 1 January 2015.

As we can clearly understand from the above, both EU and IMO develop and apply policies to reduce greenhouse gases (GHG) emissions from ships and as a result of this, two similar data collection schemes have been introduced:

- EU MRV—Monitoring, Reporting and Verification of **CO_2 emissions** (data collection started 1 January 2018)
- IMO DCS—Data Collection System on **fuel consumption** (data collection started 1 January 2019)

EU MRV and IMO DCS requirements are mandatory, and intend to be the first steps in a process to collect and analyse emission data related to the maritime sector.

The European Union regulation 2015/757 on the MRV of CO_2 emissions from maritime transport was adopted by the European Council and entered into force on 1 July 2015 (EU Regulation 601 2012). This regulation is viewed as the first step of a staged approach for the inclusion of maritime transport emissions into the EU's GHG reduction commitment, alongside the other sectors (energy production, transportation, etc). MRV Regulation covers shipboard CO_2 emissions and requires ships to monitor data on cargo carried and transportation work. The objective of the MRV is to develop a better understanding of fuel consumption and CO_2 emissions from shipping activities within the EU, which could then be used to shape and inform any future GHG monitoring, controlling and mitigation initiatives.